Reflection Asymmetry in Atomic Nuclei

Reflection Asymmetry in Atomic Nuclei

Edited by

Pengwei Zhao

Peking University, China

World Scientific

NEW JERSEY • LONDON • SINGAPORE • BEIJING • SHANGHAI • HONG KONG • TAIPEI • CHENNAI • TOKYO

Published by

World Scientific Publishing Co. Pte. Ltd.
5 Toh Tuck Link, Singapore 596224
USA office: 27 Warren Street, Suite 401-402, Hackensack, NJ 07601
UK office: 57 Shelton Street, Covent Garden, London WC2H 9HE

British Library Cataloguing-in-Publication Data
A catalogue record for this book is available from the British Library.

REFLECTION ASYMMETRY IN ATOMIC NUCLEI

ISBN 978-981-98-0595-2 (hardcover)
ISBN 978-981-98-0596-9 (ebook for institutions)
ISBN 978-981-98-0597-6 (ebook for individuals)

For any available supplementary material, please visit
https://www.worldscientific.com/worldscibooks/10.1142/14120#t=suppl

Typeset by Stallion Press
Email: enquiries@stallionpress.com

Preface

As a complex self-bound quantum many-body system, atomic nuclei have different shapes because of sizable long-range multipole-multipole interactions among nucleons. The studies of these shapes, and of the associated nuclear moments, have not only deepen our understanding of the origin of simple patterns in complex nuclear systems, but also connected to the answers of fundamental questions in the field of particle physics, astrophysics and so on. In the past few decades, with the development of worldwide facilities and the detectors, the study of the reflection asymmetry in atomic nuclei has attracted lots of attention both experimentally and theoretically.

The collection of research and review articles presented in this Special Issue encompasses a comprehensive exploration of various sub-topics, including reflection asymmetric shapes, nuclear shape phase transitions, nuclear masses, nuclear fission, clustering structures, high-spin states, octupole collective motions, electric dipole moments, and CP violation. These contributions not only highlight the progress made thus far but also pave the way for future investigations and advancements in our understanding of the subject.

We extend our heartfelt gratitude to all the authors, as well as the dedicated reviewers and editorial team who have ensured the rigor and quality of the published works. Their invaluable contributions have made this collection a comprehensive and authoritative resource for researchers and enthusiasts alike.

It is our sincere hope that this Special Issue will serve as a catalyst for further interdisciplinary collaborations and inspire researchers to embark on new journeys of exploration within the subject.

Pengwei Zhao
Peking University

Contents

Reflection-asymmetric mode in the structure of heavy nuclei*

R. V. Jolos,† E. A. Kolganova, E. V. Mardyban and T. M. Shneidman

Bogoliubov Laboratory of Theoretical Physics,
Joint Institute for Nuclear Research,
Dubna 141980, Russia
†*jolos@theor.jinr.ru*

The properties and description of the reflection-asymmetric mode in heavy nuclei are discussed. Two approaches based on the collective octupole degree of freedom and the mass asymmetry mode are presented. Angular momentum dependence of the parity splitting, phase transition in alternating parity rotational band, stabilization of the reflection-asymmetric deformation with angular momentum increase and the rigidity of the reflection-asymmetric deformation are considered.

Keywords: Reflection asymmetry; alternating parity band; parity splitting.

1. Introduction

There is a strong evidence of existence of the collective reflection asymmetry mode in nuclei belonging to different regions of the nuclide chart[1]: Ba–Ce neutron-rich isotopes, rare-earth quadrupole deformed nuclei and actinides. Since the first observation of negative parity states near the ground state,[2,3] a large amount of the experimental data has been accumulated. Among recent experimental findings, one can specify an observation of alternating parity bands in ^{123}Ba,[4] enhanced $E1$ and $E3$ strengths in ^{144,146}Ba,[5,6] $^{220-226}$Rn[7–10] and ^{222,224,228}Ra,[7,11] and also strong octupole correlations for excited 0^+ state in ^{240}Pu.[12,13] An extensive review of experimental achievements can be found in Refs. 1 and 14.

*To cite this article, please refer to its earlier version published in the *International Journal of Modern Physics E*, Volume 32(10), (23 pages), 2340002 (2023), DOI: 10.1142/S0218301323400025.

†Corresponding author.

The experimental data allow to established some of the main properties of the collective reflection-asymmetric mode. For the even–even nuclei they are in the following:

- low-lying rotational bands of odd parity with $I^{\pi} = 1^{-}, 3^{-}, 5^{-}, \ldots$ which are characterized by a shift in excitation energy with respect to the positive parity ground state band. Both positive and negative parity states form the so-called alternating parity band;
- relatively strong $E1$ transition rates between the members of the alternating parity bands;
- very low hindrance factors in α-decay observed between ground states and the very low-lying 1^{-} states in daughter even–even nuclei in the region of actinides. These hindrance factors reach a minimum in the same region where a minimum in the excitation energies of the 1^{-} states is located.[15,16]

The reflection-asymmetric correlations can be associated with an appearance of the p–h orbital pairs with $\Delta j = \Delta l = 3$ near the Fermi surface. For nuclei in the actinide region, just above the double spherical closed shells in ^{208}Pb, the $g_{9/2}$ and $j_{15/2}$ neutron orbits, and the $f_{7/2}$ and $i_{13/2}$ proton orbits are energetically very close together and very close to the Fermi surface. In the Ba–Ce region just above the double magic nucleus ^{132}Sn, the $f_{7/2}$ and $i_{13/2}$ neutron orbits, and the $d_{5/2}$ and $h_{11/2}$ proton orbits are also very close energetically. The combination of these low-lying orbitals gives rise to appearance of the low-energy $K^{\pi} = 3^{-}$ two quasiparticle configurations which form a microscopic basis for stable reflection-asymmetric deformation.

Early calculations within the random phase approximation,[17–20] shell-corrected liquid drop models[21,22] and mean-field models[23–25] show that nuclei in these mass regions are indeed soft with respect to the octupole deformation. However, the fact that in actinides α-decay hindrance factor reaches its minimum in the same region, where a minimum of the excitation energies of the 1^{-} states is located, indicates a possibility of the alternative explanation which is suggested by the cluster approach. In this case, long-range multipole correlations might lead to the formation of light clusters on the surface of a heavy nucleus. The contribution of such bicluster configurations to the nuclear wave functions naturally leads to appearance of the reflection-asymmetric deformation. There exists evidence which supports the idea of clustering. Nuclei whose spectra exhibit a strong reflection-asymmetric correlations are good α-emitters. Thus, there is a significant probability to form an α-cluster in the surface region of these nuclei. It is also known from the Nilsson–Strutinsky-type calculations for light nuclei that nuclear configurations corresponding to the minimum of the potential energy are related to certain cluster structures.[26,27] Several calculations performed for heavy nuclei[28,29] have shown that configurations with large quadrupole deformation and low-lying collective negative parity states are strongly related to clustering.

It was indicated recently that both views on the physical grounds for the appearance of the collective reflection-asymmetric correlations are not mutually exclusive as it seems at first sight.[30]

This paper is organized as follows. In Sec. 2, the parity splitting phenomenon is considered. Description of the parity splitting in the cluster model is presented in Sec. 3. Section 4 is devoted to consideration of the phase transition in alternating parity bands. Stabilization of the reflection-asymmetric deformation with angular momentum increase is analyzed in Sec. 5. In Sec. 6, the problem of rigidity of the reflection-asymmetric deformation is considered. Section 7 is devoted to analytical description of the excited state phase transition in alternating parity bands.

2. Parity Splitting

The presence of strong reflection-asymmetric correlations in nuclei is indicated by a significant lowering of the excitation energies of the negative parity states. In these nuclei, low-lying negative parity states form alternating parity bands together with the states of the ground state band. Such rotational bands with $I^\pi = 0^+, 1^-, 2^+, \ldots$ exist in molecules. However, in nuclei, in contrast to atomic molecules, these bands are disturbed: the negative parity states are shifted up in energy with respect to the positive parity ones. Especially interesting is the angular momentum dependence of the parity splitting. At the beginning of the rotational band the positive parity and negative parity states are split in energy by several hundreds of keV. Even in nuclei with largest reflection asymmetry the parity splitting is 200–300keV. At higher values of the angular momentum the parity splitting decreases and the energies of both positive and negative parity states start to have the rotational angular momentum dependence typical for the $K = 0$ band. Thus, a unified parity band is formed, however, it is not perfect.

To explain angular momentum dependence of the parity splitting $\Delta E(I)$ at low to moderate angular momenta the following model has been suggested.[31] It was assumed that the parity splitting is connected to the mode, describing reflection-asymmetric deformation in a one-dimensional well which keeps the axial symmetry. In nuclei with sufficiently strong reflection-asymmetric correlations the corresponding collective potential has two minima symmetrically located at positive and negative values of this variable. These two minima are separated by the barrier and the barrier transition frequency determines the parity splitting. The height of the barrier depends on the angular momentum via a rotational energy term. Due to the dependence of the moment of inertia on collective variable (which can be the octupole deformation parameter β_{30} in the case of the Bohr–Mottelson-type model or mass asymmetry parameter in the case of the cluster model), the barrier height between two minima increases with angular momentum I proportionally to $I(I+1)$, thus decreasing the probability of the barrier penetration. As a result, the reflection asymmetric deformation stabilizes and the parity splitting goes to zero.

The Hamiltonian of the model used can be presented as

$$H_I^{(1)} = -\frac{\hbar^2}{2B}\frac{d^2}{d\beta_{30}^2} + V_I^{(1)}(\beta_{30}), \tag{1}$$

where instead of β_{30} a mass asymmetry variable can be used. The subscript I indicates that the shape of the potential depends on the angular momentum I through the presence of the rotational energy term. The parity splitting is calculated as a difference between the energies of the lowest negative and positive parity states for the same I. Concerning the experimental determination of the parity splitting, we mention that at every value of I in the $K=0$ band there is only one state with the parity $\pi=(-1)^I$. Thus, the energy of the state with the opposite parity but the same I can be introduced only by interpolation using the energies of the states with neighboring I values and opposite parities $\pi=-(-1)^I$. For even I, the interpolated energy of the unphysical negative parity state is determined using the experimental energies of the negative parity states with angular momenta $(I-1)$ and $(I+1)$. For odd I, the interpolated energy of the unphysical positive parity state is determined using the experimental energies of the positive parity states with angular momenta $(I-1)$ and $(I+1)$.

The numerical solution of the Schrödinger equation with the Hamiltonian (1) with different variants of the two-center potential having very small and very large distances between two minima have shown[32] that the wave function of the state belonging to the ground state alternating parity band with even I can be approximated with a good accuracy by a sum of two Gaussians as

$$\begin{aligned}\Psi_I^{(+)}(\beta_{30}) = & \left(\frac{\xi^2}{\pi\beta_m^2(I)}\right)^{\frac{1}{4}} \left[2\left(1+\exp\left(-\xi^2(I)\right)\right)\right]^{-1/2} \\ & \times\left(\exp\left(-\frac{1}{2}\xi^2(I)(\beta_{30}/\beta_m(I)-1)^2\right)\right. \\ & \left.+\exp\left(-\frac{1}{2}\xi^2(I)(\beta_{30}/\beta_m(I)+1)^2\right)\right). \end{aligned} \quad (2)$$

We see from Eq. (2) that $\Psi_I^{(+)}$ is a function of the variable $\beta_{30}/\beta_m(I)$ which depends only on one parameter $\xi(I)$. In Eq. (2), $\beta_m(I)$ is a scaling parameter. The convenience of the ansatz (2) for the positive parity wave functions is related to the fact that both, the limit of octupole vibrations and the limit of stable octupole deformation, are described equally well by this ansatz.[33] The parameter $\xi(I)$ gives the ratio of a distance between the centers of the Gaussians to the sum of their widths. If $\xi(I) \ll 1$, the overlap of the components in (2) is large and the wave function $\Psi_I^{(+)}$ corresponds to the case of octupole vibrations. If $\xi(I) \gg 1$, the situation is opposite and the components in (2) are well separated. The later corresponds to the static octupole deformation.

Following the procedure of the supersymmetric quantum mechanics,[33–36] which can be used for solution of the considered problem, we substitute the wave function (2) into the Schrödinger equation with the Hamiltonian (1) and obtain the following expression for the potential $V_I^{(1)}$:

$$V_I^{(1)} = \frac{\hbar^2}{2B}\frac{d^2\Psi_I^{(+)}}{d\beta_{30}^2} \Big/ \Psi_I^{(+)} + E_I^*, \quad (3)$$

which gives us

$$\begin{aligned}V_I^{(1)}(\beta_{30}) = & \frac{\hbar\omega}{2}\left(1+\xi^2(I)\left(1+\beta_{30}^2/\beta_m^2(I)\right)-2\xi^2(I)\frac{\beta_{30}}{\beta_m(I)}\right. \\ & \left.\times\frac{\exp\left(\xi^2(I)\beta_{30}/\beta_m(I)\right)-\exp\left(-\xi^2(I)\beta_{30}/\beta_m(I)\right)}{\exp\left(\xi^2(I)\beta_{30}/\beta_m(I)\right)+\exp\left(-\xi^2(I)\beta_{30}/\beta_m(I)\right)}\right)+E_I^*, \end{aligned} \quad (4)$$

where E_I^* is the excitation energy of the lowest state of the Hamiltonian H_I with angular momentum I and positive parity. For even I, E_I^* is given by the experimental data. For odd I, this energy is determined by the interpolation using the energies of the neighboring even-I states

$$E^{*}_{\text{inter}}(I) = \frac{1}{2}(E_{\text{exp}}(I+1) + E_{\text{exp}}(I-1)) \\ - \frac{1}{8}(E_{\text{exp}}(I+3) - 2E_{\text{exp}}(I+1) + E_{\text{exp}}(I-1)). \tag{5}$$

In (4), the parameter $\hbar\omega$ is used instead of the inertia coefficient B.

The following problem which must be solved is to find the wave function of the negative parity states of the alternating parity band. In order to do this and to calculate the parity splitting, it is convenient to follow the prescription of the supersymmetric quantum mechanics.[34–36] First, we determine the supersymmetric partner potential of the potential $V_I^{(1)}(\beta_{30})$ which we denote by $V_I^{(2)}(\beta_{30})$. It is presented as

$$V_I^{(2)}(\beta_{30}) = (W_I(\beta_{30}))^2 + \frac{\hbar}{\sqrt{2B}} \frac{dW_I(\beta_{30})}{d\beta_{30}} + E_I^{*}, \tag{6}$$

where

$$W_I(\beta_{30}) = -\frac{\hbar}{\sqrt{2B}} \frac{\left(\frac{d\Psi_I^{(+)}(\beta_{30})}{d\beta_{30}}\right)}{\Psi_I^{(+)}(\beta_{30})}. \tag{7}$$

Both potentials $V_I^{(1)}$ and $V_I^{(2)}$ have been calculated and it was shown that $V_I^{(2)}$ is a one-center potential contrary to $V_I^{(1)}$. This fact makes the solution of the Schrödinger equation with the potential $V_I^{(2)}$ simpler than the solution of the Schrödinger equation with the potential $V_I^{(1)}$. The reason is that in the case of the one-center potential the harmonic oscillator basis being used provides a good convergence.[33]

According to supersymmetric quantum mechanics the energy of the lowest state of the Hamiltonian $H^{(2)}$ with the potential $V_I^{(2)}$ is equal to the energy of the first excited state of $H_I^{(1)}$. The energy of the lowest state of $H_I^{(1)}$ is fixed to be equal to zero just by definition of the constant term in $V_I^{(1)}$. Thus, the difference of the energies is $\Delta E(I)$ by the definition. The energy of the lowest state of $H_I^{(2)}$ is given by the average of $H_I^{(2)}$ in the lowest state $\Psi_I^{(2)}$

$$\Delta E(I) = \langle \Psi_{0,I}^{(2)} | H_I^{(2)} | \Psi_{0,I}^{(2)} \rangle. \tag{8}$$

As it is seen from (1), (2) and (4), the Hamiltonian and the shape of the wave function are determined by two parameters: $\hbar\omega(I)$ which has a dimension of energy and dimensionless $\xi(I)$. Then the parity splitting is given

Table 1. The calculated and the experimental values of $\Delta E(I)$ for ^{240}Pu. The experimental data are taken from Ref. 37. The following values of the parameters are used: $\hbar\omega = 0.585$MeV, $c_0 = 0.02$, $c_1 = 0.057$.

	^{240}Pu			^{240}Pu	
	$\Delta E(I)$ in keV			$\Delta E(I)$ in keV	
I	exp	cal	I	exp	cal
1	583	581	16	262	257
2	575	574	17	238	233
3	563	564	18	216	211
4	548	550	19	194	190
5	531	533	20	173	171
6	511	513	21	152	153
7	488	491	22	134	137
8	465	467	23	115	122
9	440	442	24	98	109
10	415	416	25	81	98
11	388	389	26	66	88
12	363	361	27	51	80
13	337	334	28	38	72
14	311	308	29	22	67
15	286	281	30	8	63

by the expression

$$\Delta E(I) = \hbar\omega(I) f[\xi(I)], \tag{9}$$

where the function $f(\xi)$ is determined numerically. As an example, the results of calculation of $\Delta E(I)$ for ^{240}Pu are presented in Table 1.

Comparing the results of calculation of $\Delta E(I)$ with the experimental data, we have found that $\zeta(I)$ can be approximated by a linear function of I

$$\xi(I) = c_0 + c_1 I, \tag{10}$$

where c_0 and c_1 are the fit parameters.

3. Description of Parity Splitting in Cluster Model

Instead of parameterizing the nuclear shape in terms of multipole deformation parameters we can use degrees of freedom related to the dinuclear system, or cluster system.[38] Dinuclear system is considered as a system of two nuclei (A_1, Z_1) and (A_2, Z_2) kept together in touching configuration by the nucleus–nucleus potential. As a mononucleus, we consider a unified system with the charge $Z = Z_1 + Z_2$ and mass number

$A = A_1 + A_2$. The degrees of freedom describing the collective excitations are related to the rotation of a dinuclear system as a whole, to the relative rotation of a dinuclear fragments, and to the transfer of nucleons between dinuclear fragments. The later process is described by the mass asymmetry $\eta = A_2/A$ and charge asymmetry $\eta_Z = Z_2/Z$ coordinates. The values of $\eta = 0$ and $\eta = 1$ correspond to mononucleus configuration.

The motion in η destroys reflection-symmetric shape of the nuclear system. It is assumed in this approach that the nuclear wave function can be described as a superposition of the mononucleus and different dinuclear configurations. In the considered cases, the mononucleus is taken to be quadrupole deformed. The mass asymmetry is described as a continuous variable. We presume that the main source of the reflection-asymmetric deformation is a contribution of a dinuclear system with α-particle as a light cluster. This idea is supported by the observation discussed in Introduction that actinides are good α-emitters. Thus, there is a significant probability to form α-cluster in a surface region of a nucleus. Moreover, our calculations show that the potential energy of a dinuclear system with α-cluster ($\eta = \eta_\alpha$) is close or even lower than that given by the binding energy of these nuclei, while the energies of the configurations with a cluster heavier than α-particle rapidly increase with η. Therefore, in order to treat the ground state properties of actinides, one can restrict the region of variation of mass asymmetry by the vicinity of mononucleus and α-cluster dinuclear configuration. The contribution of the configuration with η larger than η_α is negligibly small. Thus, dinuclear configuration under consideration consists of a light spherical fragment and a deformed or a spherical heavy fragment.

In addition to motion in η, we consider a relative rotation of the fragments described by the angles $\Omega_R = (\theta_R, \varphi_R)$ and rotation of the heavy fragment described by the angles $\Omega_h = (\theta_h, \varphi_h)$ (Fig. 1). We introduce also plane angle ϵ between the symmetry axis of the heavy fragment and the vector of the relative distance $\boldsymbol{R}$.

The Hamiltonian of the model is the following[39]:

$$\begin{aligned} H &= H_\eta + H_{\text{rot}}, \\ H_\eta &= -\frac{\hbar^2}{2B}\frac{1}{\eta}\frac{\partial}{\partial\eta}\eta\frac{\partial}{\partial\eta} + U(\eta,\epsilon), \\ H_{\text{rot}} &= \frac{\hbar^2}{2\mu(\eta)R_m^2(\eta)}L_R^2 + \frac{\hbar^2}{2\Im_h(\eta)}L_h^2, \end{aligned} \tag{11}$$

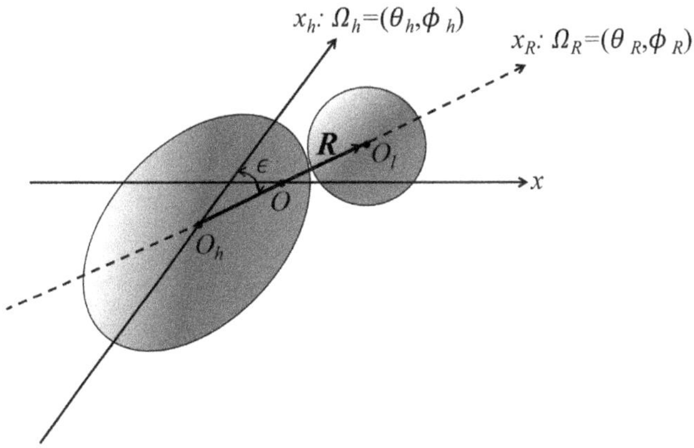

Fig. 1. The schematic picture of a dinuclear system with the indicated degrees of freedom used. The orientation of the vector $\boldsymbol{R}$ connecting the centers of nuclei is defined by the angles $\Omega_R = (\theta_R, \varphi_R)$ in the laboratory frame Ox_R. The orientation of the intrinsic coordinate system $O_h x_h$ related to the quadrupole deformed heavy fragment is defined by the angles $\Omega_h = (\theta_h, \varphi_h)$.

where H_η describes a motion in mass asymmetry and a coupling of the rotational degrees of freedom, $H_{\rm rot}$ describes the rotational motion of the heavy fragment and the relative rotation of the fragments. The ingredients of the Hamiltonian (11) are calculated within the dinuclear model. For dinuclear system with the α-particle or Li as light fragments the potential energy is taken as

$$U(\eta_i, \epsilon) = B_1(\eta_i) + B_2(\eta_i) + V(R = R_m(\epsilon), \beta_0, \eta_i, \epsilon) + \frac{1}{2}\hbar\omega_R(\eta_i), \qquad (12)$$

where $(i = \alpha, \mathrm{Li})$. Here, B_1 and B_2 are the binding energies of the fragments which are taken from the experimental data. The nucleus nucleus potential

$$V(R_m, \eta, \beta_0, \epsilon) = V_C(R_m, \eta, \beta_0, \epsilon) + V_N(R_m, \eta, \beta_0, \epsilon), \qquad (13)$$

is the sum of the Coulomb potential $V_C(R_m, \eta, \beta_0, \epsilon)$ and the nuclear interaction potential

$$V_N(R_m, \eta, \beta_0, \epsilon) = \int \rho_1(\boldsymbol{r}_1)\rho_2(\boldsymbol{R}_m - \boldsymbol{r}_2)F(\boldsymbol{r}_1 - \boldsymbol{r}_2)d\boldsymbol{r}_1 d\boldsymbol{r}_2, \qquad (14)$$

where $F(\boldsymbol{r}_1 - \boldsymbol{r}_2)$ is the density-dependent effective nucleon–nucleon interaction taken as the Migdal forces.[40] The nucleus–nucleus potential is calculated at $R = R_m$ that corresponds to the minimum of the potential

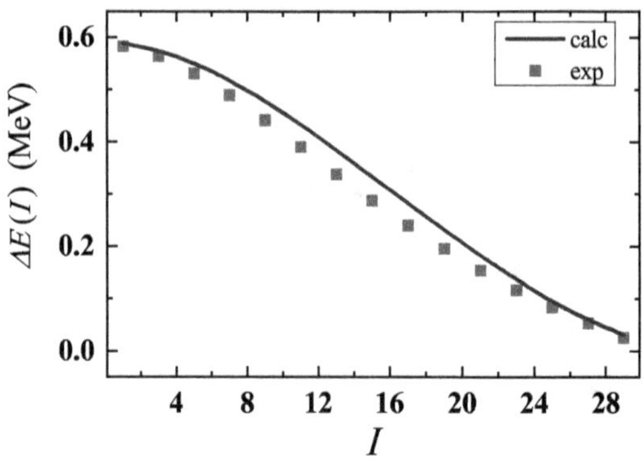

Fig. 2. Calculated (lines) and experimental (squares) values of parity splitting for ^{240}Pu as a function of the angular momentum I. Experimental data are taken from Ref. 42.

pocket in relative distance R. The last term in (12) represents the energy of the zero-point vibration in this pocket.

Another quantity required to determine the energy spectrum with the Hamiltonian (11) is the moment of inertia of a heavy fragment. This moment of inertia is approximated by the expression

$$\Im_h(\eta) = c(\eta)\Im_h^{(\text{rig})}(\eta), \tag{15}$$

where $\Im_h^{(\text{rig})}(\eta)$ is the rigid-body moment of inertia. The value of $c(\eta = 0)$ is fixed by calculation of the energy of the 2^+ state of the ground state band. For the pure cluster configuration ($\eta = \eta_\alpha$ or η_{Li}) we take $\Im_h(\eta) = 0.85\Im_h^{(\text{rig})}(\eta)$ as for the highly deformed state.[41] This relation is used because of the evolution of the single particle orbit of a heavy fragment under the influence of the light cluster in spite of the fact that the heavy fragment has a normal deformation. For $0 \leq \eta \leq \eta_\alpha$ $c(\eta)$ smoothly increases.

The example of calculation of the parity splitting in the cluster approach is shown in Fig. 2.

4. Phase Transition in Alternating Parity Bands

Let us consider evolution with angular momentum of the potential energy given by (4). As an example we consider the results for ^{240}Pu which is investigated experimentally better than the other nuclei.[43] However, similar results are obtained for other nuclei. Figure 3 shows the octupole deformation potentials calculated for ^{240}Pu for three values of the angular

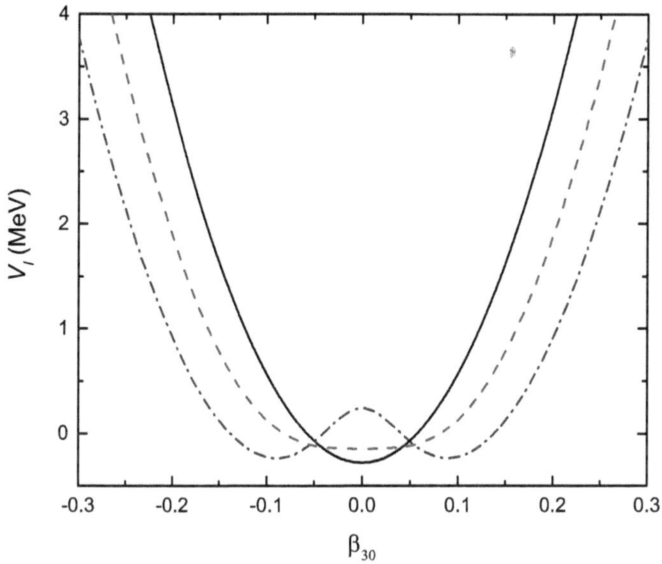

Fig. 3. The potential $V_I^{(1)}$ as a function of β_{30} calculated with $I = 0$ (solid line), 12 (dashed line) and 24 (dash-dotted line). The potential energy is counted from the ground state energy for $I = 0$ and from the excitation energy for $I = 12, 24$.

momentum: $I = 0$, 12 and 24. It is seen that at $I = 0$ the potential $V_I^{(1)}$ has a minimum at $\beta_{30} = 0$. At $I = 24$, the potential has two minima symmetrically located with respect to maximum at $\beta_{30} = 0$. At $I = 12$, we have a potential corresponding to transition from reflection-symmetric to reflection-asymmetric shape. It means that with angular momentum increase, the barrier in potential energy develops and the shape phase transition takes place in alternating parity bands. As it is seen from Fig. 3, the value of the potential at $\beta_{30} = 0$ increases with I as $\hbar\omega/2\xi^2(I)$. This follows from Eq. (4) and explains an increase of the minimum of the potential for $I \leq I_{\rm cr}$. For $I > I_{\rm cr}$, however, the minimum of the potential is shifted to $\beta_{30} \neq 0$ and the barrier starts to grow with I.

Consider this transition in detail applying the Landau theory of phase transitions.[44,45] In our case, the role of the free energy in the Landau theory is played by the potential $V_I^{(1)}$ taken at the minimum. Then β_{30} is an order parameter, and I is a control parameter. The next step is a determination of the equilibrium value of the order parameter $\beta_{30}((\beta_{30})_{\rm min})$ which is the position of the minimum of the potential. The value of $(\beta_{30})_{\rm min}$ is a function of the control parameter I.

The equilibrium value of the order parameter β_{30} is determined by the condition[46]

$$0 = \left.\frac{dV_I^{(1)}(\beta_{30})}{d\beta_{30}}\right|_{\beta_{30}=(\beta_{30})_{\min}}$$
$$= \hbar\omega\cdot\xi^2(I)\frac{(\beta_{30})_{\min}}{\beta_m(I)^2}\left(1-\frac{(\beta_{30})_{\min}}{\beta_m(I)}\tanh(\xi^2(I)(\beta_{30})_{\min}/\beta_m(I))\right.$$
$$\left.-\frac{\xi^2(I)}{(\cosh(\xi^2(I)(\beta_{30})_{\min}/\beta_m(I)))^2}\right). \tag{16}$$

The minimum of $V_I^{(1)}$ at $\beta_{30}=0$ takes place only if $\xi(I)<1/\sqrt{2}$ when the second derivative of $V_I^{(1)}$ over β_{30} is positive. If $\xi(I)>1/\sqrt{2}$ then at $\beta_{30}=0$ potential has a maximum and the minimum is smoothly shifted to nonzero value of β_{30}. This value is determined by the numerical solution. If $\xi(I)$ is larger but very close to $1/\sqrt{2}$ then the second root of Eq. (16) is given approximately by

$$(\beta_{30})_{\min} = \beta_m(I)\sqrt{\frac{3}{2\xi^6(I)}\left(\xi^2(I)-\frac{1}{2}\right)}. \tag{17}$$

Let us find the value of the potential $V_I^{(1)}$ at the minimum. If $\xi(I)<1/\sqrt{2}$ the minimum of the potential is located at $\beta_{30}=0$ and

$$V_I((\beta_{30})_{\min}) \equiv V_I(\min) = \frac{1}{2}\hbar\omega(-1+\xi^2(I)) + E_I^*. \tag{18}$$

For $\xi(I)>1/\sqrt{2}$ but close to $1/\sqrt{2}$

$$V_I(\min) \approx \frac{1}{2}\hbar\omega\left(-1+\xi^2(I)-\frac{3}{\xi^4(I)}\left(\xi^2(I)-\frac{1}{2}\right)^2\right)+E_I^*. \tag{19}$$

In the last equation, the terms of the higher order in $(\xi^2(I)-\frac{1}{2})^2$ are omitted. Remember that $\xi(I)$ is a linear function of I. Comparing (18) and (19), we see that $V_I(\min)$ is a continuous function of I at $\xi(I)=1/\sqrt{2}$. Taking the first-order derivatives of the expressions (18) and (19), we see that both derivatives coincide at $\xi=1/\sqrt{2}$. However, there is a discontinuity of the second-order derivative over I at $\xi(I)=1/\sqrt{2}$. Indeed, for $\xi(I)<1/\sqrt{2}$

$$\frac{d^2V_I(\min)}{dI^2} = \hbar\omega c_1^2 + \frac{d^2E_I^*}{dI^2}. \tag{20}$$

For $\xi(I) > 1/\sqrt{2}$

$$\frac{d^2V_I(\min)}{dI^2} = \hbar\omega c_1^2\left(1 - \frac{12}{\xi^2(I)} + \frac{42}{\xi^4(I)}\left(\xi^2(I) - \frac{1}{2}\right)\right)$$
$$- \frac{30}{\xi^6(I)}\left(\xi^2(I) - \frac{1}{2}\right)^2 + \cdots + \frac{d^2E_I^*}{dI^2}. \qquad (21)$$

At the critical point $\xi(I) = 1/\sqrt{2}$, the expressions (20) and (21) differ by the term $-12\hbar\omega c_1^2/\xi^2(I)$. Thus, we have a second-order phase transition. This is a transition from reflection-symmetric to reflection-asymmetric shape in the alternating parity band with angular momentum, i.e., with excitation energy increase.

Let us introduce a special notation for the critical value of ξ, namely, $\xi_{\text{crit}} = 1/\sqrt{2}$. The corresponding critical value of the angular momentum I_{crit} is determined by the relation

$$c_0 + c_1 \cdot I_{\text{crit}} = 1/\sqrt{2}. \qquad (22)$$

Now we can rewrite the expression (17) in the following way:

$$(\beta_{30})_{\min} = \beta_m(I_{\text{crit}})(12\sqrt{2}c_1(I - I_{\text{crit}}))^{1/2}. \qquad (23)$$

We see that the critical exponent is equal to 1/2.

The critical value of the angular momentum for such nuclei as ^{232}Th, ^{238}U and ^{238}Pu are approximately equal to 12. In the lighter Ra and Th isotopes phase transition takes place at lower angular momenta.

Thus, we have shown that the experimental data on the spectra of the ground state alternating parity bands of some nuclei indicates a second-order phase transition from reflection-symmetric to reflection-asymmetric bands.

5. Stabilization of the Reflection-Asymmetric Deformation with Angular Momentum Increase

Experimental data on the spectra and $E1$ transition probabilities characterizing alternating parity bands show that effective reflection-asymmetric deformation changing with increasing angular momentum until its value stabilizes.

It is useful to introduce a nondimensional characteristic of the spectra of the ground state alternating parity band, and to analyze the experimental data on the energies of these states in different nuclei based on this characteristic. Then it will be possible to investigate stabilization of the reflection-asymmetric deformation with angular momentum increase.

In the case of the quadrupole mode, the quantity which characterizes closeness of the nucleus to the limit of the rigid quadrupole deformation is the ratio of the energies $R(4/2) \equiv E(4_1^+)/E(2_1^+)$. If $R(4/2)$ is close to 10/3, then the corresponding nucleus is axially symmetric and well deformed. For other values of I, we can use the energy ratio $E(I_1^+)/E(2_1^+)$ which should be close to $I(I+1)/6$ in the case of the well-deformed nuclei.

By analogy with the quadrupole case, we suggest using a dimensionless quantity to characterize a closeness of a nucleus to the limit of the stable reflection-asymmetric deformation. This is a ratio of two energies. The nominator is the energy that the negative parity state with odd angular momentum I would have in the limit of the stable octupole deformation. The denominator is the experimental energy of the negative parity state with the same angular momentum.

How to determine the energy that the rotational negative parity states would have in the limit of the stable octupole deformation? It is natural to assume that in this limit the energies of the positive and negative parity states are described by a monotonously increasing function of the angular momentum I.

The function describing the angular momentum dependence of the excitation energies of the positive parity states of the ground state band can be found fitting the experimental spectra. Let us denote this function by $E_{\rm exp}^{\rm even}(I)$. The value of this function taken at the odd value of I is the energy that a negative parity state would have in the limit of the stable octupole deformation. Indeed, it is assumed by definition that in this limit we can neglect fluctuations of the octupole deformation and, therefore, the penetration through the barrier separating two symmetrically located reflection-asymmetric minima. For this reason, parity splitting disappears and the rotational energy becomes a smooth function of the angular momentum. Then the energy of the state with a given angular momentum can be determined by a smooth interpolation between the energies of the neighboring positive parity states. We denote this energy as $E_{\rm inter}(I_{\rm odd})$.

The expression for $E_{\rm inter}(I_{\rm odd})$ can be presented in the form similar to that suggested in Ref. 14

$$E_{\rm inter}(I_{\rm odd}) = W(I_{\rm odd})E_{\rm exp}^{\rm even}(I_{\rm odd}+1) + (1-W(I_{\rm odd}))E_{\rm exp}^{\rm even}(I_{\rm odd}-1), \quad (24)$$

where $W(I_{\rm odd})$ depends on the type of the spectra of the positive parity states. In fact, the final results are only slightly sensitive to the concrete choice of $W(I_{\text{odd}})$ and in the calculations we use the expression[47]

$$W(I) = \frac{I}{2I+1}, \tag{25}$$

corresponding to the rotational limit since the majority of nuclei under consideration are close to the rotational limit for the even-I states.

Thus, we introduce the following dimensionless indicator of a closeness of nuclei to the limit of the stable reflection-asymmetric deformation

$$R_{\text{oct}}(I_{\text{odd}}) = \frac{E_{\text{inter}}(I_{\text{odd}})}{E_{\text{exp}}(I_{\text{odd}})}. \tag{26}$$

If a nucleus approaches a limit of a stable reflection-asymmetric deformation the parity splitting disappears, $E_{\text{exp}}(I_{\text{odd}})$ approaches $E_{\text{inter}}(I_{\text{odd}})$ and R_{oct} $(I_{\text{odd}}) \to 1$ corresponding to the limit of a stable deformation.[47] In this limit, we obtain a molecular-type rotational band known for two-atomic asymmetric molecules such as CH. For instance, for CH molecule $R_{\text{oct}}(1) = R_{\text{oct}}(3) = R_{\text{oct}}(5) = 0.9996$.

The results obtained are shown in Figs. 4 and 5. As it is seen from these figures, at small values of I, R_{oct} is small in all considered nuclei. A large

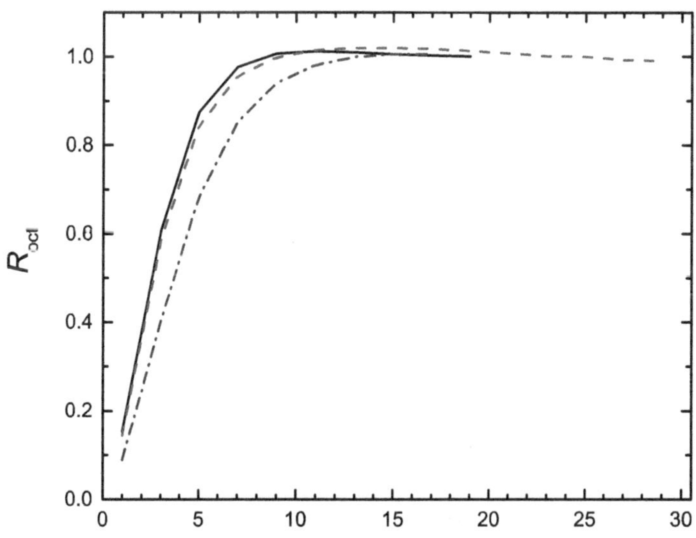

Fig. 4. The values of R_{oct} as a function of the angular momentum I derived from the experimental data. The dashed line corresponds to ^{220}Ra ($R_{4/2} = 2.30$). The solid line corresponds to ^{222}Ra ($R_{4/2} = 3.13$). The dash-dotted line corresponds to ^{226}Ra ($R_{4/2} = 2.71$).

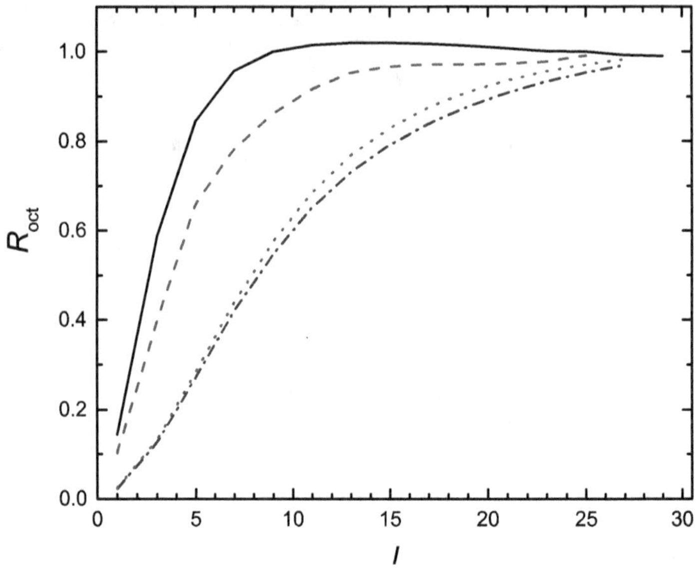

Fig. 5. The values of $R_{\rm oct}$ as a function of the angular momentum I derived from the experimental data. The solid line corresponds to ^{220}Ra ($R_{4/2} = 2.30$). The dashed line corresponds to ^{152}Gd ($R_{4/2} = 2.19$). The dash-dotted line corresponds to ^{232}Th ($R_{4/2} = 3.28$). The short-dashed line corresponds to ^{238}Pu ($R_{4/2} = 3.31$).

value of the parity splitting, i.e., a large probability for penetration of the barrier separating two physically equivalent reflection-asymmetric minima, is the reason of it. With increase of the angular momentum, $R_{\rm oct}$ increases, coming nearer to unity, and keeps this value for larger value of I, fluctuating in certain cases around unity. This happens because barrier penetration probability decreases with increasing I. The plateau is clearly seen in light Ra isotopes at the highest known values of I. This behavior of $R_{\rm oct}$, at least at low and intermediate values of I, is typical for all heavy nuclei. In the case when the experimental value of $R_{\rm oct}$ is close to 1, we can say that reflection-asymmetric deformation becomes stable.

However, there is a clear difference in angular momentum dependence of $R_{\rm oct}$ between the well-deformed nuclei ^{156}Gd, ^{232}Th, ^{238}U, ^{240}Pu and the transitional nuclei ^{144}Ba, ^{152}Gd, ^{220}Ra, ^{224}Th. This difference is especially well illustrated by the results presented in Fig. 5. In Fig. 5, two groups of lines, with one group representing nuclei with $R_{4/2}$ smaller than 3.0 and the other representing rotational nuclei with $R_{4/2} = 3.3$, are quite well separated from each other. For transitional nuclei, the value of $R_{\rm oct}$ at $I = 1, 3$ is higher than for the well-deformed nuclei. The slope of $R_{\rm oct}$ at the intermediate values of I at which R_{oct} is smaller than one is also larger for transitional nuclei compared to the well-deformed ones.

We mention that the function $R_{\text{oct}}(I)$ looks similar to the function describing establishing of the polarization with time when an external field is applied to the complex materials. The classical Debye law of relaxation is described by the exponential function

$$1 - \exp\left(-t/T_{\text{rel}}\right). \tag{27}$$

It is interesting that R_{oct} can be approximated by a similar function. For instance, for ^{222}Ra

$$R_{\text{oct}}(I) \approx 1 - \exp\left(-(I/I_{\text{crit}})^{\nu}\right), \tag{28}$$

with $I_{\text{crit}} = 3.2$ and $\nu = 1.63$. For ^{240}Pu $I_{\text{crit}} = 9.8$ and $\nu = 1.50$.

6. How Rigid is Reflection-Asymmetric Deformation?

The fluctuations of the collective variables around the equilibrium values are very important as they determine accuracy of the mean-field approximation, which is the first step in the nuclear structure calculations. Its accuracy depends on the rigidity of the deformation. Information about a rigidity of the quadrupole and octupole deformation can be obtained from the excitation spectra and the $E\lambda$ reduced transition probabilities. In the last case, a difference between the quadrupole and octupole modes is essential.

In the case of the quadrupole mode, a diagonal matrix element of the quadrupole operator can be measured, thus giving information on the static quadrupole deformation. In contrast, the octupole moment operator has negative parity and, therefore, its diagonal matrix elements are equal to zero by definition.

If we apply the Bohr–Mottelson collective model to the well-deformed axially symmetric nuclei the quadrupole operator can be divided in two parts

$$Q_{2\mu} = D^2_{\mu 0}(\langle\beta_{20}\rangle + (\beta_{20} - \langle\beta_{20}\rangle)) + \cdots , \tag{29}$$

where $\langle\beta_{20}\rangle$ is an equilibrium deformation. The term depending on γ is not presented in (29). The first term in (29) describes intraband transitions. The second term in (29) is responsible for the $E2$ transitions between the ground and the β-vibrational band. Thus, comparing the strength of the $E2$ transitions between the states of the ground state band and between the ground and the β-vibrational bands, we obtain information on the value of the amplitude of the fluctuations of β_{20} compared to the equilibrium value of β_{20}. Namely, $\sqrt{\langle(\beta_{20} - \langle\beta_{20}\rangle)^2} / \langle\beta_{20}\rangle \approx \sqrt{B(E2; 0^+_{\text{gs}} \to 2^+_{\beta}) / B(E2; 0^+_{\text{gs}} \to 2^+_1)}$.

Table 2. The calculated values of the quantity $\sqrt{B(E2;0^+_{\rm gs}\to 2^+_\beta)/B(E2;0^+_{\rm gs}\to 2^+_1)}$ for some rare-earth well-deformed nuclei. The experimental data are taken from Ref. 48.

Nucleus	$\sqrt{\frac{B(E2;0^+_{\rm gs}\to 2^+_\beta)}{B(E2;0^+_{\rm gs}\to 2^+_1)}}$	Nucleus	$\sqrt{\frac{B(E2;0^+_{\rm gs}\to 2^+_\beta)}{B(E2;0^+_{\rm gs}\to 2^+_1)}}$
^{154}Sm	0.074	^{158}Dy	0.107
^{156}Gd	0.058	^{160}Dy	0.058
^{158}Gd	0.040	^{162}Er	0.092

The typical values of this quantity for the well-deformed rare-earth nuclei are shown in Table 2. We compare them below with the results for the reflection-asymmetric mode. Of course this way of consideration of the quadrupole rigidity is not model-independent.

The octupole operator cannot be divided into two parts in a similar way since in the case of the octupole deformation there are two physically equivalent values of the equilibrium octupole deformation which are differed by the sign. Rigidity of the octupole mode can be analyzed only based on the theoretical model suggested for description of the experimental data on the properties of the octupole excitations. Below, we use the wave function (2) and the Hamiltonian introduced in Sec. 2.

As the characteristic of the octupole rigidity, we can use the value of the mean-square fluctuations of β_{30} around the minimum. Let us determine this quantity in the following way:

$$\begin{aligned}\mathrm{Fluc}(I) &\equiv \sqrt{\frac{\langle(\beta_{30}-\beta_{30,\min}(I))^2\rangle}{\beta^2_{30,\min}(I)}}\\ &= \sqrt{\int_0^\infty d\beta_{30}(\beta_{30}/\beta_{30,\min}-1)^2(\Psi_I^{(+)})(\beta_{30})^2}. \end{aligned} \tag{30}$$

The integration in (30) is taken over the region of the positive values of β_{30} in order to calculate a mean-square deviation from the positive value of β_{30}.

The results of calculations of the quantity Fluc(I) are shown in Table 3. From the definition of Fluc(I), it follows that the octupole deformation is rigid if the quantity Fluc(I) is small compared to unity, i.e., if Fluc(I) $\ll 1$. As an example we consider two nuclei, namely, ^{240}Pu and ^{224}Ra.

It is seen that a rigidity of the octupole deformation is larger in the case of ^{224}Ra than in the case of ^{240}Pu.

Table 3. The calculated values of the quantity Fluc (I) determined by Eq. (30) for ^{224}Ra and ^{240}Pu.

	^{224}Ra		^{240}Pu		^{240}Pu
I	Fluc (I)	I	Fluc(I)	I	Fluc(I)
3	0.90	12	0.82	22	0.50
4	0.71	13	0.77	23	0.49
5	0.59	14	0.72	24	0.47
6	0.52	15	0.68	25	0.45
7	0.46	16	0.65	26	0.44
8	0.41	17	0.62	27	0.43
9	0.37	18	0.59	28	0.41
10	0.34	19	0.56	29	0.40
11	0.31	20	0.54	30	0.39
12	0.29	21	0.52		

The quantity Fluc(I) in the case of the reflection-asymmetric mode plays the same role as that presented in Table 2 for the quadrupole case. Comparing the results presented in Tables 2 and 3, we see that the octupole mode is characterized by significantly larger fluctuations around the minimum than the quadrupole one.

7. Analytical Description of the Excited State Phase Transition to Reflection-Asymmetric Deformed Shape in Alternating Parity Bands

The experimental data show that the dependence of the parity splitting and transitional dipole moment on angular momentum in different nuclei looks quite universal. Therefore, based on the general idea on the role of the reflection-asymmetric mode one can try to find simple analytical description of the angular momentum dependence of these quantities containing a small number of the parameters having a clear physical meaning. These parameters can be further fitted to the experiment. It is the aim of this section to find such an analytical description of the angular momentum dependence of the main physical characteristics of the alternating parity bands.

Following the discussion in Sec. 2, we assume that the nucleus under consideration has a static quadrupole deformation $\langle\beta_{20}\rangle$ and is soft with respect to the axially symmetric ($K = 0$) reflection-asymmetric vibrations. These vibrations can be generated either by the octupole or by the mass asymmetry degree of freedom. In both cases, we denote the corresponding dynamical variable by β_{30}, although in the case of the mass asymmetry degree of freedom, the contribution of the higher-order odd multipolarity modes is effectively included. The intrinsic Hamiltonian describing collective motion in β_{30} for a given angular momentum I is presented in Sec. 2. The potential energy $V_I^{(1)}$ is an even function of β_{30}.

As it is formulated in Sec. 2, the wave functions of the states of the ground state alternating parity band can be approximated as a sum of two Gaussians centered at $\beta_{30} = \pm\beta_m(I)$ if we restrict consideration by the lowest positive parity states.

We are interested in the calculation of the parity splitting which is determined as a difference between the energies of the negative parity $E_I^{(-)}$ and the positive parity $E_I^{(+)}$ states having the same angular momentum I. Since, due to the $K = 0$ selection rules, there exists only one physical excited state for a given I (positive parity for even I and negative parity for odd I), an experimental parity splitting can be determined as a difference between the experimental excitation energy for one parity and the energy obtained by interpolation between the energies of the neighboring states of the opposite parity.

The ansatz (2) for the wave function $\Psi_I^{(+)}(\beta_{30})$ yields the expression for the potential energy of the axially symmetric reflection-asymmetric mode determined up to I-dependent constant by (4). Using the dimensionless variable $x = \beta_{30}/\beta_m(I)$ and the parameter ξ defined above, one can rewrite the Hamiltonian H_I and the potential energy $V_I^{(1)}$ in a convenient form

$$\begin{aligned} H_I &= \hbar\omega(I)h(\xi(I)), \\ h(\xi) &= -\frac{1}{2\xi^2}\frac{d^2}{dx^2} + v_\xi(x), \\ v_\xi(x) &= \frac{1}{2}(\xi^2 - 1) + \frac{1}{2}\xi^2 x^2 - \xi^2 x \tanh(\xi^2 x). \end{aligned} \tag{31}$$

From Eq. (31), it follows that the parity splitting can be parametrized as it is done in Sec. 2 (see Eq. (9))

$$\Delta E(I) \equiv E_I^{(-)} - E_I^{(+)} = \hbar\omega(I)f(\xi_I), \tag{32}$$

where $f(\xi(I))$ is the energy of the first-excited state of the Hamiltonian $h(\xi)$ (the ground state energy of this Hamiltonian is zero). All nucleus specific information is contained in the dependence of $\xi(I)$ on angular momentum and enters the Hamiltonian h_ξ and the function $f(\xi(I))$ implicitly. Due to its universal character, it make sense to find an approximate analytical expression for $f(\xi(I))$.

As it was shown in Ref. 49, $f(\xi(I))$ can be approximated with a good accuracy by the expression

$$f(\xi) = \frac{\xi^2 e^{-\xi^2}}{2\left[1 + (1 - e^{-\alpha\xi^2})\frac{\sqrt{\pi}}{4}\xi\right]}\coth\left(\frac{\xi^2}{2}\right), \quad \text{for } \alpha = 0.053, \tag{33}$$

where the value of the parameter α has been obtained by fitting the numerical results for the $f(\xi)$. We have found that in the limiting cases of very small and very large ξ, the difference between the approximate and exact values of $f(\xi)$ is negligible and vanishes asymptotically. The maximum deviation reaches about 2% at $\xi \approx 2$.

It is worth to note, that for the actual description of the experimental data, one can set $\alpha = 0$, which yields simpler expression for the parity splitting as

$$\Delta E(I) = \hbar\omega(I)\xi_I^2 e^{-\xi_I^2}\coth(\xi_I^2/2). \tag{34}$$

Since in the region of $\xi \gg 1$, the values of parity splitting are small and are influenced by many effects which are not included in the model (as for example, band crossing), we can neglect the deviations of (34) from (33) at large values of ξ and use the expression (34).

From Eq. (34), it follows that the angular momentum dependence of the parity splitting is determined by $\xi(I)$. All information on the nucleus is contained in actual dependence of ξ on angular momentum, while the function $f(\xi)$ is universal.

Our calculations have shown that with sufficiently good accuracy the angular momentum dependence of $\hbar\omega_I$ and ξ_I can be fitted as

$$\hbar\omega(I) = \text{const.}, \quad \xi(I) = c \cdot I. \tag{35}$$

Here, we have taken into account that c_0 in (10) is small. This parametrization contains very small number of parameters. If we use this parametrization of the potential, we can see that the value of the frequency $\hbar\omega$ is immediately determined by the value of the parity splitting at zero angular momentum, $\Delta E_{\text{exp}}(0)$. Indeed, if $I = 0$ then $\xi(0) = 0$ and $V_I^{(1)}(\beta_{30})$ reduces to the oscillator potential. Interval between the ground state and first excited state is then given by the frequency $\hbar\omega$. Therefore, we obtain that $\hbar\omega = \Delta E_{\text{exp}}(0)$. The function $f(\xi(I))$ depends only on the parameter c. Moreover, if we use the results of Ref. 46, we can connect the value of c to the value of the critical angular momentum I_{crit}, at which the phase transition from the octupole nondeformed to the octupole deformed shape takes place, namely

$$cI_{\text{crit}} = \frac{1}{\sqrt{2}}. \tag{36}$$

Finally, we obtain

$$\Delta E(I) = \Delta E_{\text{exp}}(0) f\left(\frac{I}{\sqrt{2}I_{\text{crit}}}\right). \tag{37}$$

The critical angular momentum I_{crit} is fitted to give the best overall description of the parity splitting in the range of the angular momenta $0 \leq I \leq 20$. The obtained values of I_{crit} are presented in Table 4.

In addition to the appearance of the low-lying negative parity states, a common property of nuclei exhibiting strong reflection-asymmetric correlations is the large values of the electric dipole transition probabilities.[48,50] While the absolute values of dipole moment for the transitions between negative and positive parity states depend on nucleus, its angular momentum dependence can be described, as it is shown below, by the universal function, similar to the situation with the parity splitting.

In the case of the well-deformed axially symmetric nuclei, the operator of the collective electric dipole moment can be written in the intrinsic system as

$$D \sim C\beta_{20}\beta_{30}, \tag{38}$$

where C is the dipole polarizability determined by the asymmetry between neutron and proton densities.[51]

Table 4. The values of the parameters $\Delta E(0)$ (keV) and I_{crit} used to describe the parity splitting in the alternating parity bands of various actinide nuclei.

Nucleus	$\Delta E(0)$	I_{crit}	Nucleus	$\Delta E(0)$	I_{crit}
^{222}Ra	209	2.8	^{238}Pu	584	13.3
^{224}Ra	192	3.4	^{240}Pu	585	12.1
^{226}Ra	235	4.7	^{242}Pu	767	11.8
^{228}Ra	456	7.5	^{244}Pu	888	14.9
^{224}Th	226	2.9	^{230}U	351	11.2
^{226}Th	209	4.9	^{232}U	548	16.2
^{228}Th	311	7.5	^{234}U	772	22.9
^{230}Th	492	10.2	^{236}U	674	15.3
^{232}Th	699	14.5	^{238}U	669	12.7
^{234}Th	685	11.7	^{240}U	789	12.1

Therefore, we see that the angular momentum dependence of the transitional dipole moment is determined by the matrix element $\langle i||\beta_{30}||f\rangle$, where vectors $|i\rangle$ and $|f\rangle$ denote the initial and final states, respectively. The lowest negative parity eigenfunction of the Hamiltonian (1) can be found by the numerical solution of the Schrödinger equation with potential (4). However, having the aim to get the result in the analytical form we supplemented the ansatz (2) for the positive parity ground state wave function by the expression for the lowest negative parity wave function as

$$\begin{aligned}\Psi_I^{(-)}(\beta_{30}) = \left(\frac{\xi^2}{\pi\beta_m^2(I)}\right)^{1/4} & (2\{1 - \exp[-\xi(I)^2]\})^{-1/2} \\ & \times \left(\exp\left(-\frac{1}{2}\xi^2(I)(\beta_{30}/\beta_m - 1)^2\right)\right. \\ & \left. - \exp\left(-\frac{1}{2}\xi^2(I)(\beta_{30}/\beta_m + 1)^2\right)\right). \end{aligned} \tag{39}$$

This form is confirmed by the numerical calculation. Note that in Ref. 33, the methods based on supersymmetric quantum mechanics have been used.

Using the ansatz (2) and (39) for the intrinsic wave functions of the positive and negative parity members of the alternating parity bands we can find an analytical expression for the angular momentum dependence of the matrix element of β_{30}, namely

$$\langle i||\beta_{30}||f\rangle = \sqrt{\frac{\hbar}{B\omega}} \frac{\sqrt{2}e^{\frac{(\xi_f^2-\xi_i^2)^2}{2(\xi_f^2+\xi_i^2)}}(\xi_f\xi_i)^{1/2}}{\sqrt{(e^{\xi_f^2}-1)(e^{\xi_i^2}+1)}} \times \frac{(\xi_f^2 - \xi_i^2 + e^{\frac{2\xi_f^2\xi_i^2}{\xi_f^2+\xi_i^2}}(\xi_f^2+\xi_i^2))}{(\xi_f^2+\xi_i^2)^{3/2}}. \tag{40}$$

The last expression can be simplified if we assumed that approximately $\xi_i \approx \xi_f = \xi$. Then

$$\langle i||\beta_{30}||f\rangle = \sqrt{\frac{\hbar}{B\omega}} \frac{\xi e^{\xi^2}}{\sqrt{e^{2\xi^2} - 1}}. \tag{41}$$

From Eq. (41), we see that in the vibrational limit of the octupole motion ($\xi \ll 1$), we have

$$\langle i||\beta_{30}||f\rangle \approx \sqrt{\frac{\hbar}{2B\omega}}. \tag{42}$$

For large values of ξ, the dipole moment is an increasing function of ξ. This increase is almost linear for $\xi > 1$ and we obtain

$$\langle i|\beta_{30}|f\rangle \approx \beta_m, \quad (\xi \gg 1). \tag{43}$$

As follows from the definition of the dipole operator in Eq. (38), the behavior of matrix element $\langle i||\beta_{30}||f\rangle$ determines the angular momentum dependence of the dipole reduced transition probability

$$B(E1, i \to f) = B(E1, 0^+ \to 1^-)\frac{\xi^2 e^{2\xi^2}}{e^{2\xi^2} - 1}, \tag{44}$$

where $\xi = \sqrt{\xi_i \xi_f}$.

Using the obtained values of the critical angular momenta I_{crit}, one can calculate the angular momentum dependence of the reduced matrix element for $E1$ transitions. In Fig. 6, the results for the ^{226}Ra are presented. The calculated values are compared with the experimental data on dipole moment obtained in Ref. 48. As seen, matrix elements $\langle I+1||M(E1)||I\rangle$ increase with angular momentum, following the increase of $\langle i||\beta_{30}||f\rangle$, because of their proportionality. As we can see, the calculation with use of Eq. (44) agree well with the experimental data.

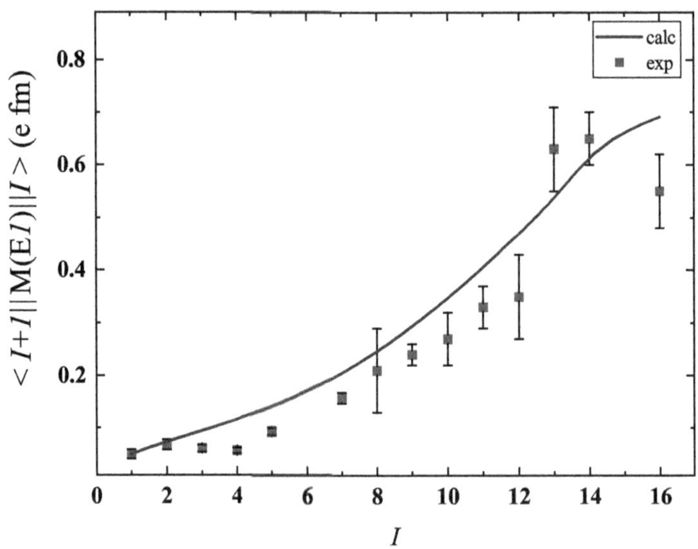

Fig. 6. Dependence of the calculated and experimental values of the $E1$ matrix element on the angular momentum I for ^{226}Ra.

8. Conclusion

The interpretation of the experimental data on the properties of the alternating parity bands in heavy nuclei is suggested. This interpretation is based on the assumption that the main role plays the reflection-asymmetric mode which preserves the axial symmetry. This mode can be related to the octupole deformation of the nuclear shape or to the mass asymmetry parameter of a dinuclear system. The collective potential corresponding to the reflection-asymmetric mode has at low angular momentum I one minimum located at the zero value of the reflection-asymmetric deformation. However, with increasing I the potential becomes more and more flat and, finally, at I exceeding some critical value the barrier appears which separates two minima symmetrically located at nonzero deformation. It is shown that this approach provides a good description of the experimental data on parity splitting and electric dipole transitions both below and above the critical angular momentum.

It is shown that the experimental data on the spectra of the ground state alternating parity bands indicate a second-order phase transition from reflection-symmetric to reflection-asymmetric shape in these bands. The phase transition takes place at some critical value of the angular momentum which in our consideration plays the role of the control parameter.

The nondimensional characteristic of the ground state alternating parity band ($R_{\text{oct}}(I)$) is introduced to characterize stabilization of the reflection-asymmetric deformation with angular momentum increase. The analysis of the experimental data has shown that in all considered nuclei R_{oct} is close to zero at small I and asymptotically comes nearer to unity at large I. This indicates that at large I positive and negative parity states form a unified molecular-type band. The angular momentum dependence of R_{oct} characterizes a phase transition from nondeformed shape at low I to reflection-asymmetric deformed shape at high I. It is also shown that transitional nuclei become reflection-asymmetrically deformed at smaller values of the angular momentum than axially symmetric nuclei with static quadrupole deformation.

Based on the suggested approach the approximate analytical expressions for the angular momentum dependence of the parity splitting and the electric dipole transitional moment are obtained. These analytical expressions contain a small number of the parameters with the clear physical meaning. These parameters can be fitted according to the experimental data or calculated using microscopical model.

References

1. P. Butler and W. Nazarewicz, *Rev. Mod. Phys.* **68** (1996) 349.
2. F. Asaro, F. S. Stephens, Jr. and I. Perlman, *Phys. Rev.* **92** (1953) 1495.
3. F. S. Stephens, Jr., F. Asaro and I. Perlman, *Phys. Rev.* **100** (1955) 1543.
4. X. C. Chen *et al.*, *Phys. Rev. C* **94** (2016) 021301(R).
5. B. Bucher *et al.*, *Phys. Rev. Lett.* **116** (2016) 112503.
6. B. Bucher *et al.*, *Phys. Rev. Lett.* **118** (2017) 152504.
7. L. P. Gaffney *et al.*, *Nature* **497** (2013) 199.
8. P. Spagnoletti *et al.*, *Phys. Rev. C* **105** (2022) 024323.
9. P. A. Butler *et al.*, *Nat. Commun.* **10** (2019) 2473.
10. P. A. Butler *et al.*, *Nat. Commun.* **11** (2020) 3560.
11. P. A. Butler *et al.*, *Phys. Rev. Lett.* **124** (2020) 042503.
12. M. Spieker *et al.*, *Phys. Rev. C* **88** (2013) 041303(R).
13. M. Spieker *et al.*, *Phys. Rev. C* **97** (2013) 064319.
14. I. Ahmad and P. A. Butler, *Annu. Rev. Nucl. Part. Sci.* **43** (1993) 71.
15. R. K. Sheline and B. B.-M. Rossinga, *Phys. Rev. C* **44** (1991) 218.
16. R. K. Sheline and M. A. Riley, *Phys. Rev. C* **61** (2000) 057301.
17. V. G. Soloviev and P. Vogel, *Phys. Lett.* **6** (1963) 126.
18. L. A. Malov, V. G. Soloviev and P. Vogel, *Phys. Lett.* **22** (1966) 441.
19. K. Neergard and P. Vogel, *Nucl. Phys. A* **145** (1070) 33.
20. R. G. Nazmitdinov, I. N. Mikhailov and C. Briancon, *Phys. Lett. B* **188** (1987) 171.
21. G. A. Leander *et al.*, *Nucl. Phys. A* **388** (1982) 452.
22. W. Nazarewicz *et al.*, *Nucl. Phys. A* **429** (1984) 269.
23. P. Bonche, P. H. Heenen, M. Flocard and D. Vautherin, *Phys. Lett. B* **175** (1986) 387.
24. L. M. Robledo, J. L. Egido, J. F. Berger and M. Girod, *Phys. Lett. B* **187** (1987) 223.
25. J. Egido and L. Robledo, *Nucl. Phys. A* **494** (1989) 85.
26. W. D. M. Rae, *Int. J. Mod. Phys. A* **3** (1988) 1343.
27. M. Freer and A. C. Merchant, *J. Phys. G* **23** (1997) 261.
28. W. Nazarewicz *et al.*, *Phys. Lett. B* **322** (1994) 304.
29. S. Aberg and L.-O. Jonsson, *Z. Phys. A* **349** (1994) 205.
30. A. V. Afanasjev, H. Abusara and S. E. Agbemava, *Phys. Scr.* **93** (2018) 034002.
31. R. V. Jolos and P. von Brentano, *Phys. Rev. C* **49** (1994) R2301.
32. R. V. Jolos, N. Minkov and W. Scheid, *Phys. Rev. C* **72** (2005) 064312.
33. R. V. Jolos and P. von Brentano, *Phys. Rev. C* **84** (2011) 024312.
34. H. Kalka and G. Soff, *Supersymmetrie*, *Teubner Studienbücher Physik* (Teubner, Stuttgart, 1997).
35. A. Frank, P. Van Isacker and J. Jolie, *Symmetries in Atomic Nuclei: From Isospin to Supersymmetry*, *Springer Tracks in Physics*, Vol. **230** (Springer, Cham, 2009).
36. F. Cooper, A. Khare and U. Sukhatme, *Supersymmetry in Quantum Mechanics* (World Scientific, Singapore, 2004).
37. X. Wang *et al.*, *Phys. Rev. Lett.* **102** (2009) 122501.

38. G. G. Adamian, N. V. Antonenko and W. Scheid, *Clusters in Nuclei, Vol. 2*, ed. C. Beck, *Lecture Notes in Physics*, Vol. **848** (Springer-Verlag, Berlin, 2012), p. 165.
39. T. M. Shneidman, G. G. Adamian, N. V. Antonenko, R. V. Jolos and S.-G. Zhou, *Phys. Rev. C* **92** (2015) 034302.
40. A. B. Migdal, *Theory of Finite Fermi Systems and Applications to Atomic Nuclei* (Wiley, New York, 1967).
41. T. M. Shneidman, G. G. Adamian, N. V. Antonenko, S. P. Ivanova and W. Scheid, *Nucl. Phys. A* **671** (2000) 119.
42. B. Singh and E. Browne, *Nucl. Data Sheets* **109** (2008) 2439.
43. R. V. Jolos, P. von Brentano and R. F. Casten, *Phys. Rev. C* **88** (2013) 034306.
44. L. Landau, *Phys. Z. Sowjetunion* **11** (1937) 26.
45. L. D. Landau and E. M. Lifshitz, *Statistical Physics, Part 1*, *Course of Theoretical Physics*, Vol. **V** (Butterworth-Heinemann, Oxford, 2001).
46. R. V. Jolos, P. von Brentano and J. Jolie, *Phys. Rev. C* **86** (2012) 024319.
47. R. V. Jolos and P. von Brentano, *Phys. Rev. C* **92** (2015) 044318.
48. H. J. Wollersheim *et al.*, *Nucl. Phys. A* **556** (1993) 261.
49. E. V. Mardyban, T. M. Shneidman, E. A. Kolganova, R. V. Jolos and S.-G. Zhou, *Chin. Phys. C* **42** (2018) 124104.
50. I. Wiedenhöver *et al.*, *Phys. Rev. Lett.* **83** (1999) 2143.
51. V. M. Strutinsky, *J. Nucl. Energy* **4** (1957) 523.

Ground state and fission properties of even-A uranium isotopes from multidimensionally-constrained relativistic mean field model*

Xiang-Quan Deng[†,‡] and Shan-Gui Zhou[†,‡,§,¶,‖,**]

[†]*CAS Key Laboratory of Theoretical Physics,*
Institute of Theoretical Physics,
Chinese Academy of Sciences,
Beijing 100190, P. R. China

[‡]*School of Physical Sciences,*
University of Chinese Academy of Sciences,
Beijing 100049, P. R. China

[§]*School of Nuclear Science and Technology,*
University of Chinese Academy of Sciences,
Beijing 100049, P. R. China

[¶]*Synergetic Innovation Center for Quantum*
Effects and Application,
Hunan Normal University,
Changsha 410081, P. R. China

[‖]*Peng Huanwu Collaborative Center*
for Research and Education,
Beihang University,
Beijing 100191, P. R. China
[**]*sgzhou@itp.ac.cn*

The multidimensionally-constrained covariant density functional theories (MDC-CDFTs) have been developed to study the influence of octupole and triaxial deformations on the ground state and fission properties. In this paper, we present a brief review of the applications of MDC-CDFTs and discuss the results of a systematical study of even-A uranium isotopes with the multidimensionally-constrained relativistic mean field (MDC-RMF) model which is one of the MDC-CDFTs with pairing correlations treated by using the Bardeen-Cooper-Schrieffer (BCS) approach. We examine in detail the two-dimensional potential energy

*To cite this article, please refer to its earlier version published in the *International Journal of Modern Physics E*, Volume 32(10), (20 pages), 2340004 (2023), DOI: 10.1142/S0218301323400049.

**Corresponding author.

surfaces $E(\beta_{20}, \beta_{30})$ of these U isotopes and discuss the ground state and fission properties as well as the third and fourth minima on the potential energy surfaces. The emphasis is put on the effects of octupole and triaxial deformations.

Keywords: Multidimensionally-constrained relativistic mean field model; fission barrier; octupole deformation; triaxial deformation.

1. Introduction

Atomic nuclei are quantum many-body systems consisting of nucleons, i.e., protons and neutrons. The ground states of nuclei are characterized by structural properties such as the spin and parity I^π, mass excess or binding energy, size and shape, which are governed by both the nucleon-nucleon interaction and many-body features from which various nuclear phenomena emerge.[1,2]

In the intrinsic frame, many nuclear shapes may appear and manifest themselves in the low-lying collective spectra: The rotational spectrum $E(I) \sim I(I+1)$ corresponds to an axial quadrupole deformation;[3,4] a static octupole deformation results in the parity doublet bands[1,5–8]; the triaxial quadrupole deformation is characterized by the chiral doublet bands[9,10] or the wobbling motion[1] in certain nuclei. More exotic intrinsic nuclear shapes have been explored extensively, e.g., the hyperdeformed shapes, see Ref. 11 and references therein, the rod shapes,[12] the tetrahedral or octahedral shapes,[13–15] the triangle shape[16] and the toroidal or ring shapes.[17,18]

Nuclear fission — the large amplitude collective motion — can be described as the evolution of nuclear shape in a multidimensional deformation space.[19–23] The potential energy surface (PES) in such a deformation space is crucial for the study of nuclear fission.[11,23–30] In particular, the characteristics of fission barrier, e.g., the height and width are important inputs for theoretical models of nuclear fission.[31–38] For instance, in the statistical models for calculating the survival probability of hot compound superheavy nuclei (SHN), the competition between the neutron evaporation and fission is mainly determined by the neutron separation energy and fission barrier height.[39–43] It has been revealed that, besides the axial quadrupole deformation which is the most important shape degree of freedom in describing nuclear fission, the nonaxial and reflection asymmetric deformations are crucial as well.[22,23]

Many theoretical models have been developed and used to study the potential energy surfaces (PESs) and barrier heights, including the macroscopic-microscopic (MM) models,[44–54] the extended Thomas–Fermi plus Strutinsky intergral (ETFSI) method[55] and the density functional theories (DFTs).[56–63] The multidimensionally-constrained (MDC) covariant density functional theories (CDFTs) have been developed for the study of ground state and fission properties as well as the PESs.[15,27,60] In MDC-CDFTs, both nonaxial and reflection asymmetric deformations are considered self-consistently. The MDC-CDFTs include the MDC relativistic mean field (MDC-RMF) model in which the pairing correlations are treated by using the Bardeen-Cooper-Schrieffer (BCS) approach and the MDC relativistic Hartree–Bogoliubov (MDC-RHB) model.

In the present work, by using the MDC-RMF model, we study systematically the PESs of even-A U isotopes from the two-proton drip line up to the two-neutron drip line. We study the ground state properties and the primary barrier heights and examine the effects of octupole and triaxial deformations. The third and fourth minima appearing on the PESs of some U isotopes are also discussed.

The paper is organized as follows. The applications of the MDC-CDFTs are briefly reviewed in Sec. 2. In Sec. 3, we present numerical results of the PESs and discuss in detail the ground state and fission properties of even-A U isotopes. A summary of this work is given in Sec. 4.

2. Brief Review of MDC-CDFTs

The CDFT has been very successful in self-consistent descriptions of atomic nuclei throughout almost the whole chart of nuclides.[60,64–74] In the CDFT, one usually derives the equations of motion of nucleons from the Lagrangian incorporating the nucleon fields and the interaction between nucleons which is realized either via meson exchanges (ME) or through point couplings (PC).

In the MDC-CDFTs, the equations of motion for nucleons are solved with the basis expansion method. For axially deformed nuclei with the reflection asymmetry, a two-center basis would be more appropriate; e.g., the reflection asymmetric RMF (RAS-RMF) model[75] has been developed in a two-center harmonic-oscillator basis and used to study octupole deformations in ^{226}Ra[75] and Sm,[76] Ba,[77] Th[78] and Dy[79] isotopes. However, in the MDC-CDFTs, the reflection symmetric, axially deformed harmonic oscillator (ADHO) basis[80] was adopted for convenience. In Ref. 80, the nuclei in question were assumed to be spherical or axially deformed with the reflection symmetry, i.e., in the following multipole expansion of the nuclear surface, only even λ is considered and $\mu = 0$,

$$R(\theta, \varphi) = R_0 \left[1 + \sum_{\lambda=2}^{\infty} \sum_{\mu=-\lambda}^{\lambda} \beta_{\lambda\mu}^{*} Y_{\lambda\mu}(\theta, \varphi) \right]. \tag{1}$$

In the MDC-CDFTs, the V_4 symmetry was imposed when solving the equations of motion for nucleons. Thus all deformations $\beta_{\lambda\mu}$ with even μ, including nonaxial and reflection asymmetric deformations, are included self-consistently. The pairing correlations are treated by using the BCS method in the MDC-RMF model[27] and by implementing the Bogoliubov transformation in the MDC-RHB model.[15] For the details of the MDC-CDFTs, the readers are referred to Refs. 15, 27 and 60.

The MDC-CDFTs have been extensively used to study normal and hypernuclei, with emphasis on the properties concerning various deformation effects on nuclear properties including the ground state properties, PESs, superdeformed and hyperdeformed states and fission properties. Next, we present a brief review of the applications and further developments of the MDC-CDFTs.

Fission barriers of actinide nuclei The MDC-CDFTs have been applied to the study of fission barriers of actinide nuclei.[23,27] The typical double-humped fission barriers in actinides were well reproduced. Besides the lowering effects of the triaxial distortion on the first barrier and the large influence of the octupole deformation on the second barrier, it was found that the triaxial deformation also lowers the second barrier considerably. When both triaxial and octupole deformations are included, the calculated fission barriers heights agree satisfactorily with the available empirical values.

Third minima in PESs of light actinides The PESs of light actinides were carefully examined and the third minima were investigated.[11] The origin of these minima, corresponding to hyperdeformed states, has been attributed to the $Z = 90$ proton shell gap at very large deformations.

Shapes and PESs of superheavy nuclei One- and two-dimensional PESs have been obtained for ^{270}Hs — a deformed doubly magic SHN.[81] The influences of the nonaxial and reflection asymmetric distortions on the fission barrier and fission pathway have been investigated: When the axial symmetry is imposed, the reflection symmetric and asymmetric fission paths both show a double-humped structure and the latter is energetically favored; when nonaxial shapes are allowed, the reflection symmetric fission pathway becomes favorable. The higher-order deformation effects on the ground state properties of SHN of and near ^{270}Hs have been studied.[82] It was found that among higher-order deformations β_λ ($\lambda = 4, 6, 8$ and 10), β_6 has the greatest impact.

Nonaxial octupole Y_{32} correlations The nonaxial reflection-asymmetric β_{32} shapes in transfermium nuclei with $N = 150$ have been studied.[83] It was found that in these nuclei, the origin of the Y_{32} correlations is mainly from a pair of neutron orbitals, $[734]9/2(\nu j_{15/2})$ and $[622]5/2(\nu g_{9/2})$, and a pair of proton orbitals, $[521]3/2(\pi f_{7/2})$ and $[633]7/2(\pi i_{13/2})$. The tetrahedral shapes in neutron-rich even-even Zr isotopes have been investigated.[15] The tetrahedral ground states are mainly caused by large shell gaps around $Z = 40$ and $N = 70$.

Axial octupole Y_{30} correlations The MDC-CDFTs have been used to investigate the coexistence of chirality and octupole correlations in ^{76}Br[84] and ^{78}Br[85,86]; the latter is the first example of chiral geometry in octupole soft nuclei. The octupole correlations in ^{123,125}Ba,[87] ^{73}Br[88] and ^{71}Ge[89] have been revealed with the MDC-CDFTs.

Structure of hypernuclei The structure of hypernuclei has been studied by using the MDC-CDFTs.[90–95] For brevity, we only briefly mention two works concerning the deformation effects. In the study of the shapes of light normal and Λ hypernuclei, it was found that the shape polarization effect of Λ is so strong that the shapes of some Λ hypernuclei, e.g., $^{13}_{\Lambda}$C, $^{23}_{\Lambda}$C and $^{31}_{\Lambda}$Si, are drastically different from the corresponding cores.[90] The superdeformed (SD) states and corresponding SD Λ hypernuclei of Ar isotopes were examined. A strong localization effect with a ring structure was predicted in density distributions of the SD states, resulting in a larger Λ separation energy in SD states than in the ground states.[91]

Fission dynamics and α decays Based on the PESs from the MDC-CDFTs, the dynamics of spontaneous and induced fissions in actinide nuclei has been studied extensively.[96–105] For example, in Ref. 97, the spontaneous fission dynamics of ^{264}Fm and ^{250}Fm was explored and it was concluded that the inclusion of pairing correlations in the space of collective coordinates favors axially symmetric shapes along the dynamic pathway of the fissioning system. Such dynamic studies tell us more about the role played by the various deformations in fission than static studies focusing only on PESs. Recently, microscopic investigations of half-lives for α decays in ^{108}Xe and ^{104}Te[106] and α and 2α decays in ^{212}Po and ^{224}Ra[107] have also been performed based on the PESs calculated from the MDC-CDFTs.

Angular momentum and parity projections The angular momentum projection (AMP) and parity projection (PP) have been implemented in the MDC-RHB model to restore the rotational and parity symmetries which are both broken in the mean-field level. Such a projected MDC-RHB (p-MDCRHB) model was presented in Ref. 16. With the p-MDCRHB model, one may study nuclear spectra corresponding to exotic intrinsic shapes, e.g., triangle or tetrahedron. In Ref. 108, an anatomy of octupole correlations in ^{96}Zr has been performed with the p-MDCRHB model. It was found that the PESs of this nucleus are strongly dependent on the angular momentum and parity and both triaxial and octupole deformations should be included in order to give a decent description of the structure of ^{96}Zr.

3. Results and Discussions

In this section, we systematically study the properties of even-A U isotopes with the MDC-RMF model. First, we present and discuss two-dimensional PESs of several selected U isotopes. Second, we focus on the ground state properties of the isotopes and show the density distribution profiles of several typical nuclei with deformed ground state shapes. Third, for each isotope, we perform calculations in the vicinity of each saddle point with the reflection symmetry and the axial symmetry broken. The primary barrier heights with octupole and triaxial deformations are compared with the available empirical values and/or calculation results of other models. Finally, we discuss hyperdeformed third and even fourth minima on the PESs of several U isotopes.

In our calculations, the effective interaction PC-PK1[109] is used. The truncation in the expansion of the large component of the Dirac spinor $N_{\mathrm{f}} = 20$ and that for the small component $N_{\mathrm{g}} = 21$. The quadrupole deformation parameter of the ADHO basis is chosen to be half of the desired β_{20} value. The pairing correlations are treated with the BCS approach and the separable pairing force of finite-range[110–112] is used. The pairing strength and effective range are taken as $G = 1.1G_0$ with $G_0 = 728\,\mathrm{MeV\,fm^3}$ and $a = 0.644\,\mathrm{fm}$.

3.1. *Two-dimensional PESs*

We calculate 2D PESs $E(\beta_{20}, \beta_{30})$ of the U isotopes, from the two-proton drip line ^{214}U, the lightest known uranium isotope,[113] to the two-neutron drip line ^{350}U. In these calculations the nuclei are assumed to be axially symmetric. For each isotope, β_{20} runs from -0.20 to 2.00 and β_{30} from 0.00 to 1.00, both with a step size of 0.05. The PESs with negative β_{30} value are obtained through the relation $E(\beta_{20}, \beta_{30}) = E(\beta_{20}, |\beta_{30}|)$.

The PESs of several selected isotopes are shown in Fig. 1 where the ground states and the highest saddle points are shown by red dots and red

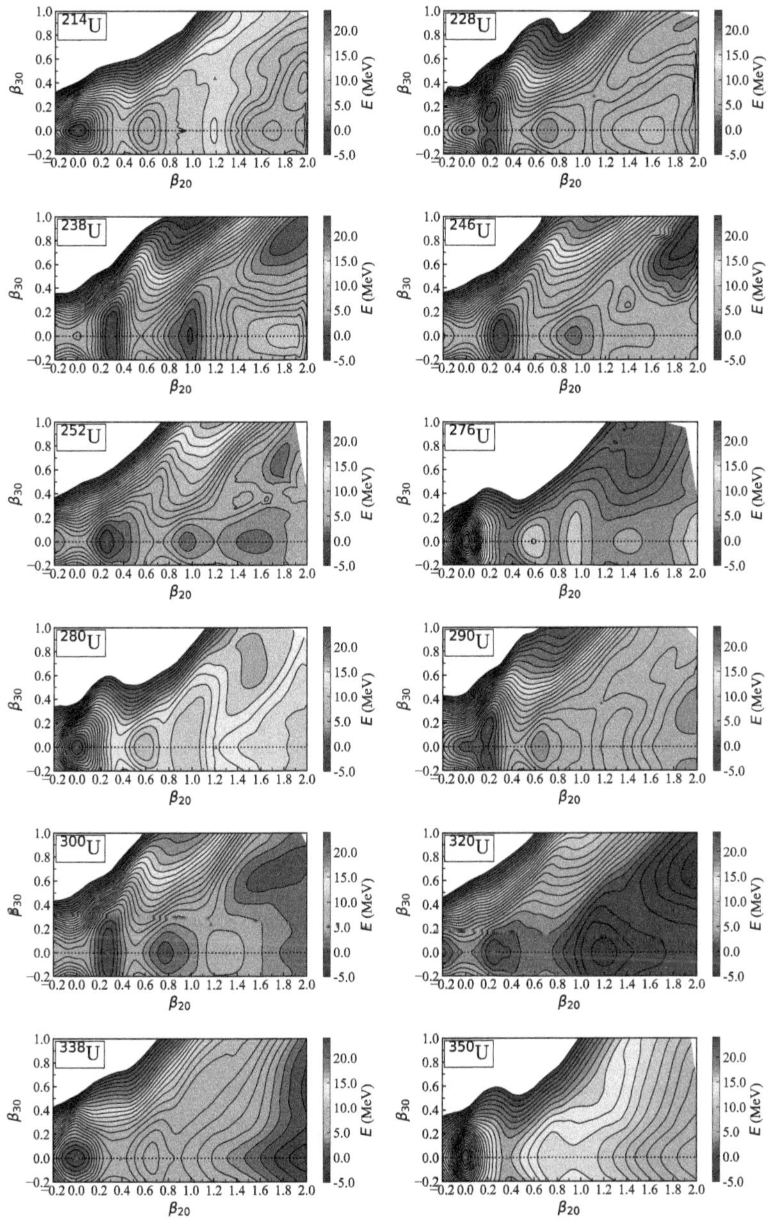

Fig. 1. (Color online) 2D PESs of selected U isotopes obtained in the MDC-RMF calculations with the effective interaction PC-PK1.[109] The ground states are indicated by red dots. The saddle points corresponding to primary barriers are shown by red triangles. The contour interval is 1.0MeV.

triangles. The 2D PESs of U isotopes are mostly smooth and continuous with obvious minima and saddle points and the static fission pathways can be easily identified. First let us take ^{214}U as an example. There is a reflection-symmetric (RS) fission pathway and a reflection-asymmetric (RA) one on the PES, both starting from a spherical ground state. There are three saddle points on the RS pathway and two on the RA pathway. Two shallow third minima can be seen, meaning that ^{214}U may fission through both RS and RA pathways.

By examining the PESs of $^{214-250}$U, it is found that the RA fission pathways are energetically favorable, as can be clearly seen in Fig. 1. For instance, the PES of ^{238}U is featured by a RS inner barrier, a low RA outer barrier and a high RS outer barrier which strongly hinder the nucleus from fissioning through the RS pathway. Specially, for $^{242-248}$U there are two RA fission pathways starting from the second minimum; an example, ^{246}U, is shown in Fig. 1. Similar results were discussed for ^{248}Cm in Ref. 23 (Fig. 5). For $^{252-260}$U, we can find two saddle points on the PESs, corresponding to the typical double-humped barriers in actinide nuclei. In these U isotopes, besides the lowest static RS fission path, there is also a RA one, as seen in the PES of ^{252}U in Fig. 1. For $^{268-282}$U, the fission barriers are very high. There are three or four RS barriers on the PESs and the potential wells between the barriers may correspond to hyperdeformed fission isomers. The hyperdeformed third or fourth minima will be discussed in Sec. 3.4. The RS and RA pathways also appear on the PESs of the isotopes $^{286-334}$U. From the PESs of $^{336-350}$U, it can be expected that these isotopes fission mainly through the RS pathway.

We can readily locate the ground states of the U isotopes from the 2D PESs. For most of the isotopes, the ground states are reflection symmetric. But for $^{226-232}$U and $^{288-294}$U, the β_3 values are nonzero and are around 0.2 for ^{228,290}U as seen in Fig. 1. Detailed ground state properties will be discussed in the next subsection.

3.2. *Ground state properties*

The two-neutron separation energies S_{2n} and root-mean-square (rms) radii of neutron, proton, matter and charge distributions for U isotopes are shown in Fig. 2.

In Fig. 2(a), the two-neutron separation energies are compared with available experimental or evaluated values from the AME2020[114,115] and the results calculated from the deformed relativistic Hartree–Bogoliubov theory in continuum (DRHBc).[116] Since the reflection symmetry is imposed in the

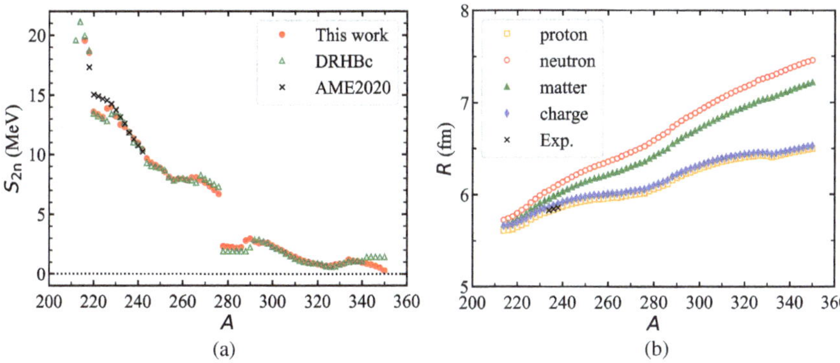

Fig. 2. (Color online) Ground state properties of U isotopes obtained in the MDC-RMF calculations with the effective interaction PC-PK1[109]: (a) Two-neutron separation energies S_{2n} compared with available values from the AME2020[114,115] and the DRHBc mass table[116]; (b) root-mean-square (rms) radii of neutron, proton, matter and charge distributions as functions of the mass number A. Experimental data for charge radii[117] are shown by the black crosses in (b).

DRHBc theory, the octupole deformations in the ground states of some U isotopes would certainly result in differences in the binding and separation energies between the MDC-RMF model and the DRHBc theory. The pairing approach and pairing interaction used in these two models are different, which would also lead to some differences in the ground state properties. For ^{218}U, both the MDC-RMF model and the DRHBc theory overestimate the experimental S_{2n} values and for $^{220-226}$U, they underestimate the experiments. In other cases, the MDC-RMF and the DRHBc S_{2n} values are in agreement with the AME2020 whenever available. As A increases, the S_{2n}'s from both models share the same trend. For $342 \leq A \leq 350$, the MDC-RMF model predicts that S_{2n} decreases monotonically with A increasing and ^{350}U is the last bound even-A U isotope. However, in the DRHBc mass table, $S_{2n} = 1.4\,\text{MeV}$ for these five U isotopes though ^{350}U is also predicted to be the last bound even-A U isotope. Such differences concerning S_{2n} between these two models and the different predictions concerning the lightest bound U isotopes, i.e., ^{212}U by the DRHBc theory[116] and ^{214}U by the MDC-RMF model, may stem from that the DRHBc theory can provide a proper description of exotic nuclei by considering the deformation and continuum effects.[118–120]

The root-mean-square (rms) radii of neutron, proton, matter and charge distributions of the ground states are shown in Fig. 2(b). The available experimental values for the charge radii[117] are also included for comparison. The neutron radii grows faster than proton as the nucleus gets heavier, leading to thicker neutron skins. The charge radius $R_{\rm ch}$ is a significant observable characterizing the size of a nucleus. For ^{234}U, ^{236}U and ^{238}U, the experimental values for charge radii are 5.829 fm, 5.843 fm and 5.857 fm, respectively. The charge radii calculated with the MDC-RMF model for the three isotopes are 5.864 fm, 5.882 fm and 5.900 fm and are agree with the data within 1%. However, it is clear that the MDC-RMF model overestimates the data systematically. Note that such systematic overestimation of the charge radii for ^{234,236,238}U ($R_{\rm ch} = 5.863$ fm, 5.882 fm and 5.897 fm) also exists in the DRHBc calculations.[116] Although an accuracy of 1% is acceptable for microscopic models like MDC-RMF and DRHBc, such systematic discrepancy may hint some insights concerning the possible improvements of these models.

As we have mentioned in the previous subsection, quite a large number of U isotopes have nonspherical ground state shapes. We plot the quadrupole and octupole deformation parameters (β_{20} and β_{30}) of the ground states in Fig. 3. It can be seen that spherical, prolate RA and prolate RS ground state shapes appear alternately as the mass number increases. For $^{214-224}$U, $^{264-286}$U and $^{334-350}$U, the ground state shapes are spherical. For $^{226-262}$U, β_{20} increases to a maximum then decreases gradually, corresponding to various prolate shapes. In this mass region, β_{20} peaks at ^{242}U with the maximal value of 0.30. Similarly, in the region of $^{288-324}$U, ^{306}U and ^{308}U

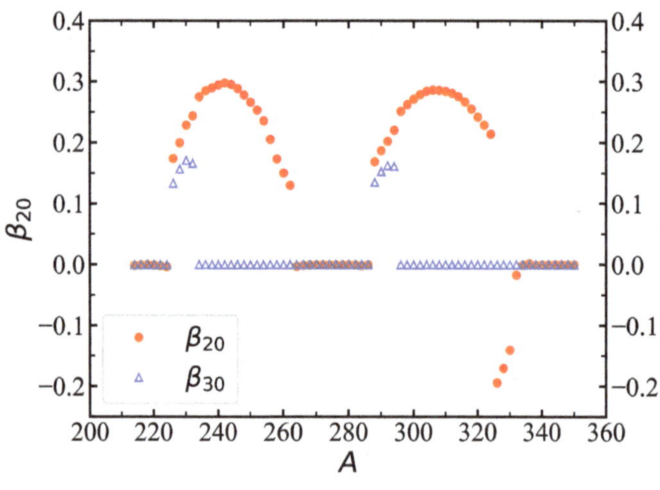

Fig. 3. (Color online) Quadrupole and octupole deformation parameters of even-A U isotopes as functions of the mass number A obtained in the MDC-RMF calculations with the effective interaction PC-PK1.[109]

both have prolate ground state shapes with $\beta_{20} = 0.29$. From ^{326}U to ^{332}U, the ground state evolves from a largely deformed oblate shape to a nearly spherical one, with $\beta_{20} = -0.19$ and -0.02 for ^{326}U and ^{332}U, respectively. From Fig. 3, we find that for eight isotopes, i.e., $^{226-232}$U and $^{288-294}$U, both the quadrupole and octupole deformations appear in their ground states. These isotopes are featured by pear-shaped ground states.

In Fig. 4, we display density distributions of four typical nuclei with different kinds of ground state shapes. The ground states of ^{224}U, ^{242}U and ^{326}U have a spherical shape, an axially symmetric prolate shape and an axially symmetric oblate shape, respectively. The ground state of ^{230}U has a pear shape with $\beta_{20} = 0.23$ and $\beta_{30} = 0.17$.

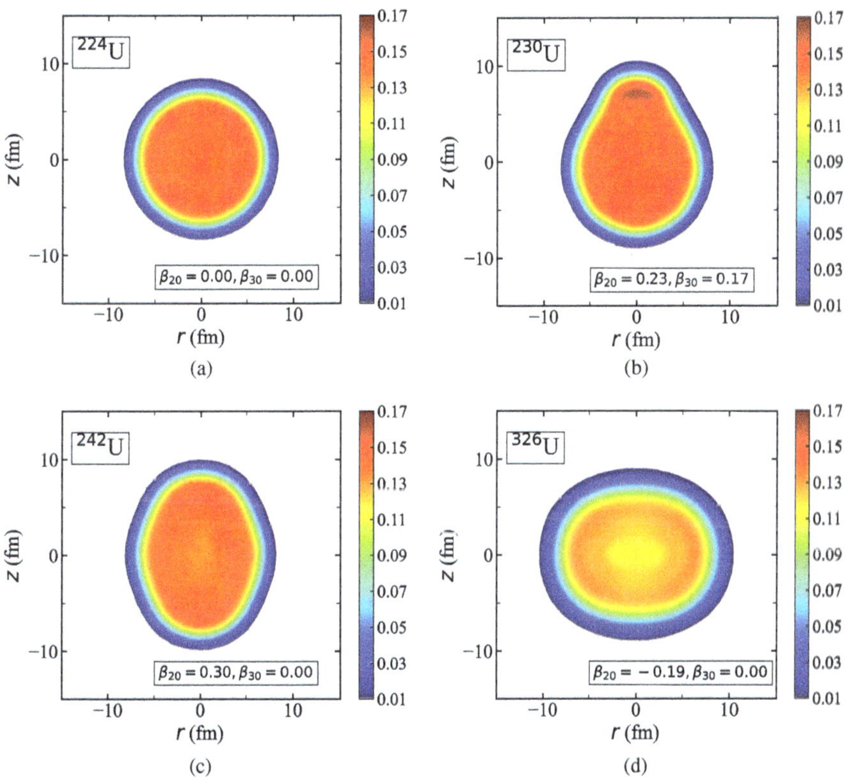

Fig. 4. (Color online) Density distribution profiles of the ground states of (a) ^{224}U, (b) ^{230}U, (c) ^{242}U and (d) ^{326}U obtained in the MDC-RMF calculations with the effective interaction PC-PK1.[109] The z-axis is the symmetry axis. The quadrupole and octupole deformation parameters of the nuclear shapes are shown.

3.3. *Primary barrier heights calculated with octupole and triaxial deformations considered*

Next, we focus on the primary barrier, the highest one, in each even-A U isotope. The height of the primary barrier is defined as the energy difference between the highest saddle point and the ground state. With the calculation results of the 2D PESs we can find all saddle points for each isotope and determine the primary barrier height. As seen in the PESs, for many isotopes, the primary barrier in the RA pathway is obviously lower than the primary barrier in the RS pathway, which means that the primary barrier height of the nuclei is lowered when the octupole distortion is allowed.

As we have mentioned, the triaxiality may further lower the height of a barrier.[22,23] So in addition to the octupole deformation, we must take the triaxial deformation into consideration to give a more accurate description of the primary barrier heights. Since the MDC-RMF calculations are very time-consuming in the whole $(\beta_{20}, \beta_{22}, \beta_{30})$ deformation space, we break the axial symmetry only in the vicinity of each saddle point to take the lowering effect of triaxiality on the barriers into account. With the octupole deformation considered, the primary barrier heights of U isotopes with and without triaxiality are shown in Fig. 5(a). For most of the U isotopes, the triaxiality lowers the height of the primary barrier. The primary barrier may "shift" from one barrier to another when the triaxiality is considered because the lowering effects of the triaxial distortion for different barriers may be different. But the general trend of the primary barrier heights versus A is unchanged. No matter whether we break the axial symmetry or not, the barrier height peaks at three isotopes with $N = 126$, 150 and 184, corresponding to neutron magic numbers and neutron subshells. For $N = 184$, i.e., ^{276}U, the primary barrier height is as large as 17.48 MeV, leading to a relatively high stability against fission.

We compare the primary barrier heights with available empirical values taken from RIPL-3[121] in Fig. 5(b). The calculated primary barrier heights agree with the RIPL-3 data for ^{232}U, ^{236}U and ^{238}U; in both our calculation and RIPL-3, for ^{232}U and ^{236}U the primary barrier is the outer one and for ^{238}U the inner one. The barrier height for ^{234}U is reproduced by our calculation, but we predict the inner barrier as the primary one for this isotope.

Based on the FUNF program,[122,123] the fission properties of some actinides have been investigated in detail.[124] For each isotope, the empirical heights of the inner, middle and outer barriers are given. The empirical

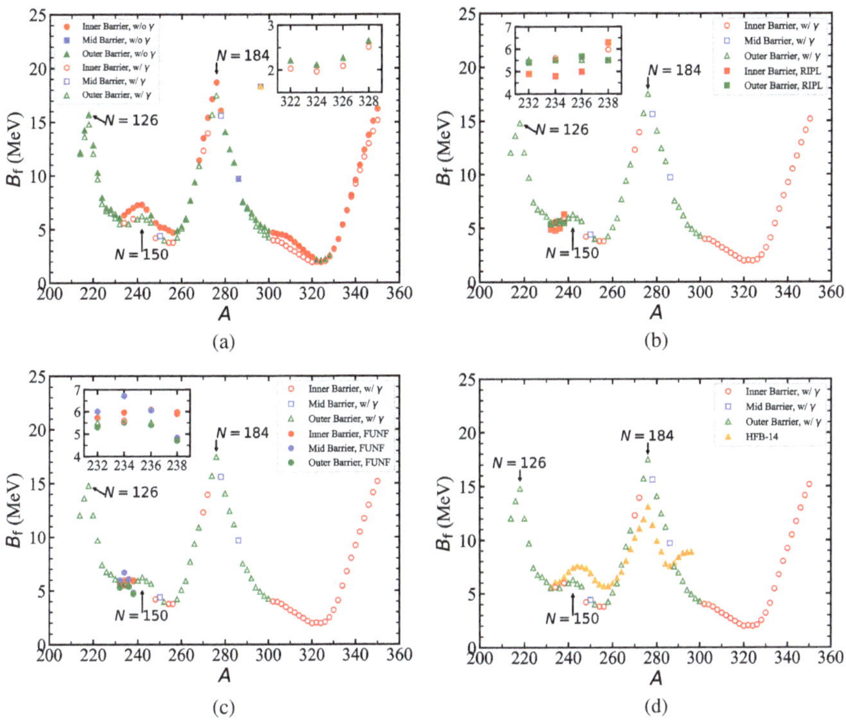

Fig. 5. (Color online) (a) Primary barrier heights with and without triaxiality obtained in the MDC-RMF calculations with the effective interaction PC-PK1[109] and comparisons of the barrier heights (b) with empirical values from RIPL-3,[121] (c) with the results given by the FUNF program[122–124] and (d) with HFB-14 calculations.[57] For the MDC-RMF calculation results, we symbolize the barrier height with a red circle, blue square or green triangle if the primary barrier of the isotope is the inner barrier, one of the middle barriers or the outer barrier, respectively. The barrier heights without and with triaxiality are shown by full symbols and hollow symbols in (a). The empirical values of the barrier heights and the FUNF and the HFB-14 results are shown by full symbols in (b), (c) and (d), respectively.

barrier heights of $^{232-238}$U and our calculation results are shown together in Fig. 5(c). The heights of outer barriers for ^{232,236}U and the inner barrier for ^{238}U given by the MDC-RMF model are in agreement with the FUNF results. The inner barrier for ^{234}U from our calculation is about 0.5 MeV lower than that from the FUNF program. However, in most cases the primary barriers predicted by the MDC-RMF model are different from the FUNF results. Only for ^{238}U, both the MDC-RMF model and the FUNF program show that the inner barrier is the primary one. The FUNF calculations predict that the middle barrier is the primary one for ^{232}U and ^{234}U. For ^{236}U, the FUNF program shows that the inner barrier is the highest among the investigated barriers. It is worth noting that the FUNF program predicts the appearance of a middle barrier with a considerable height for $^{232-236}$U. Such results for the middle barriers are different from those given by the MDC-RMF calculations.

Based on other microscopic models, the barrier heights of the U isotopic chain are also investigated. In Fig. 5(d), the comparison of the MDC-RMF results with fission barrier heights for $^{232-296}$U given by HFB-14 calculations[57] is shown. For $^{232-286}$U, the barrier heights from the MDC-RMF model and the HFB-14 method show the same trend as the mass number increases. For $^{232-262}$U and $^{288-296}$U, the HFB-14 method gives higher barriers in comparison with the MDC-RMF model; while in $^{264-286}$U, our calculation results predict higher barriers. A drastic difference between the two model calculation results can be seen for the mass region $286 \leq A \leq 296$: The HFB-14 barrier heights increase with A while the MDC-RMF values decrease. The reason behind such a different behavior is yet not clear to us and should be studied in the future.

3.4. *Hyperdeformed third and fourth minima on the PESs*

By investigating the potential energy curves or surfaces of some actinides, the appearance of third minima corresponding to hyperdeformed nuclear shapes has been reported and many experiment efforts have also been made.[11,56,125–136] From our MDC-RMF calculation results for the even-A U isotopes, we can find obvious RS third and even fourth minima in PESs of $^{268-282}$U. As examples, we show the potential energy curves $E(\beta_{20})$ at $\beta_{30} = 0$ for ^{276}U and ^{280}U in Fig. 6. ^{276}U, the isotope with neutron magic number $N = 184$, has the highest fission barriers among the U isotopes. The height of the inner barrier is 18.8 MeV and the value is lowered by 1.64 MeV when the triaxial deformations are considered. The depths (the energy difference between the minimum and the lower one of the two saddle points around it) of the second, third and fourth potential wells for ^{276}U are 2.29 MeV, 1.87 MeV and 0.80 MeV. The positions and depths of these wells are indicated in Fig. 6.

The depths of the third well of ^{226,228,230,232}Th and ^{232,234,238}U have been studied by using the MDC-RMF model with functionals DD-ME2 and PC-PK1.[11] With PC-PK1, the third well only appears in ^{226,228,230}Th and the depths are 1.29 MeV, 0.78 MeV and 0.44 MeV, respectively. By investigating the deformed single-particle levels of proton and neutron, the appearance of the third minimum has been attributed to a proton shell gap at large deformations which stems from several pairs of single-proton states in the

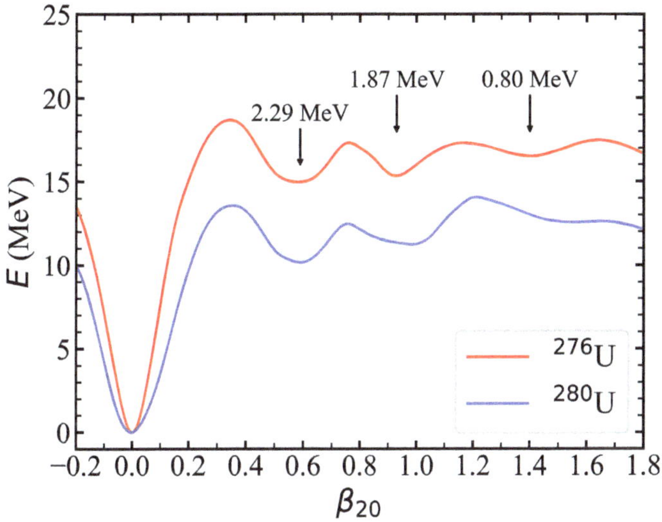

Fig. 6. (Color online) One-dimensional potential energy curves of ^{276}U and ^{280}U obtained in the MDC-RMF calculations with the effective interaction PC-PK1.[109] For ^{276}U, the positions and depths (the energy difference between the minimum and the lower one of the two saddle points around it) of the second, third and fourth potential wells are indicated. The curves represent fission pathways with the axial symmetry and the reflection symmetry imposed.

vicinity of the Fermi surface.[11] In comparison with the results given in Ref. 11, the third well of ^{276}U with a depth of 1.87 MeV is deeper than those of the previously investigated Th isotopes. The fourth well of ^{276}U is even deeper than the third wells of ^{228}Th and ^{230}Th. On the potential energy curve of ^{280}U in Fig. 6, the second and third minima can be identified, but the fourth well no longer exists due to the absence of a fourth saddle point.

The density distribution profiles of ^{276}U at the ground state and the second, third and fourth minima are shown in Fig. 7. The ground state of ^{276}U is spherical. At the second minimum, the nucleus is in a prolate shape. The nucleus further elongates when it is in the third well. At the fourth minimum, the neck appears and the nucleus shows a tendency to split up. With $\beta_{20} = 0.95$ and $\beta_{20} = 1.40$ at the third and fourth minima, ^{276}U is highly deformed, corresponding to hyperdeformed fission isomers.

4. Summary

In summary, we have systematically studied the properties of even-A U isotopes by using the MDC-RMF model. The PESs of ^{214}U to ^{350}U in the

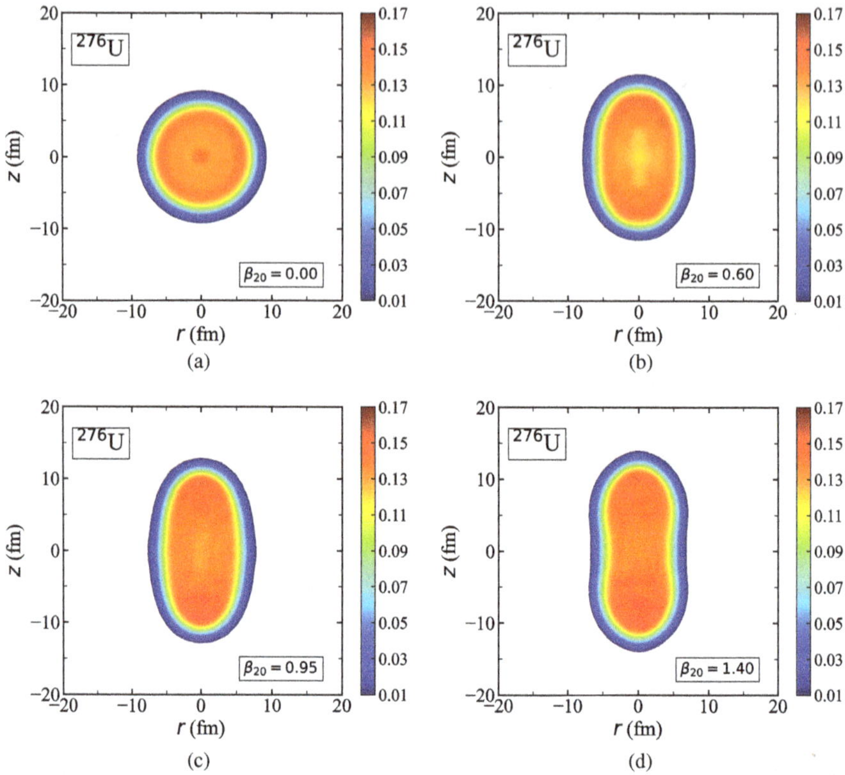

Fig. 7. (Color online) Density distribution profiles of ^{276}U at (a) the ground state and (b) the second, (c) third and (d) fourth minima obtained in the MDC-RMF calculations with the effective interaction PC-PK1.[109] The nucleus is in axial-symmetric shapes at these minima and the quadrupole deformation parameters are shown.

(β_{20}, β_{30}) plane are obtained and examined. Most of the PESs are characterized by a double-humped barrier structure. But for some isotopes, three or four saddle points appear. Detailed results concerning the ground state properties, including two-neutron separation energies, root-mean-square radii for the neutron, proton, matter and charge distributions and quadrupole and octupole deformations are presented. It can be seen from the calculation results that many of these U isotopes have nonspherical ground states and several of them, e.g., ^{230}U, have a pear-shaped ground state with large β_{20} and β_{30}.

By considering both the triaxiality and the reflection asymmetry in the vicinities of the saddle points, the primary barriers of these U isotopes have been investigated in detail. For many of them, the octupole deformations lower the primary barrier. The triaxiality may further lower the barriers and in some cases the primary barrier may shift from one barrier to another because the lowering effects of the triaxial distortion for the inner, middle and outer barriers may be different. The primary barrier heights with octupole and triaxial deformations can reproduce available empirical values. The hyperdeformed third and fourth minima on the PESs are discussed. Taking ^{276}U as an example, the depths of the third and the fourth potential wells are 1.87 MeV and 0.80 MeV, respectively.

Acknowledgments

Helpful discussions with Bao-Ge Deng, Bing-Nan Lu, Yu-Ting Rong, Xiang-Xiang Sun, Zhong-Hao Tu, Kun Wang, Xiao-Qian Wang and Zhen-Hua Zhang are gratefully acknowledged. We thank Xiao-Jun Sun for providing us the FUNF results prior to publication. This work has been partly supported by the Strategic Priority Research Program of Chinese Academy of Sciences (Grant Nos. XDB34010000 and XDPB15), the National Key R&D Program of China (Grant No. 2018YFA0404402), the National Natural Science Foundation of China (Grant Nos. 11525524, 12070131001, 12047503 and 11961141004), the Inter-Governmental S&T Cooperation Project between China and Croatia and the IAEA Coordinated Research Project "F41033". The results described in this paper are obtained on the High-performance Computing Cluster of ITP-CAS and the ScGrid of the Supercomputing Center, Computer Network Information Center of Chinese Academy of Sciences.

References

1. A. Bohr and B. R. Mottelson, *Nuclear Structure* (World Scientific, Singapore, 1998).
2. P. Ring and P. Schuck, *The Nuclear Many Body Problem* (Springer-Verlag, New York, Heidelberg, Berlin, 1980).
3. A. Bohr and B. R. Mottelson, *Phys. Rev.* **89** (1953) 316.
4. A. Bohr and B. R. Mottelson, *Phys. Rev.* **90** (1953) 717.
5. K. Alder, A. Bohr, T. Huus, B. Mottelson and A. Winther, *Rev. Mod. Phys.* **28** (1956) 432.
6. K. Alder, A. Bohr, T. Huus, B. Mottelson and A. Winther, *Rev. Mod. Phys.* **30** (1958) 353.
7. A. Bohr and B. R. Mottelson, *Nucl. Phys.* **4** (1957) 529.
8. A. Bohr and B. R. Mottelson, *Nucl. Phys.* **9** (1958) 687.
9. S. Frauendorf and J. Meng, *Nucl. Phys. A* **617** (1997) 131.
10. J. Meng, J. Peng, S. Q. Zhang and S.-G. Zhou, *Phys. Rev. C* **73** (2006) 037303.

11. J. Zhao, B.-N. Lu, D. Vretenar, E.-G. Zhao and S.-G. Zhou, *Phys. Rev. C* **91** (2015) 014321.
12. P. W. Zhao, N. Itagaki and J. Meng, *Phys. Rev. Lett.* **115** (2015) 022501.
13. J. Dudek, D. Curien, N. Dubray, J. Dobaczewski, V. Pangon, P. Olbratowski and N. Schunck, *Phys. Rev. Lett.* **97** (2006) 072501.
14. Y. Chen and Z.-C. Gao, *Nucl. Phys. A* **834** (2010) 378c.
15. J. Zhao, B.-N. Lu, E.-G. Zhao and S.-G. Zhou, *Phys. Rev. C* **95** (2017) 014320.
16. K. Wang and B.-N. Lu, *Commun. Theor. Phys.* **74** (2022) 015303.
17. C. Wong, *Ann. Phys.* **77** (1973) 279.
18. X. G. Cao, E. J. Kim, K. Schmidt, K. Hagel, M. Barbui, J. Gauthier, S. Wuenschel, G. Giuliani, M. R. D. Rodriguez, S. Kowalski, H. Zheng, M. Huang, A. Bonasera, R. Wada, N. Blando, G. Q. Zhang, C. Y. Wong, A. Staszczak, Z. X. Ren, Y. K. Wang, S. Q. Zhang, J. Meng and J. B. Natowitz, *Phys. Rev. C* **99** (2019) 014606.
19. N. Bohr and J. A. Wheeler, *Phys. Rev.* **56** (1939) 426.
20. V. Strutinsky, *Nucl. Phys. A* **95** (1967) 420.
21. V. Strutinsky, *Nucl. Phys. A* **122** (1968) 1.
22. P. Möller and J. R. Nix, Calculation of Fission Barriers, in *Proc. Third IAEA Symp. Physics and Chemistry of Fission*, Rochester, New York, 13–17 August 1973 (International Atomic Energy Agency, Vienna, Austria, 1974), pp. 103–140.
23. B.-N. Lu, E.-G. Zhao and S.-G. Zhou, *Phys. Rev. C* **85** (2012) 011301(R).
24. M. Brack, J. Damgaard, A. S. Jensen, H. C. Pauli, V. M. Strutinsky and C. Y. Wong, *Rev. Mod. Phys.* **44** (1972) 320.
25. H. Abusara, A. V. Afanasjev and P. Ring, *Phys. Rev. C* **82** (2010) 044303.
26. H. Abusara, A. V. Afanasjev and P. Ring, *Phys. Rev. C* **85** (2012) 024314.
27. B.-N. Lu, J. Zhao, E.-G. Zhao and S.-G. Zhou, *Phys. Rev. C* **89** (2014) 014323.
28. N. Schunck and L. M. Robledo, *Rep. Prog. Phys.* **79** (2016) 116301.
29. K.-H. Schmidt and B. Jurado, *Rep. Prog. Phys.* **81** (2018) 106301.
30. N. Schunck and D. Regnier, *Prog. Part. Nucl. Phys.* **125** (2022) 103963.
31. J. A. Sheikh, W. Nazarewicz and J. C. Pei, *Phys. Rev. C* **80** (2009) 011302.
32. M. Kowal, P. Jachimowicz and A. Sobiczewski, *Phys. Rev. C* **82** (2010) 014303.
33. M. Eichler, A. Arcones, A. Kelic, O. Korobkin, K. Langanke, T. Marketin, G. Martinez-Pinedo, I. Panov, T. Rauscher, S. Rosswog, C. Winteler, N. T. Zinner and F.-K. Thielemann, *Astrophys. J.* **808** (2015) 30.
34. A. A. Sonzogni, E. A. McCutchan, T. D. Johnson and P. Dimitriou, *Phys. Rev. Lett.* **116** (2016) 132502.
35. S. E. Agbemava, A. V. Afanasjev, D. Ray and P. Ring, *Phys. Rev. C* **95** (2017) 054324.
36. D. Brown, M. Chadwick, R. Capote, A. Kahler, A. Trkov, M. Herman, A. Sonzogni, Y. Danon, A. Carlson, M. Dunn, D. Smith, G. Hale, G. Arbanas, R. Arcilla, C. Bates, B. Beck, B. Becker, F. Brown, R. Casperson, J. Conlin, D. Cullen, M.-A. Descalle, R. Firestone, T. Gaines, K. Guber, A. Hawari, J. Holmes, T. Johnson, T. Kawano, B. Kiedrowski, A. Koning, S. Kopecky, L. Leal, J. Lestone, C. Lubitz, J. Márquez Damián, C. Mattoon, E. McCutchan, S. Mughabghab, P. Navratil, D. Neudecker, G. Nobre, G. Noguere, M. Paris,

M. Pigni, A. Plompen, B. Pritychenko, V. Pronyaev, D. Roubtsov, D. Rochman, P. Romano, P. Schillebeeckx, S. Simakov, M. Sin, I. Sirakov, B. Sleaford, V. Sobes, E. Soukhovitskii, I. Stetcu, P. Talou, I. Thompson, S. van der Marck, L. Welser-Sherrill, D. Wiarda, M. White, J. Wormald, R. Wright, M. Zerkle, G. Žerovnik and Y. Zhu, *Nucl. Data Sheets* **148** (2018) 1.
37. Z. Ge, R. Xu, H. Wu, Y. Zhang, G. Chen, Y. Jin, N. Shu, Y. Chen, X. Tao, Y. Tian, P. Liu, J. Qian, J. Wang, H. Zhang, L. Liu and X. Huang, *EPJ Web Conf.* **239** (2020) 09001.
38. A. J. M. Plompen, O. Cabellos, C. De Saint Jean, M. Fleming, A. Algora, M. Angelone, P. Archier, E. Bauge, O. Bersillon, A. Blokhin, F. Cantargi, A. Chebboubi, C. Diez, H. Duarte, E. Dupont, J. Dyrda, B. Erasmus, L. Fiorito, U. Fischer, D. Flammini, D. Foligno, M. R. Gilbert, J. R. Granada, W. Haeck, F.-J. Hambsch, P. Helgesson, S. Hilaire, I. Hill, M. Hursin, R. Ichou, R. Jacqmin, B. Jansky, C. Jouanne, M. A. Kellett, D. H. Kim, H. I. Kim, I. Kodeli, A. J. Koning, A. Yu. Konobeyev, S. Kopecky, B. Kos, A. Krása, L. C. Leal, N. Leclaire, P. Leconte, Y. O. Lee, H. Leeb, O. Litaize, M. Majerle, J. I. Márquez Damián, F. Michel-Sendis, R. W. Mills, B. Morillon, G. Noguère, M. Pecchia, S. Pelloni, P. Pereslavtsev, R. J. Perry, D. Rochman, A. Röhrmoser, P. Romain, P. Romojaro, D. Roubtsov, P. Sauvan, P. Schillebeeckx, K. H. Schmidt, O. Serot, S. Simakov, I. Sirakov, H. Sjöstrand, A. Stankovskiy, J. C. Sublet, P. Tamagno, A. Trkov, S. van der Marck, F. Álvarez-Velarde, R. Villari, T. C. Ware, K. Yokoyama and G. Žerovnik, *Eur. Phys. J. A* **56** (2020) 181.
39. A. Zubov, G. Adamian, N. Antonenko, S. Ivanova and W. Scheid, *Phys. Rev. C* **65** (2002) 024308.
40. C. J. Xia, B. X. Sun, E. G. Zhao and S. G. Zhou, *Sci. China-Phys. Mech. Astron.* **54** (2011) 109.
41. N. Wang, E.-G. Zhao, W. Scheid and S.-G. Zhou, *Phys. Rev. C* **85** (2012) 041601(R).
42. G. G. Adamian, N. V. Antonenko, H. Lenske, L. A. Malov and S.-G. Zhou, *Eur. Phys. J. A* **57** (2021) 89.
43. X.-Q. Deng and S.-G. Zhou, *Phys. Rev. C* **107** (2023) 014616.
44. W. Howard and P. Möller, *Atom. Data Nucl. Data Tables* **25** (1980) 219.
45. W. Myers and W. Swiatecki, *Nucl. Phys. A* **601** (1996) 141.
46. P. Jachimowicz, M. Kowal and J. Skalski, *Phys. Rev. C* **85** (2012) 034305.
47. A. Staszczak, A. Baran and W. Nazarewicz, *Phys. Rev. C* **87** (2013) 024320.
48. K. Pomorski, *Phys. Scripta T* **154** (2013) 014023.
49. C.-L. Zhong and T.-S. Fan, *Commun. Theor. Phys.* **62** (2014) 405.
50. P. Möller, A. J. Sierk, T. Ichikawa, A. Iwamoto and M. Mumpower, *Phys. Rev. C* **91** (2015) 024310.
51. K. Pomorski, B. Nerlo-Pomorska and J. Bartel, *Phys. Scripta* **92** (2017) 064006.
52. Q.-Z. Chai, W.-J. Zhao, M.-L. Liu and H.-L. Wang, *Chin. Phys. C* **42** (2018) 054101.
53. P. Jachimowicz, M. Kowal and J. Skalski, *Phys. Rev. C* **101** (2020) 014311.

54. X. Zhu, Z.-M. Wang, W.-J. Zhu, C.-L. Zhong, Y.-M. Zhang, L.-Y. Liao and T.-S. Fan, *Commun. Theor. Phys.* **72** (2020) 105301.
55. A. Mamdouh, J. Pearson, M. Rayet and F. Tondeur, *Nucl. Phys. A* **679** (2001) 337.
56. J.-P. Delaroche, M. Girod, H. Goutte and J. Libert, *Nucl. Phys. A* **771** (2006) 103.
57. S. Goriely, S. Hilaire, A. J. Koning, M. Sin and R. Capote, *Phys. Rev. C* **79** (2009) 024612.
58. S. A. Giuliani and L. M. Robledo, *Phys. Rev. C* **88** (2013) 054325.
59. R. Rodríguez-Guzmán and L. M. Robledo, *Phys. Rev. C* **89** (2014) 054310.
60. S.-G. Zhou, *Phys. Scripta* **91** (2016) 063008.
61. S. A. Giuliani, G. Martínez-Pinedo and L. M. Robledo, *Phys. Rev. C* **97** (2018) 034323.
62. C. Ling, C. Zhou and Y. Shi, *Eur. Phys. J. A* **56** (2020) 180.
63. A. Taninah, S. E. Agbemava and A. V. Afanasjev, *Phys. Rev. C* **102** (2020) 054330.
64. B. D. Serot and J. D. Walecka, *The Relativistic Nuclear Many-Body Problem, Advances in Nuclear Physics*, Vol. **16** (Plenum Press, New York, 1986).
65. P. G. Reinhard, *Rep. Prog. Phys.* **52** (1989) 439.
66. P. Ring, *Prog. Part. Nucl. Phys.* **37** (1996) 193.
67. D. Vretenar, A. Afanasjev, G. Lalazissis and P. Ring, *Phys. Rep.* **409** (2005) 101.
68. J. Meng, H. Toki, S. G. Zhou, S. Q. Zhang, W. H. Long and L. S. Geng, *Prog. Part. Nucl. Phys.* **57** (2006) 470.
69. T. Nikšić, D. Vretenar and P. Ring, *Prog. Part. Nucl. Phys.* **66** (2011) 519.
70. J. Meng, J. Peng, S.-Q. Zhang and P.-W. Zhao, *Front. Phys.* **8** (2013) 55.
71. H. Liang, J. Meng and S.-G. Zhou, *Phys. Rep.* **570** (2015) 1.
72. J. Meng and S. G. Zhou, *J. Phys. G: Nucl. Part. Phys.* **42** (2015) 093101.
73. J. Meng, *Relativistic Density Functional for Nuclear Structure, International Review of Nuclear Physics*, Vol. **10** (World Scientific, Singapore, 2016).
74. S. Shen, H. Liang, W. H. Long, J. Meng and P. Ring, *Prog. Part. Nucl. Phys.* **109** (2019) 103713.
75. L.-S. Geng, M. Jie and T. Hiroshi, *Chin. Phys. Lett.* **24** (2007) 1865.
76. W. Zhang, Z. P. Li, S. Q. Zhang and J. Meng, *Phys. Rev. C* **81** (2010) 034302.
77. W. Zhang, Z. P. Li and S. Q. Zhang, *Chin. Phys. C* **34** (2010) 1094.
78. J.-Y. Guo, P. Jiao and X.-Z. Fang, *Phys. Rev. C* **82** (2010) 047301.
79. Y.-T. Qiu, X.-W. Wang and J.-Y. Guo, *Phys. Rev. C* **106** (2022) 034301.
80. Y. Gambhir, P. Ring and A. Thimet, *Annals Phys.* **198** (1990) 132.
81. X. Meng, B. N. Lu and S. G. Zhou, *Sci. China-Phys. Mech. Astron.* **63** (2020) 212011.
82. X.-Q. Wang, X.-X. Sun and S.-G. Zhou, *Chin. Phys. C* **46** (2022) 024107.
83. J. Zhao, B.-N. Lu, E.-G. Zhao and S.-G. Zhou, *Phys. Rev. C* **86** (2012) 057304.
84. W. Xu, S. Wang, C. Liu, X. Wu, R. Guo, B. Qi, J. Zhao, A. Rohilla, H. Jia, G. Li, Y. Zheng, C. Li, X. Han, L. Mu, X. Xiao, S. Wang, D. Sun, Z. Li, Y. Zhang, C. Wang and Y. Li, *Phys. Lett. B* **833** (2022) 137287.

85. C. Liu, S. Y. Wang, R. A. Bark, S. Q. Zhang, J. Meng, B. Qi, P. Jones, S. M. Wyngaardt, J. Zhao, C. Xu, S.-G. Zhou, S. Wang, D. P. Sun, L. Liu, Z. Q. Li, N. B. Zhang, H. Jia, X. Q. Li, H. Hua, Q. B. Chen, Z. G. Xiao, H. J. Li, L. H. Zhu, T. D. Bucher, T. Dinoko, J. Easton, K. Juhász, A. Kamblawe, E. Khaleel, N. Khumalo, E. A. Lawrie, J. J. Lawrie, S. N. T. Majola, S. M. Mullins, S. Murray, J. Ndayishimye, D. Negi, S. P. Noncolela, S. S. Ntshangase, B. M. Nyakó, J. N. Orce, P. Papka, J. F. Sharpey-Schafer, O. Shirinda, P. Sithole, M. A. Stankiewicz and M. Wiedeking, *Phys. Rev. Lett.* **116** (2016) 112501.
86. Y. Wang, S. Zhang, P. Zhao and J. Meng, *Phys. Lett. B* **792** (2019) 454.
87. X. C. Chen, J. Zhao, C. Xu, H. Hua, T. M. Shneidman, S. G. Zhou, X. G. Wu, X. Q. Li, S. Q. Zhang, Z. H. Li, W. Y. Liang, J. Meng, F. R. Xu, B. Qi, Y. L. Ye, D. X. Jiang, Y. Y. Cheng, C. He, J. J. Sun, R. Han, C. Y. Niu, C. G. Li, P. J. Li, C. G. Wang, H. Y. Wu, Z. H. Li, H. Zhou, S. P. Hu, H. Q. Zhang, G. S. Li, C. Y. He, Y. Zheng, C. B. Li, H. W. Li, Y. H. Wu, P. W. Luo and J. Zhong, *Phys. Rev. C* **94** (2016) 021301.
88. Y. Y. Wang, Q. B. Chen and S. Q. Zhang, *Phys. Rev. C* **105** (2022) 044316.
89. C. G. Wang, R. Han, C. Xu, H. Hua, R. A. Bark, S. Q. Zhang, S. Y. Wang, T. M. Shneidman, S. G. Zhou, J. Meng, S. M. Wyngaardt, A. C. Dai, F. R. Xu, X. Q. Li, Z. H. Li, Y. L. Ye, D. X. Jiang, C. G. Li, C. Y. Niu, Z. Q. Chen, H. Y. Wu, D. W. Luo, S. Wang, D. P. Sun, C. Liu, Z. Q. Li, N. B. Zhang, R. J. Guo, P. Jones, E. A. Lawrie, J. J. Lawrie, J. F. Sharpey-Schafer, M. Wiedeking, S. N. T. Majola, T. D. Bucher, T. Dinoko, B. Maqabuka, L. Makhathini, L. Mdletshe, O. Shirinda and K. Sowazi, *Phys. Rev. C* **106** (2022) L011303.
90. B.-N. Lu, E.-G. Zhao and S.-G. Zhou, *Phys. Rev. C* **84** (2011) 014328.
91. B.-N. Lu, E. Hiyama, H. Sagawa and S.-G. Zhou, *Phys. Rev. C* **89** (2014) 044307.
92. Y.-T. Rong, P. Zhao and S.-G. Zhou, *Phys. Lett. B* **807** (2020) 135533.
93. Y.-T. Rong, Z.-H. Tu and S.-G. Zhou, *Phys. Rev. C* **104** (2021) 054321.
94. C. Chen, Q.-K. Sun, Y.-X. Li and T.-T. Sun, *Sci. China-Phys. Mech. Astron.* **64** (2021) 282011.
95. Q.-K. Sun, T.-T. Sun, W. Zhang, S.-S. Zhang and C. Chen, *Chin. Phys. C* **46** (2022) 074106.
96. J. Zhao, B.-N. Lu, T. Nikšić and D. Vretenar, *Phys. Rev. C* **92** (2015) 064315.
97. J. Zhao, B.-N. Lu, T. Nikšić, D. Vretenar and S.-G. Zhou, *Phys. Rev. C* **93** (2016) 044315.
98. J. Zhao, T. Nikšić, D. Vretenar and S.-G. Zhou, *Phys. Rev. C* **99** (2019) 014618.
99. J. Zhao, J. Xiang, Z.-P. Li, T. Nikšić, D. Vretenar and S.-G. Zhou, *Phys. Rev. C* **99** (2019) 054613.
100. J. Zhao, T. Nikšić and D. Vretenar, *Phys. Rev. C* **102** (2020) 054606.
101. J. Zhao, T. Nikšić, D. Vretenar and S.-G. Zhou, *Phys. Rev. C* **101** (2020) 064605.
102. J. Zhao, T. Nikšić and D. Vretenar, *Phys. Rev. C* **104** (2021) 044612.
103. Z. X. Ren, J. Zhao, D. Vretenar, T. Nikšić, P. W. Zhao and J. Meng, *Phys. Rev. C* **105** (2022) 044313.
104. J. Zhao, T. Nikšić and D. Vretenar, *Phys. Rev. C* **105** (2022) 054604.
105. J. Zhao, T. Nikšić and D. Vretenar, *Phys. Rev. C* **106** (2022) 054609.

106. F. Mercier, J. Zhao, R.-D. Lasseri, J.-P. Ebran, E. Khan, T. Nikšić and D. Vretenar, *Phys. Rev. C* **102** (2020) 011301.
107. F. Mercier, J. Zhao, J.-P. Ebran, E. Khan, T. Nikšić and D. Vretenar, *Phys. Rev. Lett.* **127** (2021) 012501.
108. Y.-T. Rong, X.-Y. Wu, B.-N. Lu and J.-M. Yao, *Phys. Lett. B* **840** (2023) 137896.
109. P. W. Zhao, Z. P. Li, J. M. Yao and J. Meng, *Phys. Rev. C* **82** (2010) 054319.
110. Y. Tian and Z.-y. Ma, *Chin. Phys. Lett.* **23** (2006) 3226.
111. Y. Tian, Z. Ma and P. Ring, *Phys. Lett. B* **676** (2009) 44.
112. Y. Tian, Z.-y. Ma and P. Ring, *Phys. Rev. C* **79** (2009) 064301.
113. Z. Y. Zhang, H. B. Yang, M. H. Huang, Z. G. Gan, C. X. Yuan, C. Qi, A. N. Andreyev, M. L. Liu, L. Ma, M. M. Zhang, Y. L. Tian, Y. S. Wang, J. G. Wang, C. L. Yang, G. S. Li, Y. H. Qiang, W. Q. Yang, R. F. Chen, H. B. Zhang, Z. W. Lu, X. X. Xu, L. M. Duan, H. R. Yang, W. X. Huang, Z. Liu, X. H. Zhou, Y. H. Zhang, H. S. Xu, N. Wang, H. B. Zhou, X. J. Wen, S. Huang, W. Hua, L. Zhu, X. Wang, Y. C. Mao, X. T. He, S. Y. Wang, W. Z. Xu, H. W. Li, Z. Z. Ren and S. G. Zhou, *Phys. Rev. Lett.* **126** (2021) 152502.
114. W. Huang, M. Wang, F. Kondev, G. Audi and S. Naimi, *Chin. Phys. C* **45** (2021) 030002.
115. M. Wang, W. Huang, F. Kondev, G. Audi and S. Naimi, *Chin. Phys. C* **45** (2021) 030003.
116. DRHBc Mass Table Collaboration (K. Zhang, M.-K. Cheoun, Y.-B. Choi, P. S. Chong, J. Dong, Z. Dong, X. Du, L. Geng, E. Ha, X.-T. He, C. Heo, M. C. Ho, E. J. In, S. Kim, Y. Kim, C.-H. Lee, J. Lee, H. Li, Z. Li, T. Luo, J. Meng, M.-H. Mun, Z. Niu, C. Pan, P. Papakonstantinou, X. Shang, C. Shen, G. Shen, W. Sun, X.-X. Sun, C. K. Tam, Thaivayongnou, C. Wang, X. Wang, S. H. Wong, J. Wu, X. Wu, X. Xia, Y. Yan, R. W.-Y. Yeung, T. C. Yiu, S. Zhang, W. Zhang, X. Zhang, Q. Zhao and S.-G. Zhou), *At. Data Nucl. Data Tables* **144** (2022) 101488.
117. I. Angeli and K. Marinova, *Atom. Data Nucl. Data Tables* **99** (2013) 69.
118. S.-G. Zhou, J. Meng, P. Ring and E.-G. Zhao, *Phys. Rev. C* **82** (2010) 011301.
119. L. Li, J. Meng, P. Ring, E.-G. Zhao and S.-G. Zhou, *Phys. Rev. C* **85** (2012) 024312.
120. Y. Chen, L. Li, H. Liang and J. Meng, *Phys. Rev. C* **85** (2012) 067301.
121. R. Capote, M. Herman, P. Obložinský, P. Young, S. Goriely, T. Belgya, A. Ignatyuk, A. Koning, S. Hilaire, V. Plujko, M. Avrigeanu, O. Bersillon, M. Chadwick, T. Fukahori, Z. Ge, Y. Han, S. Kailas, J. Kopecky, V. Maslov, G. Reffo, M. Sin, E. Sh. Soukhovitskii and P. Talou, *Nucl. Data Sheets* **110** (2009) 3107.
122. J. Zhang, *Nucl. Sci. Eng.* **114** (1993) 55.
123. J. Zhang, *User Manual of FUNF Code* (Atomic Energy Press, Beijing, 2012).
124. X.-J. Sun and J.-S. Zhang, Private communication (2022).
125. S. Ćwiok, W. Nazarewicz, J. Saladin, W. Płóciennik and A. Johnson, *Phys. Lett. B* **322** (1994) 304.
126. K. Rutz, J. Maruhn, P.-G. Reinhard and W. Greiner, *Nucl. Phys. A* **590** (1995) 680.

127. A. Krasznahorkay, M. Hunyadi, M. Csatlós, A. Bałanda, A. Gollwitzer, G. Graw, J. Gulyás, D. Habs, R. Hertenberger, Z. Máté, H. J. Maier, D. Rudolph, P. Thirolf, J. Timár and B. D. Valnion, *Acta Phys. Hung. A* **7** (1998) 35.
128. A. Marinov, S. Gelberg, D. Kolb and J. L. Weil, *Int. J. Mod. Phys. E* **10** (2001) 209.
129. P. G. Thirolf, S. Y. van der Werf, J. Ott, Z. Máté, J. Gulyás, M. Csatlós, M. Hunyadi, A. Krasznahorkay, T. Faestermann, A. Metz, H. J. Maier, R. Hertenberger, G. Graw, Y. Eisermann, D. Habs and D. Gassmann, *Acta Phys. Hung. A* **13** (2001) 111.
130. P. Thirolf and D. Habs, *Prog. Part. Nucl. Phys.* **49** (2002) 325.
131. A. Krasznahorkay, M. Csatlós, L. Csige, Y. Eisermann, T. Faestermann, Z. Gácsi, G. Graw, J. Gulyás, D. Habs, M. N. Harakeh, M. Heil, R. Hertenberger, F. Kappeler, A. Jr. Krasznahorkay, H. J. Maier, Z. Mate, R. Reifarth, P. G. Thirolf, J. Timár and H. F. Wirth, *Acta Phys. Hung. A* **18** (2003) 323.
132. M. Sin, R. Capote, A. Ventura, M. Herman and P. Obložinský, *Phys. Rev. C* **74** (2006) 014608.
133. M. Kowal and J. Skalski, *Phys. Rev. C* **85** (2012) 061302.
134. T. Ichikawa, P. Möller and A. J. Sierk, *Phys. Rev. C* **87** (2013) 054326.
135. P. Jachimowicz, M. Kowal and J. Skalski, *Phys. Rev. C* **87** (2013) 044308.
136. J. D. McDonnell, W. Nazarewicz and J. A. Sheikh, *Phys. Rev. C* **87** (2013) 054327.

An improved microscopic core–quasiparticle coupling model for spectroscopy of odd-mass nuclei with octupole correlations*

Xin Zhao and Zhipan Li [†]

School of Physical Science and Technology,
Southwest University, Chongqing 400715, P. R. China
[†]*zpliphy@swu.edu.cn*

Dario Vretenar

Physics Department, Faculty of Science,
University of Zagreb, Croatia

The microscopic core–quasiparticle coupling model for odd-A nuclei with octupole correlations is extended to include the monopole, quadrupole and octupole couplings between the core and the odd nucleon, based on the framework of covariant density functional theory. The model is tested in a study of low-lying excitation spectra for the odd-A nucleus ^{225}Ra. Theoretical results, e.g., the parity doublets, interband $B(E1)$ and intraband $B(E2)$, are in very good agreement with the available data. In particular, the description of signature splitting in the ground-state band has been improved. This method provides a very useful tool for a systematic study of low-energy spectra in odd-mass actinides with octupole correlations.

Keywords: Octupole deformation; parity doublets; core–quasiparticle coupling model; covariant density functional theory.

1. Introduction

Similar to other many-body systems, an atomic nucleus can be distorted from spherical shape by spontaneous symmetry breaking. Most deformed medium-heavy and heavy nuclei exhibit quadrupole, reflection-symmetric equilibrium shapes. Since the first report by a Berkeley group in the 1950s,[1–3] reflection-asymmetric octupole shapes present a recurrent theme of interest in nuclear structure physics, both in experiments and theory.[4,5] Recently,

*To cite this article, please refer to its earlier version published in the *International Journal of Modern Physics E*, Volume 32(10), (14 pages), 2340006 (2023), DOI: 10.1142/S0218301323400062.

experimental evidence for stable octupole deformation has been reported in, e.g., ^{222,224}Ra,[6,7] ^{144}Ba[8] and ^{146}Ba.[9] This type of shape deformation occurs predominantly in nuclei with neutron (proton) number $N(Z) \approx 34$, 56, 88 and 134, characterized by the presence of low-lying negative-parity bands, as well as pronounced electric octupole transitions.[10–26]

Octupole deformations are also relevant for nuclei with odd nucleon number, in which the interplay between single-particle and quadrupole and octupole collective degrees of freedom determines the structure of low-energy excitation spectra. In the past decade, considerable effort has been devoted to studies of excited states of odd-mass nuclei characterized by static and dynamic octupole deformation, and new levels, transitions and/or dipole moments have been measured for nuclei in different mass regions, e.g., $(Z, N) \sim (32, 40)$,[27] neutron-deficient Ba isotopes,[28,29] neutron-rich Ba-Nd region with $N \sim 90$,[30–37] Ra isotopes and actinides.[38–41] Very recently, the observation of strong $E1$ transitions and the well-behaved rotational sequence built on the $15/2^-$ octupole state provided the first experimental evidence of an octupole rotational band in Ge isotopes, suggesting enhanced octupole correlations around $N = 40$ in the $A \sim 70$ mass region.[27] Moreover, specific odd-A octupole deformed nuclei, such as ^{225}Ra, have attracted interest in connection with observed parity-doublet structures and enhanced Schiff moments that arise from time-reversal and parity-violating interactions in the nuclear medium.[42–47]

From a theoretical point of view, nuclear properties related to octupole degrees of freedom have been investigated using a variety of approaches, such as density functional theory (DFT) and DFT-based beyond-mean-field approaches,[48–85] macroscopic-microscopic models,[49,86–89] interacting boson model (IBM),[90–103] phenomenological collective models[104–112] and shell model.[113–116] The study of odd-A nuclei is particularly challenging due to the coupling effect between the unpaired nucleon and the core. Several approaches have been developed over the past few decades, such as the particle-rotor model,[117–125,88,126] the cluster model,[127] the reflection-asymmetric shell model[128] and the interacting boson-fermion model (IBFM).[129,55] Recently, we have introduced a microscopic core–quasiparticle coupling (CQC) model based on covariant density functional theory, which includes both quadrupole and octupole couplings between the even-mass core and odd nucleon.[130] Model predictions for low-energy excitation spectra and transition rates of odd-mass radium isotopes ^{223,225,227}Ra were shown to be in reasonable agreement with available data. However, the Fermi surface for the odd nucleon is a free parameter, and sometimes it has to be adjusted to values rather different from those in the neighboring even–even core nuclei. Additionally, some theoretical parity doublets exhibit incorrect signature splitting, such as, for instance, the ground-state band of ^{225}Ra.

In this work, we aim to extend the CQC model by including the monopole coupling between the core and odd nucleon. In this way, the Fermi surface in the odd-A nucleus is no longer an adjustable parameter as in the previous version of the model. Additionally, the monopole coupling produces a compression or expansion of the single-nucleon energy spectra. The improved CQC model is tested with an illustrative calculation of parity doublets in the odd-A ^{225}Ra, located next to the even–even nucleus ^{224}Ra, characterized by static octupole deformation. Furthermore, ^{225}Ra is a favorable case to explore for a permanent electric dipole moment. Section 2 describes the framework of the improved CQC model. Illustrative results for spectroscopic properties, such as low-lying excitation spectra, electric transitions and wave functions are discussed in Sec. 3. Section 4 concludes.

2. Theoretical Framework

Here, we briefly outline the core–quasiparticle coupling ($\mathrm{CQC_m}$) model, that includes monopole, quadrupole and octupole interactions in the particle–hole (p–h) channel. In the core–quasiparticle coupling scheme, the configuration space of the odd-mass nucleus with mass number A is composed of both a particle coupled to the lighter even–even neighbor $A-1$, and a hole coupled to the heavier even–even neighbor $A+1$. For an odd-A nucleus in which the low-energy excitation spectra are determined by octupole correlations, both the ground-state band and the lowest negative-parity band of the even–even cores are included in the configuration space. The ansatz for the wave function of the odd-mass system can therefore be written in the following form:

$$|\alpha J M_J \pi\rangle^A = \sum_{\mu\Omega} \{U_{\mu\Omega}[a_\mu^\dagger|\Omega\rangle]^{A-1}_{JM_J\pi} + V_{\mu\Omega}[a_\mu|\Omega\rangle]^{A+1}_{JM_J\pi}\}, \tag{1}$$

where α denotes all quantum numbers in addition to the total angular momentum J, its projection M_J and the parity π for the odd-A nucleus. μ is the generic notation for the angular momentum j, its projection m_j, parity π_j and additional quantum numbers n_μ of single-particle states, while Ω denotes the angular momentum R, its projection M_R, parity π_R and additional quantum numbers n_Ω of collective states of the even–even core. The coefficients $U_{\mu\Omega}$ and $V_{\mu\Omega}$ represent the probability amplitudes for the particle-like and hole-like states, respectively.

The Hamiltonian for CQC model reads

$$H = H_{\rm qp} + H_{\rm c}$$
$$= \begin{pmatrix} (\varepsilon^{A-1} - \varepsilon_f) + \Gamma^{A-1} & \Delta^{A+1} \\ \Delta^{\dagger A-1} & -(\varepsilon^{A+1} - \varepsilon_f) - \Gamma^{A+1} \end{pmatrix} + \begin{pmatrix} E^{A-1} & 0 \\ 0 & E^{A+1} \end{pmatrix}. \quad (2)$$

The matrices $(\varepsilon^{A\pm1} - \varepsilon_f)$ and $(E^{A\pm1})$ are diagonal with respect to the basis states in the decomposition of Eq. (1):

$$(\varepsilon^{A\pm1} - \varepsilon_f) = (\varepsilon_\mu^{A\pm1} - \varepsilon_f)\delta_{\mu\mu'}\delta_{\Omega\Omega'}, \quad (3)$$

$$(E^{A\pm1}) = E_\Omega^{A\pm1}\delta_{\mu\mu'}\delta_{\Omega\Omega'} \quad (4)$$

with the spherical single-particle energies $\varepsilon_\mu^{A\pm1}$, the collective excitation energies $E_\Omega^{A\pm1}$ and Fermi surface $\varepsilon_f = \frac{1}{2}[E^{A+1}(0_1^+) - E^{A-1}(0_1^+)]$.[131] $E^{A\pm1}(0_1^+)$ are the absolute ground-state energies of the neighboring even–even cores. Γ and Δ are the mean field and pairing field that correspond to the long-range particle–hole, and short-range particle–particle interactions, respectively. The monopole, quadrupole and octupole p–h interactions determine the field Γ, while a monopole force is used for the pairing field Δ:

$$\begin{aligned}(\Gamma^{A\pm1}) = & -\chi_0/\sqrt{(2j+1)(2R+1)} \\ & -\chi_2(-1)^{j+R+J}\begin{Bmatrix} j & 2 & j' \\ R' & J & R \end{Bmatrix}[\langle\mu\|\hat{q}_2\|\mu'\rangle\langle\Omega\|\hat{Q}_2\|\Omega'\rangle]^{A\pm1} \\ & -\chi_3(-1)^{j+R+J}\begin{Bmatrix} j & 3 & j' \\ R' & J & R \end{Bmatrix}[\langle\mu\|\hat{q}_3\|\mu'\rangle\langle\Omega\|\hat{Q}_3\|\Omega'\rangle]^{A\pm1}, \end{aligned} \quad (5)$$

$$\begin{aligned}(\Delta^{A+1}) = (\Delta^{A-1}) &= \langle\Omega, A-1|\hat{\Delta}|\Omega', A+1\rangle\delta_{\mu\mu'} \\ &\approx \frac{1}{2}(\Delta_\Omega^{A-1} + \Delta_\Omega^{A+1})\delta_{\mu\mu'}\delta_{\Omega\Omega'} \equiv (\Delta), \end{aligned} \quad (6)$$

where $\langle\mu\|\hat{q}_\lambda\|\mu'\rangle^{A\pm1}$ and $\langle\Omega\|\hat{Q}_\lambda\|\Omega'\rangle^{A\pm1}$ are the reduced quadrupole ($\lambda = 2$) and octupole ($\lambda = 3$) matrix elements of the spherical single-particle (hole) and even–even core states, respectively. χ_0, χ_2 and χ_3 denote the corresponding coupling strengths of the monopole, quadrupole and octupole interactions. $\Delta_\Omega^{A\pm1}$ are the average pairing gaps of the collective states $|\Omega\rangle^{A\pm1}$. The CQC Hamiltonian of Eq. (2) is determined by the energies $\varepsilon_\mu^{A\pm1}$ and $E_\Omega^{A\pm1}$, the quadrupole matrix elements $\langle\mu\|\hat{q}_2\|\mu'\rangle^{A\pm1}$ and $\langle\Omega\|\hat{Q}_2\|\Omega'\rangle^{A\pm1}$, the octupole matrix elements $\langle\mu\|\hat{q}_3\|\mu'\rangle^{A\pm1}$ and $\langle\Omega\|\hat{Q}_3\|\Omega'\rangle^{A\pm1}$ and the pairing gaps $\Delta_\Omega^{A\pm1}$. The input of the CQC model can be calculated using the quadrupole–octupole collective Hamiltonian,[81] combined with a constrained reflection-asymmetric relativistic Hartree–Bogoliubov (RHB) model.[132] A detailed description of the CQC model can be found in Ref. 130.

In this work, the coupling strengths χ_0, χ_2 and χ_3 are free parameters that can be adjusted to reproduce selected experimental results, e.g., the ground-state spin, parity and the excitation energies of few lowest levels. Finally, the excitation energies $E_{\alpha J\pi}$, and coefficients $U_{\mu\Omega}$ and $V_{\mu\Omega}$, are obtained by solving the eigen-equation

$$\begin{pmatrix} (\varepsilon^{A-1} - \varepsilon_f) + \Gamma^{A-1} + E^{A-1} & \Delta \\ \Delta & -(\varepsilon^{A+1} - \varepsilon_f) - \Gamma^{A+1} + E^{A+1} \end{pmatrix} \begin{pmatrix} U \\ V \end{pmatrix} = E_{\alpha J\pi} \begin{pmatrix} U \\ V \end{pmatrix}. \tag{7}$$

The eigenstates are used to calculate the electromagnetic multipole transition rates, as described in Ref. 130.

3. Results and Discussion

The improved CQC model is here tested in a calculation of parity doublets in the odd-A nucleus ^{225}Ra. The model input is the single-particle energy levels and the excitation spectra of ^{224,226}Ra, which are obtained by using the self-consistent RHB method with the PC-PK1 functional,[133] and the quadrupole–octupole collective Hamiltonian (QOCH). The single-particle space consists of spherical positive- and negative-parity single-neutron states with $\varepsilon_f - 2\hbar\omega \le \varepsilon_\mu \le 2$MeV, where $\hbar\omega = 41A^{-\frac{1}{3}}$MeV. The positive-parity ground-state and lowest negative-parity bands of the core nuclei with angular momenta $R \le 18\hbar$ are included in the CQC calculation.

Figure 1 displays the low-lying excitation spectrum of the odd-mass nucleus ^{225}Ra calculated with the CQC$_\mathrm{m}$ model, in comparison with the available experimental results[134] and theoretical values obtained using the standard CQC model.[130] The absolute energies of the ground state of ^{224}Ra and ^{226}Ra, calculated with the QOCH model, are $E(0_1^+) = -1720.60$ and -1731.67MeV, respectively, in very good agreement with the experimental binding energies: -1720.30 and -1731.60MeV.[135] The Fermi energy of the odd-A nucleus ^{225}Ra is obtained at -5.54MeV. The coupling strength parameters are adjusted to the following values: $\chi_0 = 7.2$MeV, $\chi_2 = 6.0\mathrm{MeV}/b^2$ and $\chi_3 = 17.2\mathrm{MeV}/b^3$ to reproduce selected experimental results, e.g., the ground-state spin, parity and the excitation energies of $5/2^+$ and $3/2^+$ in Band 1 and $3/2^-$ in Band 2. b in the units of χ_2 and χ_3

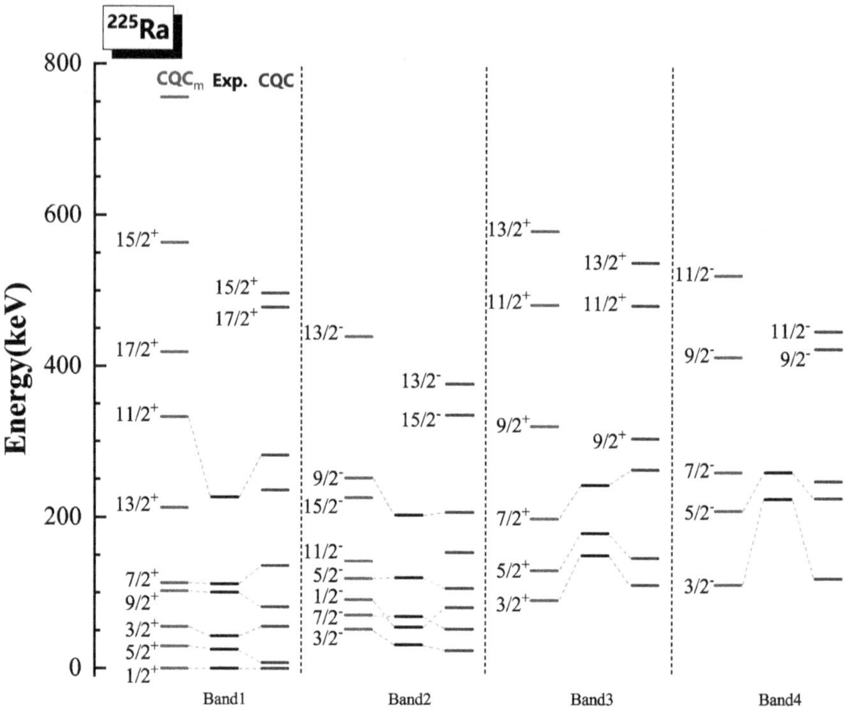

Fig. 1. (Color online) The low-lying excitation spectrum of the odd-mass nucleus ^{225}Ra calculated with the CQC$_\mathrm{m}$ model, plotted in comparison with the available data[134] and theoretical values obtained using the standard CQC model.[130]

means barn, a unit for area, and $1b = 100\,\mathrm{fm}^2$. States are grouped into different bands according to the dominant $E2$ decay rates and the similarity in the composition of their wave functions. Here, four low-lying bands with band-heads $J^\pi = 1/2^\pm$ and $J^\pi = 3/2^\pm$ are shown, and they are displayed as two parity-doublet bands. In general, the parity doublets are reproduced by both CQC$_\mathrm{m}$ and CQC models. However, the CQC$_\mathrm{m}$ model gives a better description of signature splitting for the ground-state band $K^\pi = 1/2^+$, while in the results of CQC model the sequence of $5/2^+$ and $9/2^+$ states is too low, and the sequence of $3/2^+$ and $7/2^+$ states lies too high. The improved model CQC$_\mathrm{m}$ could be more appropriate for the calculation of the Shiff moment.[43,45] For the second parity doublet, i.e., the Bands 3 and 4, the overall structure is reproduced but the calculated band-heads are approximately 90 keV lower than the corresponding experimental levels.

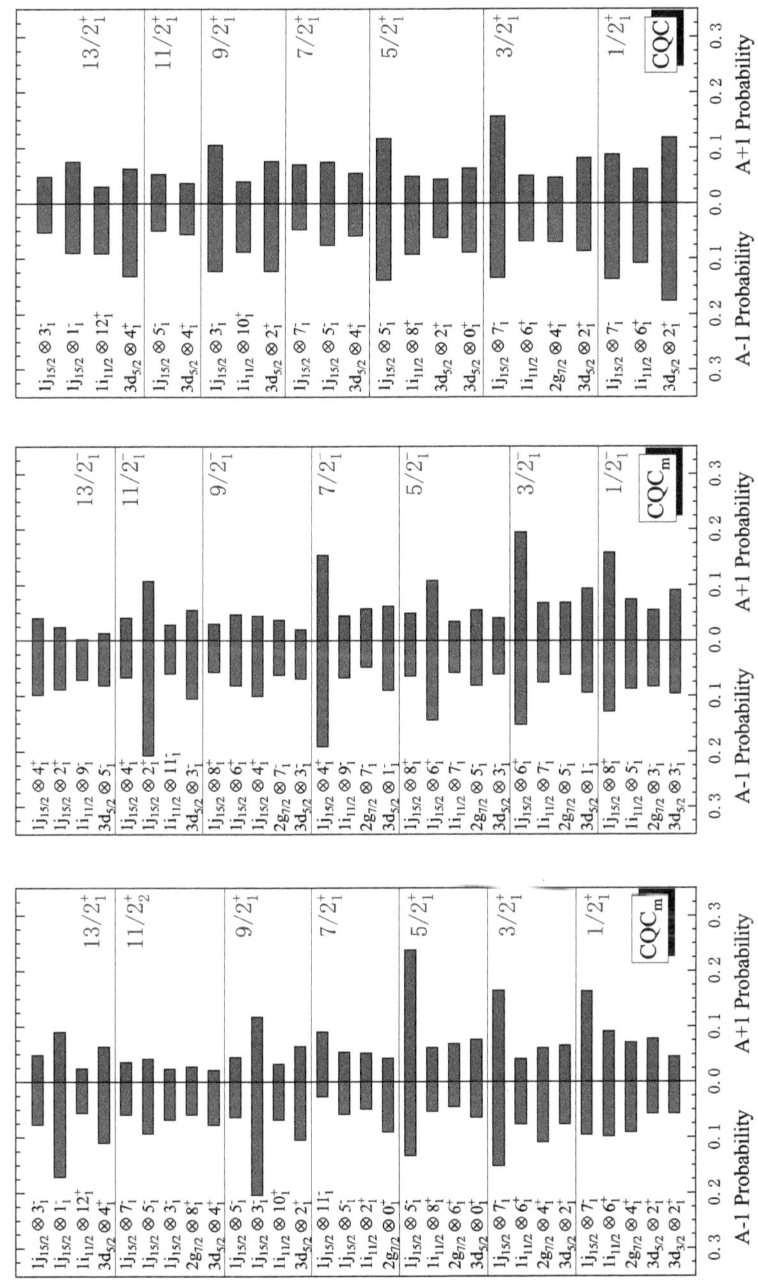

Fig. 2. (Color online) The probabilities of the dominant configurations in the wave functions of selected states in the ground-state band and its negative-parity partner band of ^{225}Ra, obtained with the CQC_m model ((a) and (b)) and the CQC model (c).

The probabilities of dominant configurations in the wave functions of selected states in the ground-state band and its partner band of ^{225}Ra, obtained by the CQC$_{\rm m}$ model, are shown in Figs. 2(a) and 2(b), respectively. The parity doublet structure is characterized by the strongly mixed configurations based on $3d_{5/2}$, $2g_{7/2}$, $1i_{11/2}$ and $1j_{15/2}$. The pronounced mixing between positive- and negative-parity single-particle states and collective core states reflects the strong octupole coupling between the odd neutron

Table 1. Theoretical interband $E1$ and $E3$ transition probabilities, and intraband $E2$ (in Weisskopf units), for the low-lying bands in ^{225}Ra calculated with the CQC$_{\rm m}$ and CQC models, in comparison with the available experimental values.[134]

$B(E\lambda; J_i^\pi \to J_f^\pi)$	CQC$_{\rm m}$	CQC	Exp.
$B(E1; 3/2^-_{\rm Band2} \to 1/2^+_{\rm Band1})$	0.0026	0.0024	0.00081(20)
$B(E1; 5/2^-_{\rm Band2} \to 3/2^+_{\rm Band1})$	0.0026	0.0031	
$B(E1; 7/2^-_{\rm Band2} \to 5/2^+_{\rm Band1})$	0.0043	0.0038	
$B(E1; 9/2^-_{\rm Band2} \to 7/2^+_{\rm Band1})$	0.0023	0.0028	
$B(E1; 11/2^-_{\rm Band2} \to 9/2^+_{\rm Band1})$	0.0039	0.0042	
$B(E1; 13/2^-_{\rm Band2} \to 11/2^+_{\rm Band1})$	0.0031	0.0038	
$B(E2; 3/2^+_{\rm Band1} \to 1/2^+_{\rm Band1})$	5.97	56.25	> 2.0
$B(E2; 5/2^+_{\rm Band1} \to 1/2^+_{\rm Band1})$	83.60	91.63	102(6)
$B(E2; 5/2^+_{\rm Band1} \to 3/2^+_{\rm Band1})$	60.49	64.14	
$B(E2; 7/2^+_{\rm Band1} \to 3/2^+_{\rm Band1})$	58.12	101.87	
$B(E2; 7/2^+_{\rm Band1} \to 5/2^+_{\rm Band1})$	18.25	37.22	
$B(E2; 9/2^+_{\rm Band1} \to 5/2^+_{\rm Band1})$	142.36	151.37	
$B(E2; 9/2^+_{\rm Band1} \to 7/2^+_{\rm Band1})$	16.98	31.24	
$B(E2; 11/2^+_{\rm Band1} \to 7/2^+_{\rm Band1})$	85.46	154.06	
$B(E2; 11/2^+_{\rm Band1} \to 9/2^+_{\rm Band1})$	22.40	22.01	
$B(E2; 3/2^-_{\rm Band2} \to 1/2^-_{\rm Band2})$	87.44	94.38	
$B(E2; 5/2^-_{\rm Band2} \to 1/2^-_{\rm Band2})$	12.65	32.85	
$B(E2; 5/2^-_{\rm Band2} \to 3/2^-_{\rm Band2})$	71.20	66.73	
$B(E2; 7/2^-_{\rm Band2} \to 3/2^-_{\rm Band2})$	127.24	130.65	
$B(E2; 7/2^-_{\rm Band2} \to 5/2^-_{\rm Band2})$	51.23	49.33	
$B(E2; 9/2^-_{\rm Band2} \to 5/2^-_{\rm Band2})$	124.76	84.59	
$B(E2; 9/2^-_{\rm Band2} \to 7/2^-_{\rm Band2})$	32.89	22.68	
$B(E2; 11/2^-_{\rm Band2} \to 7/2^-_{\rm Band2})$	159.55	164.03	
$B(E2; 11/2^-_{\rm Band2} \to 9/2^-_{\rm Band2})$	25.41	21.91	
$B(E3; 7/2^-_{\rm Band2} \to 1/2^+_{\rm Band1})$	34.98	38.26	
$B(E3; 9/2^-_{\rm Band2} \to 3/2^+_{\rm Band1})$	36.81	29.27	
$B(E3; 11/2^-_{\rm Band2} \to 5/2^+_{\rm Band1})$	65.84	71.21	
$B(E3; 13/2^-_{\rm Band2} \to 7/2^+_{\rm Band1})$	33.30	60.78	
$B(E3; 15/2^-_{\rm Band2} \to 9/2^+_{\rm Band1})$	74.51	83.75	
$B(E3; 17/2^-_{\rm Band2} \to 11/2^+_{\rm Band1})$	74.79	79.58	
$B(E3; 19/2^-_{\rm Band2} \to 13/2^+_{\rm Band1})$	93.73	90.93	
$B(E3; 21/2^-_{\rm Band2} \to 15/2^+_{\rm Band1})$	92.34	97.79	

and the even–even cores. To illustrate the improvement of signature splitting in the ground-state band, we also show the wave function for the ground-state band obtained using the original CQC model in Fig. 2(c). The CQC_m model wave functions are characterized by a more pronounced contribution of the $1j_{15/2}$ configuration, especially in the sequence of $1/2^+, 5/2^+, 9/2^+$ and $13/2^+$, which results in a stronger Coriolis coupling and brings the theoretical results in better agreement with the experimental values. Furthermore, we find that the single-particle levels, i.e., $3d_{5/2}$, $2g_{7/2}$, $1i_{11/2}$ and $1j_{15/2}$, are compressed by the monopole p–h interaction in the CQC_m model, and this leads to stronger admixtures of the $1j_{15/2}$ configuration.

Table 1 collects the results for interband electric dipole $E1$ and octupole $E3$ transition probabilities, and intraband electric quadrupole $E2$ transition probabilities for ^{225}Ra. The transitions rates obtained with both models are rather similar, except for few $E2$ transitions in the ground-state band, which may be caused by the different Coriolis coupling. For the available experimental values, both models reproduce the data, especially considering that the present calculations include the full configuration space and, therefore, do not contain adjustable parameters in the form of effective charges. Of course, the available data are quite limited at present and, therefore, further experiments are called for, especially for the $E1$ and $E3$ transitions related to the ground-state band.

4. Conclusion

In summary, the microscopic core–quasiparticle coupling model for the spectroscopy of odd-A nuclei with octupole correlations has been extended by including the monopole coupling between the core and the odd nucleon. In the new version of the model, the Fermi surface is determined as the difference between the absolute binding energies of the neighboring even–even cores, and the coupling strengths for the monopole, quadrupole and octupole interactions are adjusted to the experimental excitation energies of few selected states. The model has been tested in a study of low-lying excitation spectra for the odd-A nucleus ^{225}Ra, characterized by octupole deformation. The theoretical results, e.g., the parity doublets, interband $B(E1)$ and intraband $B(E2)$, are in agreement with the available data. In particular, the description of the signature splitting in the ground-state band has been improved. The improved CQC model can be further applied to a systematic study of odd-A Ra and actinides, a work that is currently in progress.

Acknowledgments

This work was partly supported by the National Natural Science Foundation of China (Grants No. 11875225), the Fundamental Research Funds for the Central Universities and the Fok Ying-Tong Education Foundation.

ORCID

Zhipan Li https://orcid.org/0000-0002-3309-2832

References

1. F. Asaro, F. Stephens and I. Perlman, *Phys. Rev.* **92** (1953) 1495.
2. F. S. Stephens, F. Asaro and I. Perlman, *Phys. Rev.* **100** (1955) 1543.
3. K. Lee and D. R. Inglis, *Phys. Rev.* **108** (1957) 774.
4. P. A. Butler and W. Nazarewicz, *Rev. Mod. Phys.* **68** (1996) 349.
5. P. A. Butler, *J. Phys. G.* **43** (2016) 073002.
6. P. A. Butler, L. P. Gaffney, P. Spagnoletti, K. Abrahams, M. Bowry, J. Cederkäll, G. de Angelis, H. De Witte, P. E. Garrett, A. Goldkuhle, C. Henrich, A. Illana, K. Johnston, D. T. Joss, J. M. Keatings, N. A. Kelly, M. Komorowska, J. Konki, T. Kröll, M. Lozano, B. S. Nara Singh, D. O'Donnell, J. Ojala, R. D. Page, L. G. Pedersen, C. Raison, P. Reiter, J. A. Rodriguez, D. Rosiak, S. Rothe, M. Scheck, M. Seidlitz, T. M. Shneidman, B. Siebeck, J. Sinclair, J. F. Smith, M. Stryjczyk, P. Van Duppen, S. Vinals, V. Virtanen, N. Warr, K. Wrzosek-Lipska and M. Zielińska, *Phys. Rev. Lett.* **124** (2020) 042503.
7. L. P. Gaffney, P. A. Butler, M. Scheck, A. B. Hayes, F. Wenander, M. Albers, B. Bastin, C. Bauer, A. Blazhev, S. Bnig, N. Bree, J. Cederkll, T. Chupp, D. Cline, T. E. Cocolios, T. Davinson, H. De Witte, J. Diriken, T. Grahn, A. Herzan, M. Huyse, D. G. Jenkins, D. T. Joss, N. Kesteloot, J. Konki, M. Kowalczyk, Th. Krll, E. Kwan, R. Lutter, K. Moschner, P. Napiorkowski, J. Pakarinen, M. Pfeiffer, D. Radeck, P. Reiter, K. Reynders, S. V. Rigby, L. M. Robledo, M. Rudigier, S. Sambi, M. Seidlitz, B. Siebeck, T. Stora, P. Thoele, P. Van Duppen, M. J. Vermeulen, M. von Schmid, D. Voulot, N. Warr, K. Wimmer, K. Wrzosek-Lipska, C. Y. Wu and M. Zielinska, *Nature* **497** (2013) 199.
8. B. Bucher, S. Zhu, C. Y. Wu, R. V. F. Janssens, D. Cline, A. B. Hayes, M. Albers, A. D. Ayangeakaa, P. A. Butler, C. M. Campbell, M. P. Carpenter, C. J. Chiara, J. A. Clark, H. L. Crawford, M. Cromaz, H. M. David, C. Dickerson, E. T. Gregor, J. Harker, C. R. Hoffman, B. P. Kay, F. G. Kondev, A. Korichi, T. Lauritsen, A. O. Macchiavelli, R. C. Pardo, A. Richard, M. A. Riley, G. Savard, M. Scheck, D. Seweryniak, M. K. Smith, R. Vondrasek and A. Wiens, *Phys. Rev. Lett.* **116** (2016) 112503.
9. B. Bucher, S. Zhu, C. Y. Wu, R. V. F. Janssens, R. N. Bernard, L. M. Robledo, T. R. Rodríguez, D. Cline, A. B. Hayes, A. D. Ayangeakaa, M. Q. Buckner,

C. M. Campbell, M. P. Carpenter, J. A. Clark, H. L. Crawford, H. M. David, C. Dickerson, J. Harker, C. R. Hoffman, B. P. Kay, F. G. Kondev, T. Lauritsen, A. O. Macchiavelli, R. C. Pardo, G. Savard, D. Seweryniak and R. Vondrasek, *Phys. Rev. Lett.* **118** (2017) 152504.
10. S. L. Rugari, R. H. France, B. J. Lund, Z. Zhao, M. Gai, P. A. Butler, V. A. Holliday, A. N. James, G. D. Jones, R. J. Poynter, R. J. Tanner, K. L. Ying and J. Simpson, *Phys. Rev. C* **48** (1993) 2078.
11. C. Fahlander, D. Seweryniak, J. Nyberg, Zs. Dombrdi, G. E. Perez, M. Jzsa, B. M. Nyak, A. Atac, B. Cederwall, A. Johnson, A. Kerek, J. Kownacki, L.-O. Norlin, R. Wyss, E. Adamides, E. Ideguchi, R. Julin, S. Juutinen, W. Karczmarczyk, S. Mitarai, M. Piiparinen, R. Schubart, G. Sletten, S. Trmnen and A. Virtanen, *Nucl. Phys. A* **577** (1994) 773–785.
12. J. F. Smith, C. J. Chiara, D. B. Fossan, G. J. Lane, J. M. Sears, I. Thorslund, H. Amro, C. N. Davids, R. V. F. Janssens, D. Seweryniak, I. M. Hibbert, R. Wadsworth, I. Y. Lee and A. O. Macchiavelli, *Phys. Rev. C* **57** (1998) R1037.
13. J. DeGraaf, M. Cromaz, T. E. Drake, V. P. Janzen, D. C. Radford and D. Ward, *Phys. Rev. C* **58** (1998) 164.
14. E. S. Paul, H. C. Scraggs, A. J. Boston, O. Dorvaux, P. T. Greenlees, K. Helariutta, P. Jones, R. Julin, S. Juutinen, H. Kankaanp, H. Kettunen, M. Muikku, P. Nieminen, P. Rahkila and O. Stezowski, *Nucl. Phys. A* **673** (2000) 31.
15. T. Rzaca-Urban, W. Urban, A. Kaczor, J. L. Durell, M. J. Leddy, M. A. Jones, W. R. Phillips, A. G. Smith, B. J. Varley, I. Ahmad, L. R. Morss, M. Bentaleb, E. Lubkiewicz and N. Schulz, *Eur. Phys. J. A — Hadrons Nucl.* **9** (2000) 165.
16. J. F. Smith, C. J. Chiara, D. B. Fossan, D. R. LaFosse, G. J. Lane, J. M. Sears, K. Starosta, M. Devlin, F. Lerma, D. G. Sarantites, S. J. Freeman, M. J. Leddy, J. L. Durell, A. J. Boston, E. S. Paul, A. T. Semple, I. Y. Lee, A. O. Macchiavelli and P. H. Heenen, *Phys. Lett. B* **523** (2001) 13.
17. G. de Angelis, A. Gadea, E. Farnea, R. Isocrate, P. Petkov, N. Marginean, D. R. Napoli, A. Dewald, M. Bellato, A. Bracco, F. Camera, D. Curien, M. DePoli, E. Fioretto, A. Fitzler, S. Kasemann, N. Kintz, T. Klug, S. Lenzi, S. Lunardi, R. Menegazzo, P. Pavan, J. L. Pedroza, V. Pucknell, C. Ring, J. Sampson and R. Wyss, *Phys. Lett. B* **535** (2002) 93.
18. L. Capponi, J. F. Smith, P. Ruotsalainen, C. Scholey, P. Rahkila, K. Auranen, L. Bianco, A. J. Boston, H. C. Boston, D. M. Cullen, X. Derkx, M. C. Drummond, T. Grahn, P. T. Greenlees, L. Grocutt, B. Hadinia, U. Jakobsson, D. T. Joss, R. Julin, S. Juutinen, M. Labiche, M. Leino, K. G. Leach, C. McPeake, K. F. Mulholland, P. Nieminen, D. O'Donnell, E. S. Paul, P. Peura, M. Sandzelius, J. Sarén, B. Saygi, J. Sorri, S. Stolze, A. Thornthwaite, M. J. Taylor and J. Uusitalo, *Phys. Rev. C* **94** (2016) 024314.
19. E. T. Gregor, M. Scheck, R. Chapman, L. P. Gaffney, J. Keatings, K. R. Mashtakov, D. ODonnell, J. F. Smith, P. Spagnoletti, M. Thrauf, V. Werner and C. Wiseman, *Eur. Phys. J. A* **53** (2017) 50.
20. E. T. Gregor, N. N. Arsenyev, M. Scheck, T. M. Shneidman, M. Thrauf, C. Bernards, A. Blanc, R. Chapman, F. Drouet, A. A. Dzhioev, G. de France,

M. Jentschel, J. Jolie, J. M. Keatings, Th. Krll, U. Kster, R. Leguillon, K. R. Mashtakov, P. Mutti, D. ODonnell, C. M. Petrache, G. S. Simpson, J. Sinclair, J. F. Smith, T. Soldner, P. Spagnoletti, A. V. Sushkov, W. Urban, A. Vancraeyenest, J. R. Vanhoy, V. Werner, K. O. Zell and M. Zieliska, *J. Phys. G: Nucl. Particle Phys.* **46** (2019) 075101.

21. M. M. R. Chishti, D. ODonnell, G. Battaglia, M. Bowry, D. A. Jaroszynski, B. S. Nara Singh, M. Scheck, P. Spagnoletti and J. F. Smith, *Nat. Phys.* **16** (2020) 853.
22. C. M. Petrache, N. Minkov, T. Nakatsukasa, B. F. Lv, A. Astier, E. Dupont, K. K. Zheng, P. Greenlees, H. Badran, T. Calverley, D. M. Cox, T. Grahn, J. Hilton, R. Julin, S. Juutinen, J. Konki, J. Pakarinen, P. Papadakis, J. Partanen, P. Rahkila, P. Ruotsalainen, M. Sandzelius, J. Saren, C. Scholey, J. Sorri, S. Stolze, J. Uusitalo, B. Cederwall, A. Ertoprak, H. Liu, S. Guo, M. L. Liu, J. G. Wang, X. H. Zhou, I. Kuti, J. Timár, A. Tucholski, J. Srebrny and C. Andreoiu, *Phys. Rev. C* **102** (2020) 014311.
23. A. Yagi, A. Odahara, H. Nishibata, R. Lozeva, C.-B. Moon, S. Nishimura, K. Yoshida, N. Yoshinaga, C. Watanabe, K. Higashiyama, T. Shimoda, R. Daido, Y. Fang, P. S. Lee, B. Moon, P. Doornenbal, G. Lorusso, P.-A. Söderström, T. Sumikama, H. Watanabe, T. Isobe, H. Baba, H. Sakurai, F. Browne, Z. Patel, S. Rice, L. Sinclair, J. Wu, Z. Y. Xu, R. Yokoyama, T. Kubo, N. Inabe, H. Suzuki, N. Fukuda, D. Kameda, H. Takeda, D. S. Ahn, Y. Shimizu, D. Murai, F. L. Bello Garrote, J.-M. Daugas, F. Didierjean, E. Ideguchi, S. Iimura, T. Ishigaki, H. S. Jung, T. Komatsubara, Y. K. Kwon, C. S. Lee, S. Morimoto, M. Niikura, I. Nishizuka and K. Tshoo, *Phys. Rev. C* **105** (2022) 044325.
24. B. F. Lv, C. M. Petrache, K. K. Zheng, Z. H. Zhang, W. Sun, Z. P. Li, X. T. He, J. Zhang, A. Astier, P. Greenlees, T. Grahn, R. Julin, S. Juutinen, M. Luoma, J. Ojala, J. Pakarinen, J. Partanen, P. Rahkila, P. Ruotsalainen, M. Sandzelius, J. Sarén, H. Tann, J. Uusitalo, G. Zimba, B. Cederwall, Ö. Aktas, A. Ertoprak, W. Zhang, S. Guo, M. L. Liu, H. J. Ong, Z. Y. Sun, J. G. Wang, X. H. Zhou, I. Kuti, B. M. Nyakó, D. Sohler, J. Timár, C. Andreoiu, M. Doncel, D. T. Joss and R. D. Page, *Phys. Rev. C* **105** (2022) 044319.
25. A. Mukherjee, S. Bhattacharya, T. Trivedi, R. P. Singh, S. Muralithar, D. Negi, R. Palit, S. Nag, S. Rajbanshi, M. Kumar Raju, S. Kumar, D. Choudhury, R. Kumar, R. K. Bhowmik, S. C. Pancholi and A. K. Jain, *Phys. Rev. C* **105** (2022) 014322.
26. C. J. Zhang and J. Y. Jia, *Phys. Rev. Lett.* **128** (2022) 022301.
27. C. G. Wang, R. Han, C. Xu, H. Hua, R. A. Bark, S. Q. Zhang, S. Y. Wang, T. M. Shneidman, S. G. Zhou, J. Meng, S. M. Wyngaardt, A. C. Dai, F. R. Xu, X. Q. Li, Z. H. Li, Y. L. Ye, D. X. Jiang, C. G. Li, C. Y. Niu, Z. Q. Chen, H. Y. Wu, D. W. Luo, S. Wang, D. P. Sun, C. Liu, Z. Q. Li, N. B. Zhang, R. J. Guo, P. Jones, E. A. Lawrie, J. J. Lawrie, J. F. Sharpey-Schafer, M. Wiedeking, S. N. T. Majola, T. D. Bucher, T. Dinoko, B. Maqabuka, L. Makhathini, L. Mdletshe, O. Shirinda and K. Sowazi, *Phys. Rev. C* **106** (2022) L011303.
28. X. C. Chen, J. Zhao, C. Xu, H. Hua, T. M. Shneidman, S. G. Zhou, X. G. Wu, X. Q. Li, S. Q. Zhang, Z. H. Li, W. Y. Liang, J. Meng, F. R. Xu, B. Qi, Y. L. Ye,

D. X. Jiang, Y. Y. Cheng, C. He, J. J. Sun, R. Han, C. Y. Niu, C. G. Li, P. J. Li, C. G. Wang, H. Y. Wu, Z. H. Li, H. Zhou, S. P. Hu, H. Q. Zhang, G. S. Li, C. Y. He, Y. Zheng, C. B. Li, H. W. Li, Y. H. Wu, P. W. Luo and J. Zhong, *Phys. Rev. C* **94** (2016) 021301.
29. B. Ding, Z. Liu, D. Seweryniak, P. J. Woods, H. L. Wang, J. Yang, H. L. Liu, C. N. Davids, M. P. Carpenter, T. Davinson, R. V. F. Janssens, R. D. Page, A. P. Robinson, J. Shergur, S. Sinha, S. Zhu, X. D. Tang, J. G. Wang, T. H. Huang, W. Q. Zhang, M. D. Sun, X. Y. Liu and H. Y. Lu, *Phys. Rev. C* **95** (2017) 024301.
30. B. Olaizola, A. Babu, R. Umashankar, A. B. Garnsworthy, G. C. Ball, V. Bildstein, M. Bowry, C. Burbadge, R. Cabellero-Folch, I. Dillmann, A. Diaz-Varela, R. Dunlop, A. Estradé, P. E. Garrett, G. Hackman, A. D. MacLean, J. Measures, C. J. Pearson, B. Shaw, D. Southall, C. E. Svensson, J. Turko, K. Whitmore and T. Zidar, *Phys. Rev. C* **104** (2021) 034307.
31. R. Lică, G. Benzoni, A. I. Morales, M. J. G. Borge, L. M. Fraile, H. Mach, M. Madurga, C. Sotty, V. Vedia, H. De Witte, J. Benito, T. Berry, N. Blasi, A. Bracco, F. Camera, S. Ceruti, V. Charviakova, N. Cieplicka-Oryńczak, C. Costache, F. C. L. Crespi, J. Creswell, G. Fernández-Martínez, H. Fynbo, P. Greenlees, I. Homm, M. Huyse, J. Jolie, V. Karayonchev, U. Kster, J. Konki, T. Krll, J. Kurcewicz, T. Kurtukian-Nieto, I. Lazarus, S. Leoni, M. Lund, N. Marginean, R. Marginean, C. Mihai, R. Mihai, A. Negret, A. Orduz, Z. Patyk, S. Pascu, V. Pucknell, P. Rahkila, J. M. Regis, F. Rotaru, N. Saed-Sami, V. Sánchez-Tembleque, M. Stanoiu, O. Tengblad, M. Thuerauf, A. Turturica, P. Van Duppen and N. Warr, *J. Phys. G: Nucl. Particle Phys.* **44** (2017) 054002.
32. J. Wiśniewski, W. Urban, T. Rzaca-Urban, A. G. Smith, J. F. Smith, G. S. Simpson, I. Ahmad and J. P. Greene, *Phys. Rev. C* **96** (2017) 064301.
33. E. H. Wang, W. Lewis, C. J. Zachary, J. H. Hamilton, A. V. Ramayya, J. K. Hwang, S. H. Liu, N. T. Brewer, Y. X. Luo, J. O. Rasmussen, S. J. Zhu, G. M. Ter-Akopian and Yu. Ts. Oganessian, *Eur. Phys. J. A* **53** (2017) 234.
34. E. Ruchowska, H. Mach, M. Kowal, J. Skalski, W. A. Płóciennik and B. Fogelberg, *Phys. Rev. C* **92** (2015) 034328.
35. H. J. Li, S. J. Zhu, J. H. Hamilton, E. H. Wang, A. V. Ramayya, Y. J. Chen, J. K. Hwang, J. Ranger, S. H. Liu, Z. G. Xiao, Y. Huang, Z. Zhang, Y. X. Luo, J. O. Rasmussen, I. Y. Lee, G. M. Ter-Akopian, Yu. Ts. Oganessian and W. C. Ma, *Phys. Rev. C* **90** (2014) 047303.
36. T. Rzaca-Urban, W. Urban, A. G. Smith, I. Ahmad and A. Syntfeld-Każuch, *Phys. Rev. C* **87** (2013) 031305.
37. T. Rzaca-Urban, W. Urban, J. A. Pinston, G. S. Simpson, A. G. Smith and I. Ahmad, *Phys. Rev. C* **86** (2012) 044324.
38. I. Ahmad, R. R. Chasman, J. P. Greene, F. G. Kondev and S. Zhu, *Phys. Rev. C* **92** (2015) 024313.
39. W. Reviol, R. V. F. Janssens, S. Frauendorf, D. G. Sarantites, M. P. Carpenter, X. Chen, C. J. Chiara, D. J. Hartley, K. Hauschild, T. Lauritsen, A. Lopez-Martens, M. Montero, O. L. Pechenaya, D. Seweryniak, J. B. Snyder and S. Zhu, *Phys. Rev. C* **90** (2014) 044318.

40. S. K. Tandel, M. Hemalatha, A. Y. Deo, S. B. Patel, R. Palit, T. Trivedi, J. Sethi, S. Saha, D. C. Biswas and S. Mukhopadhyay, *Phys. Rev. C* **87** (2013) 034319.
41. W. Reviol, D. G. Sarantites, C. J. Chiara, M. Montero, R. V. F. Janssens, M. P. Carpenter, T. L. Khoo, T. Lauritsen, C. J. Lister, D. Seweryniak, S. Zhu, O. L. Pechenaya and S. G. Frauendorf, *Phys. Rev. C* **80** (2009) 011304.
42. M. Bishof, R. H. Parker, K. G. Bailey, J. P. Greene, R. J. Holt, M. R. Kalita, W. Korsch, N. D. Lemke, Z. T. Lu, P. Mueller, T. P. O'Connor, J. T. Singh and M. R. Dietrich, *Phys. Rev. C* **94** (2016) 025501.
43. R. H. Parker, M. R. Dietrich, M. R. Kalita, N. D. Lemke, K. G. Bailey, M. Bishof, J. P. Greene, R. J. Holt, W. Korsch, Z.-T. Lu, P. Mueller, T. P. O'Connor and J. T. Singh, *Phys. Rev. Lett.* **114** (2015) 233002.
44. J. Dobaczewski and J. Engel, *Phys. Rev. Lett.* **94** (2005) 232502.
45. N. Auerbach, V. V. Flambaum and V. Spevak, *Phys. Rev. Lett.* **76** (1996) 4316.
46. V. V. Flambaum, *Phys. Rev. C* **99** (2019) 035501.
47. V. V. Flambaum and V. A. Dzuba, *Phys. Rev. A* **101** (2020) 042504.
48. R. Rodrguez-Guzmn, L. M. Robledo, K. Nomura and N. C. Hernandez, *J. Phys. G: Nucl. Particle Phys.* **49** (2021) 015101.
49. J. Yang, J. Dudek, I. Dedes, A. Baran, D. Curien, A. Gaamouci, A. Góźdź, A. Pędrak, D. Rouvel, H. L. Wang and J. Burkat, *Phys. Rev. C* **105** (2022) 034348.
50. K. Nomura, L. Lotina, T. Nikšić and D. Vretenar, *Phys. Rev. C* **103** (2021) 054301.
51. Y. C. Cao, S. E. Agbemava, A. V. Afanasjev, W. Nazarewicz and E. Olsen, *Phys. Rev. C* **102** (2020) 024311.
52. A. V. Afanasjev, H. Abusara and S. E. Agbemava, *Phys. Scr.* **93** (2018) 034002.
53. Y. Fu, H. Wang, L. J. Wang and J. M. Yao, *Phys. Rev. C* **97** (2018) 024338.
54. P. Marević, J.-P. Ebran, E. Khan, T. Nikšić and D. Vretenar, *Phys. Rev. C* **97** (2018) 024334.
55. K. Nomura, T. Nikšić and D. Vretenar, *Phys. Rev. C* **97** (2018) 024317.
56. W. Zhang and Y. F. Niu, *Phys. Rev. C* **96** (2017) 054308.
57. S. E. Agbemava and A. V. Afanasjev, *Phys. Rev. C* **96** (2017) 024301.
58. J. M. Yao and J. Engel, *Phys. Rev. C* **94** (2016) 014306.
59. L. S. Geng, J. Meng and H. Toki, *Chin. Phys. Lett.* **24** (2007) 1865.
60. J. Y. Guo, P. Jiao and X. Z. Fang, *Phys. Rev. C* **82** (2010) 047301.
61. L. M. Robledo, M. Baldo, P. Schuck and X. Viñas, *Phys. Rev. C* **81** (2010) 034315.
62. L. M. Robledo and G. F. Bertsch, *Phys. Rev. C* **84** (2011) 054302.
63. R. Rodríguez-Guzmán, L. M. Robledo and P. Sarriguren, *Phys. Rev. C* **86** (2012) 034336.
64. R. Rodrguez-Guzmn, Y. M. Humadi and L. M. Robledo, *Eur. Phys. J. A* **56** (2020) 43.
65. R. Rodrguez-Guzmn, Y. M. Humadi and L. M. Robledo, *J. Phys. G: Nucl. Particle Phys.* **48** (2020) 015103.
66. R. Rodríguez-Guzmán and L. M. Robledo, *Phys. Rev. C* **103** (2021) 044301.
67. L. M. Robledo and P. A. Butler, *Phys. Rev. C* **88** (2013) 051302.

68. R. N. Bernard, L. M. Robledo and T. R. Rodríguez, *Phys. Rev. C* **93** (2016) 061302.
69. L. M. Robledo and G. F. Bertsch, *J. Phys. G* **42** (2015) 055109.
70. J. Zhao, B. N. Lu, E. G. Zhao and S. G. Zhou, *Phys. Rev. C* **86** (2012) 057304.
71. S. G. Zhou, *Phys. Scr.* **91** (2016) 063008.
72. K. Nomura, D. Vretenar and B. N. Lu, *Phys. Rev. C* **88** (2013) 021303.
73. K. Nomura, D. Vretenar, T. Nikšić and B.-N. Lu, *Phys. Rev. C* **89** (2014) 024312.
74. K. Nomura, R. Rodríguez-Guzmán and L. M. Robledo, *Phys. Rev. C* **92** (2015) 014312.
75. Y. T. Qiu, X. W. Wang and J. Y. Guo, *Phys. Rev. C* **106** (2022) 034301.
76. S. E. Agbemava, A. V. Afanasjev and P. Ring, *Phys. Rev. C* **93** (2016) 044304.
77. J. M. Yao and K. Hagino, *Phys. Rev. C* **94** (2016) 011303.
78. W. Zhang, Z. P. Li, S. Q. Zhang and J. Meng, *Phys. Rev. C* **81** (2010) 034302.
79. Z. P. Li, B. Y. Song, J. M. Yao, D. Vretenar and J. Meng, *Phys. Lett. B* **726** (2013) 866.
80. Z. P. Li, T. Niksic and D. Vretenar, *J. Phys. G* **43** (2016) 024005.
81. S. Y. Xia, H. Tao, Y. Lu, Z. P. Li, T. Nikšić and D. Vretenar, *Phys. Rev. C* **96** (2017) 054303.
82. J. M. Yao, E. F. Zhou and Z. P. Li, *Phys. Rev. C* **92** (2015) 041304.
83. E. F. Zhou, J. M. Yao, Z. P. Li, J. Meng and P. Ring, *Phys. Lett. B* **753** (2016) 227.
84. H. Tao, J. Zhao, Z. P. Li, T. Nikšić and D. Vretenar, *Phys. Rev. C* **96** (2017) 024319.
85. S. Ebata and T. Nakatsukasa, *Phys. Scr.* **92** (2017) 064005.
86. H. L. Wang, J. Yang, M. L. Liu and F. R. Xu, *Phys. Rev. C* **92** (2015) 024303.
87. P. Mller, R. Bengtsson, B. G. Carlsson, P. Olivius, T. Ichikawa, H. Sagawa and A. Iwamoto, *Atom. Data Nucl. Data Tabl.* **94** (2008) 758.
88. G. A. Leander, W. Nazarewicz, P. Olanders, I. Ragnarsson and J. Dudek, *Phys. Lett. B* **152** (1985) 284.
89. W. Nazarewicz, P. Olanders, I. Ragnarsson, J. Dudek, G. A. Leander, P. Mller and E. Ruchowsa, *Nucl. Phys. A* **429** (1984) 269.
90. K. Nomura, *Phys. Rev. C* **105** (2022) 054318.
91. K. Nomura, R. Rodríguez-Guzmán and L. M. Robledo, *Phys. Rev. C* **104** (2021) 054320.
92. K. Nomura, R. Rodríguez-Guzmán, L. M. Robledo, J. E. García-Ramos and N. C. Hernández, *Phys. Rev. C* **104** (2021) 044324.
93. O. Vallejos and J. Barea, *Phys. Rev. C* **104** (2021) 014308.
94. K. Nomura, R. Rodríguez-Guzmán, L. M. Robledo and J. E. García-Ramos, *Phys. Rev. C* **103** (2021) 044311.
95. H. S. El-Gendy, *Nucl. Phys. A* **1006** (2021) 122117.
96. K. Nomura, R. Rodríguez-Guzmán, Y. M. Humadi, L. M. Robledo and J. E. García-Ramos, *Phys. Rev. C* **102** (2016) 064326.
97. A. M. Khalaf, G. S. M. Ahmed, A. Abdelsalam and M. R. M. Allam, *Nucl. Phys. A* **988** (2019) 1.

98. K. Nomura, R. Rodríguez-Guzmán and L. M. Robledo, *Phys. Rev. C* **92** (2015) 014312.
99. K. Nomura, D. Vretenar, T. Nikšić and B. N. Lu, *Phys. Rev. C* **89** (2014) 024312.
100. K. Nomura, D. Vretenar and B. N. Lu, *Phys. Rev. C* **88** (2013) 021303.
101. N. V. Zamfir and D. Kusnezov, *Phys. Rev. C* **67** (2003) 014305.
102. N. V. Zamfir and D. Kusnezov, *Phys. Rev. C* **63** (2001) 054306.
103. T. Otsuka and M. Sugita, *Phys. Lett. B* **209** (1988) 140.
104. R. Budaca, P. Buganu and A. I. Budaca, *Phys. Rev. C* **106** (2022) 014311.
105. M. S. Nadirbekov, S. N. Kudiratov, F. N. Temirov and N. Minkov, *Int. J. Mod. Phys. E* **29** (2020) 2050031.
106. A. Dobrowolski, K. Mazurek and A. Góźdź, *Phys. Rev. C* **94** (2016) 054322.
107. P. G. Bizzeti and A. M. Bizzeti-Sona, *Phys. Rev. C* **88** (2013) 011305.
108. N. Minkov, S. Drenska, M. Strecker, W. Scheid and H. Lenske, *Phys. Rev. C* **85** (2012) 034306.
109. R. V. Jolos, P. von Brentano and J. Jolie, *Phys. Rev. C* **86** (2012) 024319.
110. D. Lenis and D. Bonatsos, *Phys. Lett. B* **633** (2006) 474.
111. D. Bonatsos, D. Lenis, N. Minkov, D. Petrellis and P. Yotov, *Phys. Rev. C* **71** (2005) 064309.
112. P. G. Bizzeti and A. M. Bizzeti-Sona, *Phys. Rev. C* **70** (2004) 064319.
113. N. Yoshinaga, K. Yanase, C. Watanabe and K. Higashiyama, *Prog. Theor. Exp. Phys.* **2021** (2021) 063D01.
114. X. T. He and Y. C. Li, *Phys. Rev. C* **102** (2020) 064328.
115. N. Yoshinaga, K. Yanase, K. Higashiyama and E. Teruya, *Phys. Rev. C* **98** (2018) 044321.
116. Y. S. Chen and Z. C. Gao, *Phys. Rev. C* **63** (2000) 014314.
117. Y. Y. Wang, Q. B. Chen and S. Q. Zhang, *Phys. Rev. C* **105** (2022) 044316.
118. Y. Y. Wang, *Phys. Rev. C* **104** (2021) 014318.
119. Y. Y. Wang and S. Q. Zhang, *Phys. Rev. C* **102** (2020) 034303.
120. Y. P. Wang, Y. Y. Wang and J. Meng, *Phys. Rev. C* **102** (2020) 024313.
121. Y. Y. Wang, S. Q. Zhang, P. W. Zhao and J. Meng, *Phys. Lett. B* **792** (2019) 454.
122. N. Minkov and A. Pálffy, *Phys. Rev. Lett.* **118** (2017) 212501.
123. N. Minkov, S. Drenska, P. Yotov, S. Lalkovski, D. Bonatsos and W. Scheid, *Phys. Rev. C* **76** (2007) 034324.
124. G. A. Leander and Y. S. Chen, *Phys. Rev. C* **37** (1988) 2744.
125. G. A. Leander and Y. S. Chen, *Phys. Rev. C* **35** (1987) 1145.
126. G. A. Leander and R. K. Sheline, *Nucl. Phys. A* **413** (1984) 375.
127. G. G. Adamian, N. V. Antonenko, R. V. Jolos and T. M. Shneidman, *Phys. Rev. C* **70** (2004) 064318.
128. Y. J. Chen, Z. C. Gao, Y. S. Chen and Y. Tu, *Phys. Rev. C* **91** (2015) 014317.
129. C. E. Alonso, J. M. Arias, A. Frank, H. M. Sofia, S. M. Lenzi and A. Vitturi, *Nucl. Phys. A* **586** (1995) 100.
130. W. Sun, S. Quan, Z. P. Li, J. Zhao, T. Nikšić and D. Vretenar, *Phys. Rev. C* **100** (2019) 044319.

131. A. Klein, P. Protopapas, S. G. Rohoziński and K. Starosta, *Phys. Rev. C* **69** (2004) 034338.
132. P. Marević, J.-P. Ebran, E. Khan, T. Nikšić and D. Vretenar, *Phys. Rev. C* **97** (2018) 024334.
133. P. W. Zhao, Z. P. Li, J. M. Yao and J. Meng, *Phys. Rev. C* **82** (2010) 054319.
134. B. N. Laboratory, National Nuclear Data Center, http://www.nndc.bnl.gov/.
135. M. Wang, W. J. Huang, F. G. Kondev, G. Audi and S. Naimi, *Chin. Phys. C* **45** (2021) 030003.

Fission within dinuclear system approach*

H. Paşca,[†,‡] A. V. Andreev,[†]
G. G. Adamian[†,§] and N. V. Antonenko[†]

[†]*Joint Institute for Nuclear Research,*
6 Joliot-Curie Street, Dubna,
141980 Moscow, Russia

[‡]*Faculty of Physics,*
"Babeş-Bolyai" University,
1 Mihail Kogălniceanu Street,
400084 Cluj-Napoca, Romania
[§]*adamian@theor.jinr.edu*

The improved scission-point statistical model based on the dinuclear system approach is employed to describe spontaneous fission, electromagnetic-, neutron-, charged-particle- and heavy-ion-induced fission of even–even pre-actinides, actinides and superheavy nuclei and to analyze the correlations between various observables. The key element of the model is the calculation of potential energy surfaces. The evolution of fission observables with increasing excitation energy is shown to be related with the widening and migration of the minima in the potential energy surface. Conservation of asymmetric shapes of mass and charge distributions of the fission fragments at high enough excitation energies of fissioning nuclei Hg, Pb, Rn, Ra, Th, U, Cf, Fm and No is predicted. At some critical excitation energy, the saturation of the symmetric component of charge and mass yields is demonstrated. For fissioning ^{180,182,190,198}Hg, $^{250-258}$Fm and $^{250-258}$No, transitions from two-peaked to single-peaked mass distributions are predicted. The origin of the transition between asymmetric and symmetric fission modes with variations of neutron number and excitation energy is explored. For ^{182}Hg(i.f.), ^{198}Hg(i.f.), ^{254}Fm(i.f.), ^{257}Fm($n_{\rm th}$, f) and ^{260}Fm(s.f.), the unexpected difference (symmetric or asymmetric) between the shapes of charge and mass distributions is predicted for the first time. The dependence of the neutron excess ratio of fission fragments on the fragment charge number is studied. A method is suggested for experimental verification of the multi-chance fission assumption. A possible explanation of the anomaly in charge yield of Mo/Sn fragments in the fission reaction ^{238}U(n, f) at low excitation energies found by γ–γ coincidence spectroscopy is presented.

---

*To cite this article, please refer to its earlier version published in the *International Journal of Modern Physics E*, Volume 32(10), (90 pages), 2340005 (2023), DOI: 10.1142/S0218301323400050.

[§]Corresponding author.

Keywords: Spontaneous fission; induced fission; charge and mass distributions; total kinetic energy; neutron multiplicity; asymmetric and symmetric fission modes; scission-point fission model.

1. Introduction

The fission process remains one of the most interesting puzzles of modern physics despite several decades of experimental and theoretical investigations. As known from the measured data, the charge and mass distributions of pre-actinides are symmetric while mass- and charge-asymmetric fragmentations are favored in nuclei heavier than thorium, U–Cf.[1–3] The fissioning nuclei ^{A}Pa and ^{A}Th with $A < 226$ ($A > 226$) and $A = 226$ display single-peaked (two-peaked) and triple-peaked charge distributions, respectively.[4] In general, the distribution asymmetry is attributed to the shell effects in the fissioning nucleus or fission fragments. The closed spherical shells similar to those found in the vicinity of Sn ($Z_H = 50$, $N_H = 82$) or/and deformed neutron shells (e.g., $N_H = 88$) crucially influence the potential energy landscape creating minima that favor asymmetric divisions of the parent nucleus. There is a longstanding opinion that the competition between symmetric and asymmetric fission modes is mainly related to deformed shells.[1,5] However, in the neutron-deficient ^{180}Hg, the asymmetric mass distribution of fission fragments was unexpectedly observed.[2,3,6] This is unusual behavior since the original prediction was that the symmetric division would be even more pronounced due to the magic neutron $N_{L,H} = 50$ and semimagic proton $Z_{L,H} = 50$ shells in the ^{90}Zr+^{90}Zr fragmentation. In Ref. 6, this asymmetric fission was interpreted as a new mode, not related to any fission-fragment shells. The dynamical effects during the evolution from a relatively compact mononucleus system to a dinuclear system (DNS) may affect the competition between the symmetric and asymmetric fission modes.[7–9] In contrast, the theoretical and experimental studies[10–14] have linked this asymmetric fission to the deformed proton and/or neutron shell gaps around $Z_L = 36$, $N_H = 52$ and 56. Therefore, there is still no clear understanding of the roles of microscopic and macroscopic effects in the fission process.[15]

In the region of heavy actinides, the onset of symmetric distribution was first observed in the spontaneous fission of ^{257}Fm.[16–19] For ^{258}Fm, the spontaneous fission results in an unexpected narrow symmetric mass distribution.[20] Symmetric distributions were also observed in the spontaneous fission of ^{259}Fm, ^{259,260}Md, ^{258,262}No and ^{259}Lr.[16,17,19] The thermal-neutron-induced fission of ^{256,257}Fm[16,17,19] leads to the symmetric mass distribution but with larger width than that in the spontaneous fission of ^{258}Fm. For spontaneous fission of Fm isotopes, the evolution of the mass-yield shape occurs suddenly with increasing isospin.[20] However, a survey of the literature reveals a lack of data on the charge distributions of fission products in the spontaneous and the induced fission of Fm, No and Rf, particularly as a function of isospin and excitation energy. It is interesting to answer the question if there is a difference between the shapes (symmetric or asymmetric) of charge and mass distributions in the fission of nuclei in this trans-einsteinium region. Thus, one can focus their studies on the isospin and excitation energy dependencies of the charge yields. It is also important to study the correlations between all fission observables in the case of transition from unimodal to bimodal fission. Another question of interest is the possible existence of the correlations between the shapes of neutron multiplicity, total kinetic energy (TKE) and charge/mass distributions, for instance, in the transition from the asymmetric fission of ^{254,256}Fm (^{252}No) to the symmetric one of ^{258,260}Fm (^{258,262}No).

The evolution of mass and charge distributions of fission fragments with excitation energy is also an important aspect of fission studies. The transition from an asymmetric distribution to a symmetric one with increasing excitation energy is attributed to the weakening of shell effects. However, in the cases of neutron-induced fission of ^{232}Th and ^{238}U, recent experimental data have shown that the mass distributions maintain their asymmetric shape even at neutron energies of $E_n \sim 60–70$ MeV.[21–23] The asymmetric shapes of the mass distribution resulting from the fission of highly excited nuclei $^{237-240}$U, $^{239-242}$Np and $^{241-244}$Pu (at excitation energies $E^*_{\rm CN} \sim 60–70$ MeV), produced in the transfer reaction ^{18}O+^{238}U at $E_{\rm lab} = 157.5$ MeV, were experimentally observed in Refs. 24 and 25. The same effect was observed in Refs. 12 and 26–28. The presence of a strong asymmetric component in the fission of ^{244}Cm at $E^*_{\rm CN} = 23$ MeV and of ^{250}Cf at $E^*_{\rm CN} = 46$ MeV was recently observed in Refs. 29 and 30. Furthermore, the experimental mass distribution of fission fragments of the compound nuclei (CN) ^{180,182,190}Hg* and ^{178}Pt* formed in complete fusion reactions ^{36}Ar+^{142}Nd, ^{144,154}Sm and ^{36}Ar+^{142}Nd at different bombarding energies[31–35] showed a weak change of the asymmetric shape of respective distributions, even though large excitation energies ($E^*_{\rm CN} = 60–75$ MeV) and angular momenta ($L = 59–74$) were involved. At these large excitation energies, the shell effects are expected to be washed out, and the nucleus is supposed to leave with a dominant symmetric mode, that is in contrast to experiments. Thus, the statement that the fast melting of shell corrections with excitation energy always leaves a fissioning nucleus with a dominant symmetric mode is not completely correct. Possible reasons of that could be an overestimation of the role of shell effects and an underestimation of the role of macroscopic effects. Since the shell effects are very important for the description of fission properties of heaviest actinides, it is interesting to study the excitation energy dependence of the fission observables in these nuclei. So the competition between symmetric and asymmetric fission modes with increasing excitation energy is still a matter of study. Note that the multi-chance fission at $E^*_{\rm CN} \leq 70$ MeV cannot explain[36,37] the presence of the asymmetric mode in the experiments mentioned.

There is another longstanding question. At which point of the fission path are the mass (charge) and energy distributions of the fission fragments determined? There exist several versions of "saddle-point models" and "scission-point models" which put this crucial point closer either to the saddle- or to the scission-point, respectively.[5,8,9,15,38–57] There is also a dynamical approach[8] where the potential energy surface (PES) in the entire deformation range from the saddle point to scission point governs the final yields of the fission fragments. The detailed analysis indicates that the formation of the observed total kinetic energy and mass distributions of the fission fragments occurs in the region closer to the scission region rather than to the saddle point. As shown for the fission of actinides in Ref. 8, the mass asymmetry distribution reflects the structure of the PES in the scission region. In this region, the fissioning nucleus has well-necked-in shape and the fragments attain their individual character. With the scission-point models,[15,38–47] the experimental data on fission of actinides and pre-actinides, mass, charge, kinetic energy and neutron multiplicity distributions, are well described. The wide ranges of described fission observables and effects confirm the predictive power of the scission-point-type models. Further theoretical and experimental studies of the mechanism of scission configuration formation are required.[58–62]

This review focuses on describing the charge, mass, total kinetic energy (TKE) and neutron multiplicity distributions of fission products and on comparing the calculated results with the available experimental data. Correlations between the shapes of charge/mass, average neutron multiplicity and total kinetic energy distributions are revealed. Note that there are new reliable experimental data on the mass, charge, and TKE distributions and the neutron multiplicity.[29,30,33,63–68] One of the aims is to demonstrate new predictions of charge and mass distributions at large excitation energies. The fission observables are described using the improved scission-point model[69–78] based on the DNS approach.[44] The main critical ingredient of this model is the PES, since any changes in the PES minima and their positions control all fission observables. Ultimately, the model that describes all fission features must provide a clear and physically sound method for calculating the PES.

2. Model

The statistical scission-point model relies on the assumption that statistical equilibrium is established at scission of a fissioning nucleus where the observable characteristics of fission are formed. The scission configuration is imagined as two axially deformed and uniformly charged ellipsoids — the nascent fragments.[79] The two nuclei are fully formed and possess all the features of isolated nuclei, e.g., binding energies, according to the separability principle, and mutually interact through the nuclear, Coulomb and centrifugal forces. Their orientation is frozen to a tip-to-tip configuration that provides a minimum of the fragment–fragment interaction potential $V = V^C + V^N + V^R$ (see Fig. 1).[69–78] Owing to the repulsive nature of the Coulomb, V^C, and centrifugal, V^R, interactions and attractive nature of the nuclear interaction, V^N, a potential pocket is formed in the internuclear distance R with a minimum at $R = R_m$ which roughly corresponds to a separation of $d = 0.5–1$ fm between the tips of the light (L) and heavy (H) fragments, depending on the mass $A_{L,H}$ and charge $Z_{L,H}$ numbers and deformations $\beta_{L,H}$ (the ratios between the major and minor semiaxes of the ellipsoids). The depth B_{qf} of the potential pocket insures that the DNS is in local equilibrium over all collective coordinates before the fragmentation. Since the model assumes statistical equilibrium at a scission point, one can reduce the complexity of the problem by fixing the internuclear distance at the bottom of the potential pocket ($R = R_m$). The most important step of the scission-point model is the calculation of the potential energy (the energies of the fragments plus the energy of their interaction),

$$U = U_L^{\mathrm{LD}} + \delta U_L^{\mathrm{shell}} + U_H^{\mathrm{LD}} + \delta U_H^{\mathrm{shell}} + V, \tag{1}$$

of the DNS as a function of Z_i, A_i, β_i and R.[69–78] The index i designates a light or a heavy fragment. In Eq. (1), U_i^{LD} and $\delta U_i^{\mathrm{shell}}$ are the excitation-energy-dependent liquid-drop (LD)[80,81] and shell-correction energies, respectively. The surface tension coefficient depends on the deformation and excitation energy. The nuclear interaction potential V^N is calculated in the

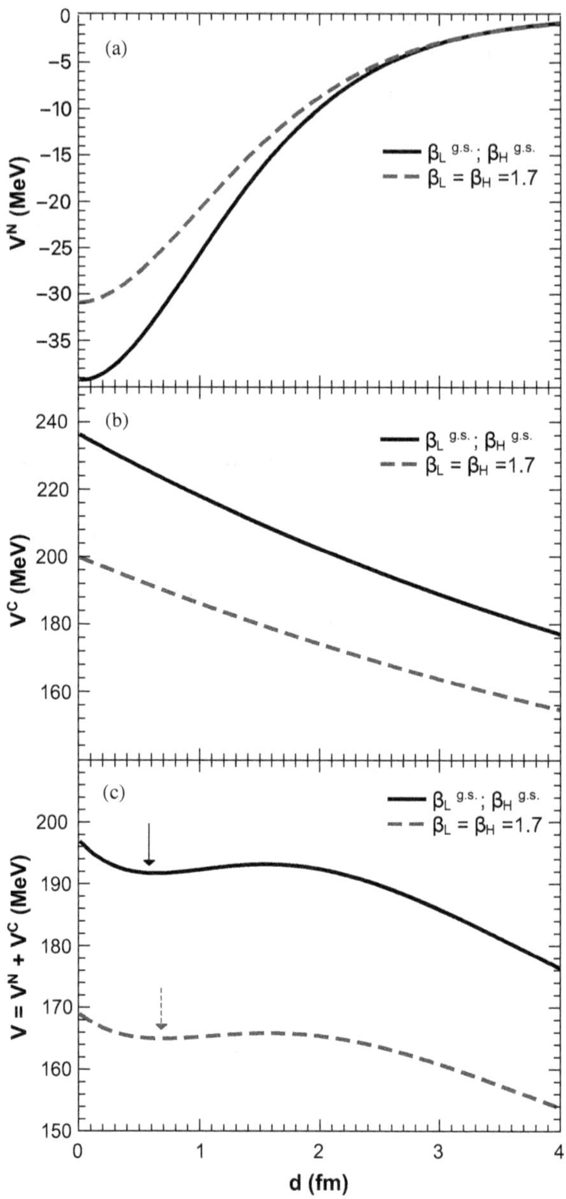

Fig. 1. The calculated nuclear V^N and Coulomb V^C interaction potentials and their sum $V = V^N + V^C$ as functions of the distance d between the tips of the fragments for the fragmentation ^{222}Th$\rightarrow$^{110}Ru+^{112}Pd. The calculations are performed at $L = 0$ or $V^R = 0$. The position of the interaction potential pocket minimum is indicated by the arrow. The deformations $\beta_{L,H}$ of the DNS nuclei are indicated.

double-folding form.[44,82] The nuclear densities are taken in the two-parameter Woods–Saxon form with the diffuseness parameter $a = 0.51$–0.56 fm depending on the charge number of the nucleus. The energy of the nuclear interaction reaches values of about $-(25-40)$ MeV for the touching nuclei depending on their deformations (Fig. 1). Antisymmetrization between the nucleons belonging to different fragments is regarded by a density dependence of the nucleon–nucleon force which gives a repulsive core in the fragment– fragment interaction potential.[82] As seen in Fig. 1, for the fragmentation ^{222}Th$\rightarrow$^{110}Ru+^{112}Pd, the interaction potential has an inner pocket and an external Coulomb barrier located at the distances d between the tips of the fragments of 0.6 fm (0.7 fm) and 1.5 fm (1.6 fm), respectively, at the ground-state deformations of fragments (at $\beta_L = \beta_H = 1.7$). The absolute value of the nuclear interaction (Coulomb interaction) at high deformations is about 9 MeV (34 MeV) less than for deformations of the ground state.

The shell corrections are calculated with the Strutinsky method for an axially deformed nucleus.[79] The damping of the shell correction with excitation energy E_i^* is introduced as

$$\delta U_i^{\text{shell}}(A_i, Z_i, \beta_i, E_i^*) = \delta U_i^{\text{shell}}(A_i, Z_i, \beta_i, E_i^* = 0)\exp\left[-E_i^*/E_D\right],$$

where $E_D = 18.5$ MeV is the damping constant. The DNS excitation energy,

$$E_{\text{DNS}}^*(A_i, Z_i, \beta_i, L) = E_{\text{CN}}^* + [U_{\text{CN}}(A, Z, \beta, L) - U(A_i, Z_i, \beta_i, R_m, L)], \quad (2)$$

(U_{CN} and E_{CN}^* are the binding and excitation energies of CN, respectively) is assumed to be distributed between the fragments proportional to their masses: $E_i^* = E_{\text{DNS}}^* A_i/A$.

The relative formation and decay probability of the DNS with particular masses, charges and deformations of the fragments is calculated within the statistical approach as follows[69–78]:

$$\begin{aligned} w(A_i, Z_i, \beta_i, L) = N_0(2L+1)\exp\left[-\frac{U(A_i, Z_i, \beta_i, R_m, L)}{T(L)}\right] \\ \times \exp\left[-\frac{B_{qf}(A_i, Z_i, \beta_i, L)}{T_{\text{DNS}}(L)}\right], \end{aligned} \quad (3)$$

where N_0 is the normalization factor and T is the temperature of the DNS (T_{DNS}) or the CN (T_{CN}) with the lowest potential energy U. The term $\exp\left[-B_{qf}/T_{\text{DNS}}\right]$ describes the decay probability of the DNS in R. The height of the barrier B_{qf} is simply taken as $B_{qf} = V(R_b, L) - V(R_m, L)$ (where $R_b \simeq R_m + 2$ fm is the position of the Coulomb barrier) and has a strong dependence on the charge numbers and deformation parameters of the interacting fragments and angular momentum L. It is smaller for mass/charge symmetric configurations than for asymmetric ones and for highly deformed configurations than for compact (small $\beta_{L,H}$) ones. Since the potential energy contains the repulsive centrifugal part, the value of B_{qf} decreases with increasing angular momentum. For the fissioning nuclei under consideration, the values of B_{qf} are quite large even for the symmetric fragmentations. In Eq. (3), the CN and, respectively, DNS temperatures are calculated as $T_{\rm CN} = \sqrt{E^*_{\rm CN}/a}$ and $T_{\rm DNS} = \sqrt{E^*_{\rm DNS}/a}$, where $a = A/12$ MeV^{-1} is the level density parameter.[44] Finally, for the calculations of mass and charge distributions, the following expressions are obtained:

$$\begin{aligned} Y(A_i) &= \sum_{L=0}^{L_{\rm max}} \int d\beta_L d\beta_H dZ_i w(A_i, Z_i, \beta_i, L), \\ Y(Z_i) &= \sum_{L=0}^{L_{\rm max}} \int d\beta_L d\beta_H dA_i w(A_i, Z_i, \beta_i, L). \end{aligned} \tag{4}$$

For the electromagnetic-, neutron- and charged-particle-induced fission, $L_{\rm max} = 0$. For the heavy-ion-induced fission,[83] the value of L is limited by either the kinematic angular momentum $L_{\rm kin}$ or the critical angular momentum $L_{\rm cr}$ in the entrance channel, depending on which is smaller: $L_{\rm max} = \min\{L_{\rm kin}, L_{\rm cr}\}$ (see Table 1). Note that in order to simulate the minimal experimental uncertainties, the mass yields are smoothed using the Gaussian function with the width $\sigma = 0.5$ u (units), unless otherwise specified.

Table 1. The excitation energies $E^*_{\rm CN}(L=0)$ at zero angular momentum as well as the maximum angular momenta of CN formed in the indicated complete fusion reactions.

Reaction	$E^*_{\rm CN}(L=0)$ (MeV)	$L_{\rm max}$
^{194}Pt(α, f)	49	30
^{36}Ar+^{144}Sm→^{180}Hg*	33.4	$L_{\rm kin} = 9$
	48	$L_{\rm kin} = 51$
	65.8	$L_{\rm cr} = 73$
^{36}Ar+^{154}Sm→^{190}Hg*	56	$L_{\rm kin} = 24$
	62.4	$L_{\rm kin} = 43$
	70.5	$L_{\rm cr} = 59$
^{40}Ca+^{142}Nd→^{182}Hg*	33	$L_{\rm kin} = 10$
	58	$L_{\rm kin} = 67$
	75	$L_{\rm cr} = 74$
^{48}Ca+^{154}Sm→^{202}Pb*	49	$L_{\rm kin} = 30$
	57	$L_{\rm kin} = 59$
	95	$L_{\rm cr} = 91$

The charge yields are not smoothed. It is also assumed that at energies considered the neutron emission prior to fission does not cause a major change in the fission yields. Thus, the role of multi-chance fission is expected to be minor.

The scission-point model is also suitable for describing the TKE of the fission fragments. Supposing that all nucleus–nucleus interaction energy transforms after fission into the TKE of the DNS primary fragments with (A_i, Z_i),

$$\begin{aligned}\mathrm{TKE}(A_i, Z_i, \beta_i, L) = V^C(A_i, Z_i, \beta_i, R_b) + V^N(A_i, Z_i, \beta_i, R_b)\\ + V^R(A_i, \beta_i, R_b, L),\end{aligned}$$

one can calculate the mean value of the TKE as a function of the fragment charge Z_i by averaging over the deformations of the primary fragments and summing over the mass numbers A_i and angular momenta L,

$$\langle \mathrm{TKE}\rangle(Z_i) = \sum_{A_i,L} \int d\beta_L d\beta_H \mathrm{TKE}(A_i, Z_i, \beta_i, L) w(A_i, Z_i, \beta_i, L). \quad (5)$$

Since the fragments are more deformed at scission than in their ground states, they are relaxed to the ground states after the DNS decays and the energies of deformations are transformed into the fragment intrinsic excitation energies. In order to calculate the neutron multiplicities from two fragments,

$$\begin{aligned}\langle n\rangle(Z_i) &= \sum_{A_i,L,\nu} \int d\beta_L d\beta_H \nu P_\nu(A_i, Z_i, \beta_i, E^*_{\mathrm{DNS}}) w(A_i, Z_i, \beta_i, L),\\ \langle n\rangle(A_i) &= \sum_{Z_i,L,\nu} \int d\beta_L d\beta_H \nu P_\nu(A_i, Z_i, \beta_i, E^*_{\mathrm{DNS}}) w(A_i, Z_i, \beta_i, L),\\ P_\nu &= \sum_{x=1}^{\nu} \int_0^{E^*_{\mathrm{DNS}}} d\epsilon^*_L P_C(\epsilon^*_L) P_{xn}(\epsilon^*_L + U^{def}_L) P_{(\nu-x)n}(E^*_{\mathrm{DNS}} + U^{def}_H - \epsilon^*_L),\end{aligned} \quad (6)$$

we should take into account the fluctuation of the excitation energy between light and heavy fragments using the microcanonical distribution $P_C(\epsilon^*_L) \sim \rho_L(\epsilon^*_L)\rho_H(E^*_{\mathrm{DNS}} - \epsilon^*_L)$ (where $\rho_i(\epsilon^*_i) \sim \exp[2a_i\epsilon^*_i)^{1/2}]$ is the Fermi-gas level density without any backshift in fragment i = L or i = H) of energy partitioned between two fragments of the DNS and the Jackson formula[84]

$$\begin{aligned}P_{xn}(\varepsilon^*_i) &= P(x) - P(x+1),\\ P(x) &= 1 - e^{-\Delta^{(i)}_x}\left[\sum_{k=0}^{2x-3} \frac{(\Delta^{(i)}_x)^k}{k!}\right],\end{aligned} \quad (7)$$

for the probability of evaporation of exactly x neutrons from the excited post-scission fragment i with excitation energy $\varepsilon_i^* = \epsilon_i^* + U_i^{def}$ (the sum of the fragment deformation energy with respect to its ground state and the fragment intrinsic excitation energy at scission point). In Eq. (7), $\Delta_x = (\varepsilon_i^* - \sum_{k=1}^{x} B_k^{(i)})/T_i$, where $B_k^{(i)}$ is the experimental neutron binding energy at the kth evaporation step and $T_i = (\varepsilon_i^*/a_i)^{1/2} = (12\varepsilon_i^*/A_i)^{1/2}$ is the temperature. Here, the quantities $P(x)$ and $P(x+1)$ are the probabilities of emission of at least x and $x+1$ neutrons, respectively. It is clear that $P(x=1) = 1$ at $\epsilon_i^* > B_1^{(i)}$. As found in Ref. 85, the values of $\langle n\rangle(Z_i)$ and $\langle n\rangle(A_i)$ weakly depends on the distribution of excitation energy between light and heavy fragments.

The main ingredient of our description within the improved scission-point fission model is the sophisticated potential energy as a function of mass (charge) asymmetry, deformations of nuclei and internuclear distance. The refinement of the scission-point model is the introduction of temperature- and isospin-dependent LD energy, which allowed us to describe and explain the mass/charge distributions resulting from spontaneous and induced fission. The knowledge of the deformations of the nascent DNS fragments is crucial at the moment of scission. Our model differs from the scission-point model of Ref. 38 by better definition of the scission configuration and the excitation energy. For each point on the PES, one can accurately calculate the excitation energy (as the difference between the total energy of the CN and the energy of the system at scission point), and hence the temperature. The distance between the tips of the fission fragments at scission and the excitation energy are not free parameters of our model but are defined from the nucleus–nucleus interaction potential and PES.

3. Simultaneous Description of Fission-Fragment Charge, Mass, Total Kinetic Energy and Neutron Multiplicity Distributions

In the calculations of mass and charge yields, the most important aspect of the PES is the existence of the minima on the PES.[15,43,69–78] These minima are the result of the interplay between the macroscopic binding energies of fragments, the microscopic shell correction energies of fragments and the fragment–fragment interaction potential. So, the minima arise either due to strong shell effects, as occurring in the fragmentation with magic and/or semimagic fragments, or due to strong macroscopic effects. In the latter case, the PES minimum is determined by the total macroscopic energy which is the sum of the macroscopic binding energies of fragments and the fragment–fragment interaction potential.[72] For the magic and semimagic fragments, the ground-state deformation is almost zero and the stiffness of the nuclear surface is large compared to the nonmagic nuclei. The effect of these two properties is the following: The minimum produced on the PES is located at relatively small deformation (deformation energy) of the fragment. In the case when the total macroscopic energy plays a crucial role, the average deformations of the fragments are relatively large.

Another important aspect is the position of the PES minima in the deformation space (β_L, β_H) (see Fig. 2). In most cases, with very rare exceptions, the minima are located at fragment deformations exceeding the ground-state deformations of fragments. As mentioned above, the large (small) deformations are associated with open-shell (magic or semimagic) primary fragments, for which the neutron binding energies are smaller (larger). Different values of the deformations $\beta_{L,H}$ of two fragments also lead to (1) different values of the fragment–fragment interaction potential, and, thus, to different values of the TKE and intrinsic excitation energies of fragments at scission; and (2) different values of the deformation energies of light and heavy fragments at scission, energies which are partly "recovered" as the internal excitation energies of two separated (after scission) fragments in the ground state. So, a part of deformation energy and the intrinsic excitation energy of fragment at scission are involved in the neutron evaporation. Thus, the positions (β_L, β_H) and widths of the minima on the PES have a strong impact on the neutron emission from fission fragments.

3.1. *Charge/mass yields, neutron multiplicities and TKE*

In Fig. 3, the calculated charge distribution of the fission fragments resulting from the thermal-ssionfission of ^{235}U is in good agreement with the experimental data.[86] The asymmetric two-peaked mass distributions are well reproduced for the 0.5-MeV neutron-induced fission of ^{239}Pu and the spontaneous fission of ^{240}Pu, respectively. It should be noted that the model also describes well the charge and mass distributions of fragments in the thermal-neutron-induced fission of ^{239}Pu.[69] The experimentally observed asymmetric two-peaked shape of the mass distribution is well reproduced in the case of spontaneous fission of ^{244}Cm (Fig. 3).

Figure 4 demonstrates the calculated dependencies of ⟨TKE⟩ on the mass number A_i of one of the fragments for the 0.5-MeV neutron-induced fission of ^{239}Pu and the spontaneous fission of ^{244}Cm. The shapes of these distributions are relatively well reproduced. The largest deviations between

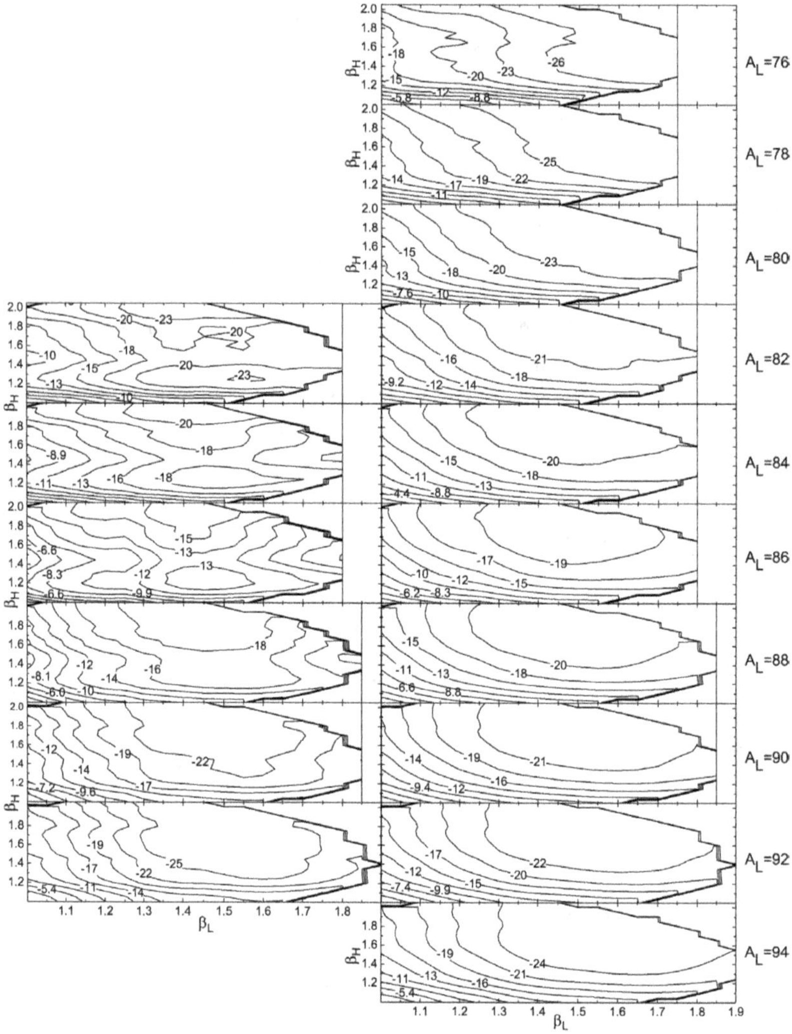

Fig. 2. The calculated PESs for different ^{A_L}Kr fragmentations at 11-MeV (left) and 60-MeV (right) excitation energies and zero spin in the case of the electromagnetic-induced fission of ^{222}Th. The value of potential energy is normalized to the energy of the DNS with spherical nuclei ($\beta_L = \beta_H = 1$). In the nuclear shape parametrization used, the value of quadrupole deformation parameter is approximately $\beta_i - 1$ ($i = L, H$).

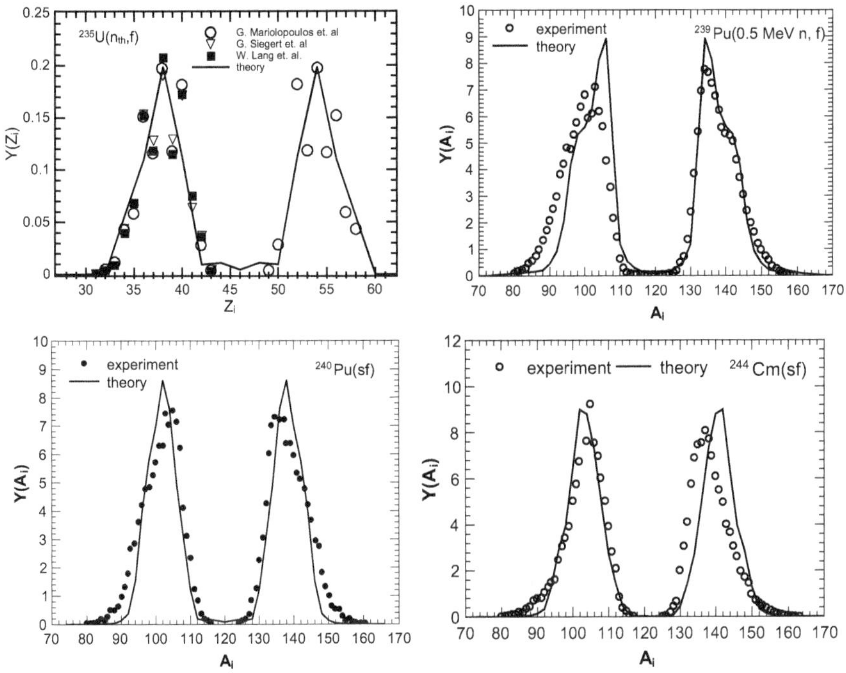

Fig. 3. The calculated (lines) and experimental (symbols)[86–89] charge and mass distributions of fission fragments for the indicated reactions. The measured mass distributions represent the distributions of secondary fission fragments, while the theoretical mass distributions refer to the distributions of primary fission products. The lines connect the calculated points for even–even fission fragments.

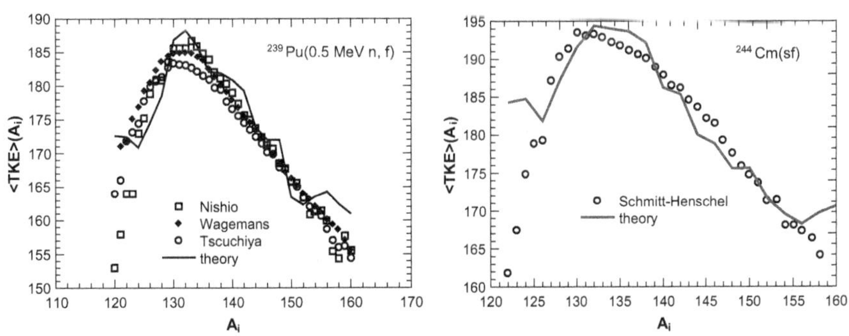

Fig. 4. The calculated (lines) and experimental (symbols)[89,90] average TKEs as functions of fragment mass for the indicated reactions. The lines connect the calculated points for even–even fission fragments.

the theoretical and experimental results are observed for the very asymmetric ($A_i > 155$) and symmetric fission events.

For the 0.5-MeV neutron-induced fission of ^{235}U and ^{239}Pu and the spontaneous fission of ^{240}Pu and ^{244}Cm, the calculated average neutron multiplicity

$$\langle n_i \rangle (A_i) = \sum_{Z_i, L, \nu_i} \int d\beta_L d\beta_H \nu_i \tilde{P}_{\nu_i}(A_i, Z_i, \beta_i, E^*_{\text{DNS}}) w(A_i, Z_i, \beta_i, L),$$

$$\tilde{P}_{\nu_i}(A_i, Z_i, \beta_i, E^*_{\text{DNS}}) = \int_0^{E^*_{\text{DNS}}} d\epsilon_i^* P_C(\epsilon_i^*) P_{\nu_i}(\epsilon_i^* + U_i^{def}),$$

of the mass number of the fission fragment (Fig. 5) and the probability

$$P_m(x) = \sum_{Z_i, A_i, L} \int d\beta_L d\beta_H P_{\nu=x}(A_i, Z_i, \beta_i, E^*_{\text{DNS}}) w(A_i, Z_i, \beta_i, L)$$

for a given neutron multiplicity x per fission act as a function of x (Fig. 6) are compared with the experimental data. The "saw-tooth" shapes of the measured neutron multiplicities are reproduced quite satisfactorily in a wide range of fission-fragment mass numbers. In these calculations, as well as in the calculations in Ref. 85, the light fragments are suggested to be disproportionately excited: the average excitation energy of the DNS light fragment is larger by the factor 1.3 than the average excitation energy of the DNS light fragment in the thermal equilibrium limit ($T_L = T_H$ or $\langle E_L^* \rangle / (E_H^*) = a_L / a_H$). The average excitation energy of a heavy fragment of the DNS is determined from the conservation of the excitation energy of the DNS. Indeed, the most probable fission-fragment has $Z_H = 50$–58 and, thus, shell effects are stronger than in the complementary light fragment. The larger the shell effects, the smaller the level density parameter and, correspondingly, the smaller average excitation energy. As found, the probability $P_m(x)$ for a given neutron multiplicity x per fission event weakly depend on this disproportionality of the distribution of the DNS excitation energy between the fragments. The multiplicity distributions $P_m(x)$ are perfectly described. The low multiplicity can be attributed to the minima on the PES with small deformation energy (correspondingly, with small intrinsic excitation energy), while the larger values of x are attributed to the structures on the PES at large deformations (correspondingly, with large intrinsic excitation energy), as the deformation energy (with respect to the deformation energy of the ground state) of the nucleus is added to its intrinsic excitation energy at scission.

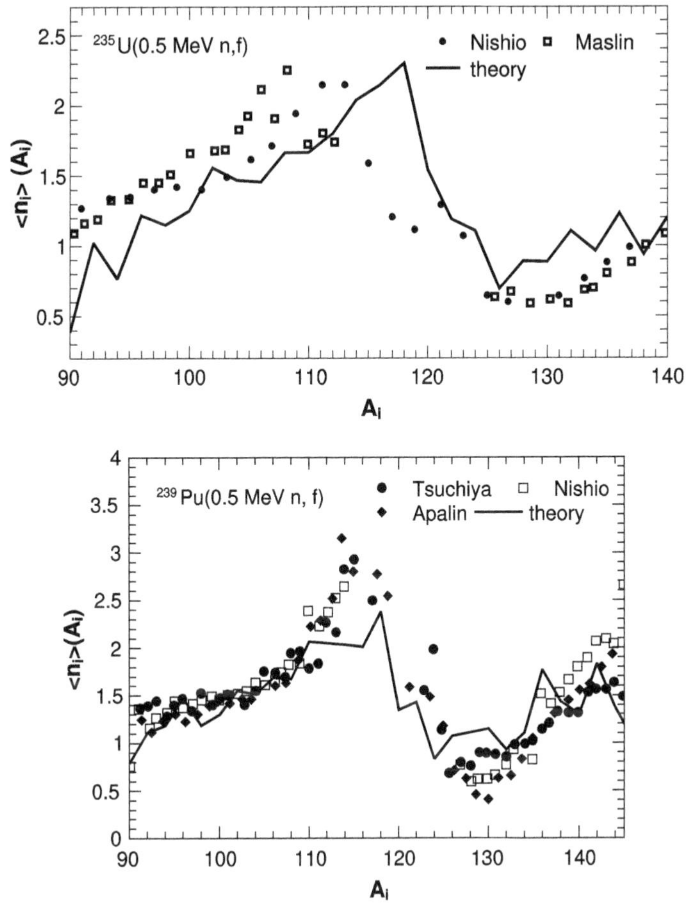

Fig. 5. The calculated (lines) and experimental (symbols)[91–94] average neutron multiplicities as functions of fragment mass A_i for the indicated reactions. The lines connect the calculated points for even–even fission fragments.

The larger excitation energies (temperatures) in each configuration lead to slightly larger values of the deformations of the fragments, which, alongside the increased excitation energy, lead to the larger deformation energy and, thus, to higher values in the neutron multiplicities. This can also be seen in the experimental $\langle \mathrm{TKE} \rangle$ as a function of the mass number of one of the fission fragments, where for the neutron-induced fission, the $\langle \mathrm{TKE} \rangle(A_i)$ are slightly lower. The calculated average TKE values are $\langle \mathrm{TKE} \rangle = 181.3\,\mathrm{MeV}$ for the ^{239}Pu(0.5-MeV n, f) reaction and $\langle \mathrm{TKE} \rangle = 177.1\,\mathrm{MeV}$ for the spontaneous fission of ^{240}Pu. The large excitation energy and deformations explain not only

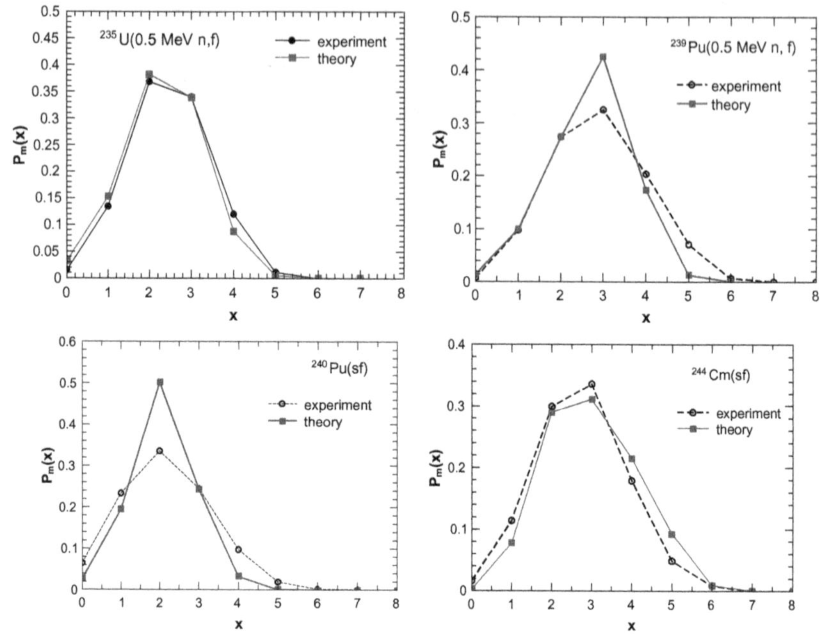

Fig. 6. The calculated (squares connected by lines) and experimental (open circles connected by lines)[95–97] probabilities $P_m(x)$ as functions of multiplicity x for the indicated fission reactions. The lines are only for an eye-guide.

the greater value of average neutron multiplicity, but also the large width of the distribution $P_m(x)$, since the large excitation energy attributed to the neutron-induced fission leads to large thermal fluctuations around the equilibrium deformations as well as the fluctuations in the excitation energies of fragments.

In Fig. 7, the calculated average numbers $\langle n \rangle$ of neutrons emitted per act of fission in the spontaneous fission of ^{230}Th, ^{236}U, ^{240}Pu, ^{244}Cm and ^{252}Cf, the thermal neutron-induced fission of ^{235}U and the 0.5-MeV neutron-induced fission of ^{239}Pu are presented in comparison with the experimental data. The experimental data are well reproduced for the aforementioned fissioning nuclei, and over a large range of mass/charge numbers of fissioning actinides. The largest differences between theory and experiment, of about 0.3 u and 0.5 u, appear for Cm and Cf nuclei, respectively. The dependence of $\langle n \rangle$ on A is explained by the fact that the average deformations and average excitation energies of fission fragments at scission increase with the mass/charge numbers of the fissioning nucleus. However, in a given isotopic chain some deviations from the global trends are possible.

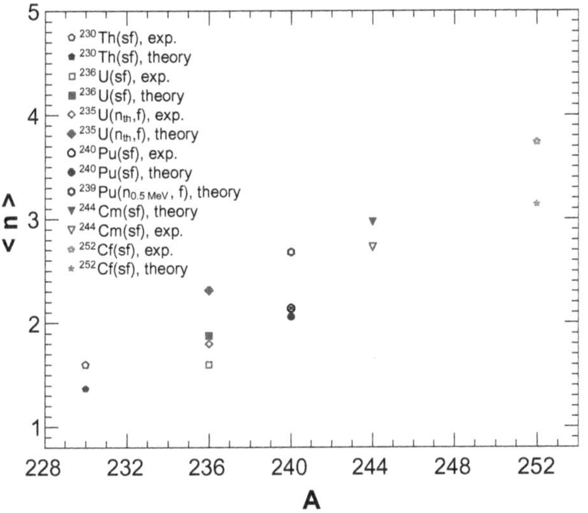

Fig. 7. The calculated (open symbols) and experimental (solid symbols)[98–101] average numbers of neutrons $\langle n \rangle$ emitted per act of fission as functions of the mass number A of the indicated fissioning nucleus.

3.2. *Correlations between fission observables*

As noted above, the mass/charge yields and all other fission observables are associated with the structures appearing on the PES. In particular, the minima on the PES of each possible binary fragmentation determine the shapes of the mass/charge distributions and the intrinsic excitation energies of the resulting DNS. So, we can expect some correlation between $\langle n \rangle$ and $\langle \text{TKE} \rangle$ distributions which is reflected in the transition from symmetric to asymmetric fission.

In the case of electromagnetic-induced ($E^*_\gamma = 11\,\text{MeV}$) fission of even Th isotopes with mass numbers $A = 222$–230, there is a transition from symmetric LD-like fission of ^{222}Th to asymmetric fission of ^{230}Th, where the shell effects in the region $A_H = 130-140$ ($Z_L = 36$, $N_L = 50$, $Z_H = 54$, $N_H = 82$ and around them) impose asymmetric charge/mass fragmentations (Fig. 8). In the case of fission of ^{226}Th, there is a mixed effect: The two asymmetric peaks are driven by the shell effects, while the symmetric peak of almost the same weight is the result of the macroscopic effects. In Figs. 8–13, the calculated charge, mass and TKE distributions of fission fragments are shown and compared with the available experimental data.[4,64,65] As can be seen, the charge, mass and TKE distributions are well reproduced. For the fissioning nuclei ^{222,224}Th and ^{226,228,230}Th, the TKE distributions have different shapes (Figs. 9–13). In the case of ^{222,224}Th, there is almost the plateau-type shape at $35 < Z_i < 55$, while in the cases of ^{226,228,230}Th the predicted TKE distributions show the pronounced asymmetric "wings" around $Z_L = 36$ and $Z_H = 54$. As the Th mass number increases, the role of

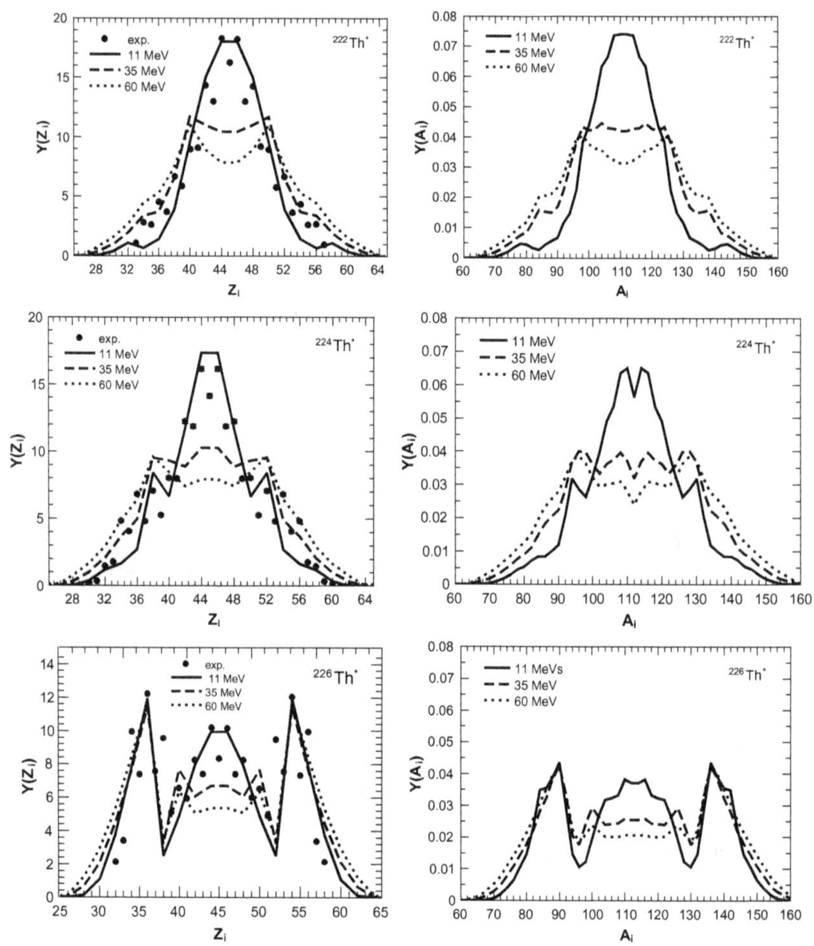

Fig. 8. The calculated (lines) charge and mass distributions for the indicated Th isotopes at 11-, 35- and 60-MeV excitation energies and zero spin of the initial CN. The lines connect the calculated yields of even–even fission fragmentations. The experimental (symbols) charge and mass distributions of fission fragments for the electromagnetic-induced fission of Th isotopes at 11-MeV excitation energy are from Refs. 4, 64 and 65.

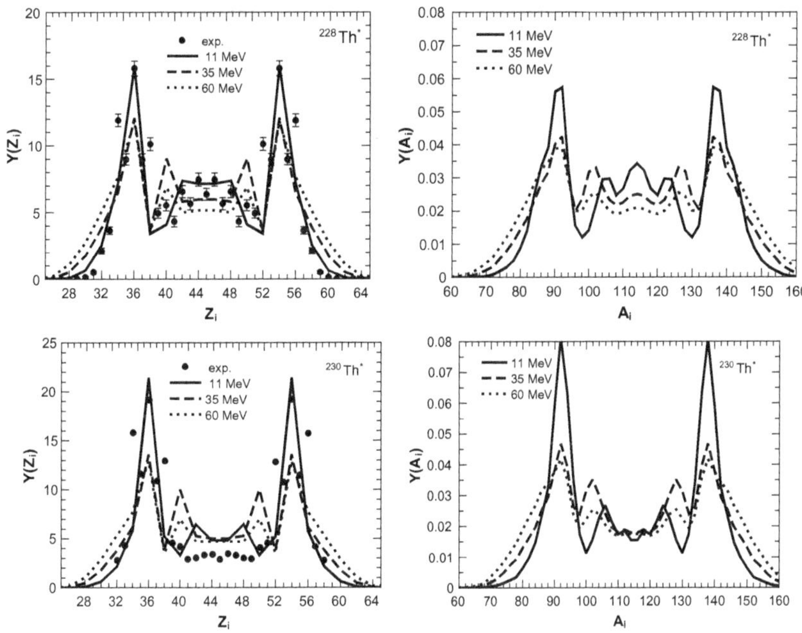

Fig. 8. (*Continued*)

shell corrections becomes dominant, and the TKE and neutron multiplicity distributions closely follow the changes occurring in the PES of the system. In the case of ^{226,228,230}Th, the asymmetric peaks in the charge/mass distributions resulting from the shell corrections are closely related to the pronounced asymmetric humps found in the TKE distribution.

For the fissioning nuclei ^{222,224,226,228,230}Th, the calculated neutron multiplicities $\langle n\rangle(Z_i)$ as functions of the charge number of one of the fragments are shown in Figs. 9–13. In the cases of ^{222,226,230}Th, the shapes of the calculated (Eq. (6)) and experimental neutron multiplicity distributions match each other. The increase of the neutron multiplicity at symmetry on going from light to heavy isotopes of thorium corresponds to the measured trend. However, there is a discrepancy in the absolute values of the calculated and experimental $\langle n\rangle(Z_i)$. In the fission of ^{222,226}Th (^{230}Th), the largest difference is about 1 (2) at symmetry. Also, there is uncertainty in $\langle n\rangle(Z_i)$ because of inaccuracy in definition of detector efficiency. It should be noted that an increase in the excitation energy $E^*_{\rm CN}$ from 11 MeV to 15 MeV leads to an increase in the neutron multiplicity by an average of about 0.5 units.

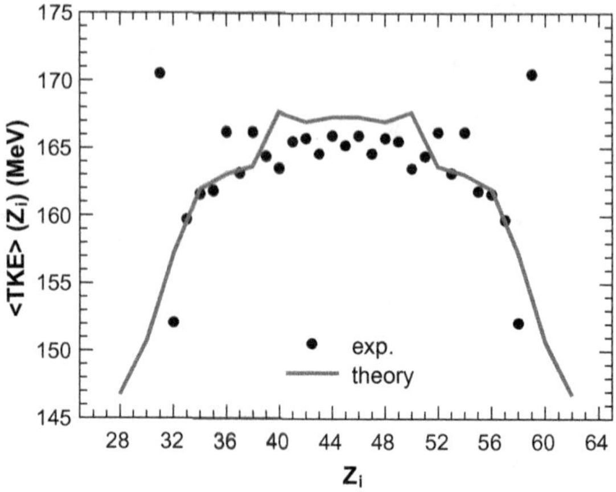

(a) ^{222}Th

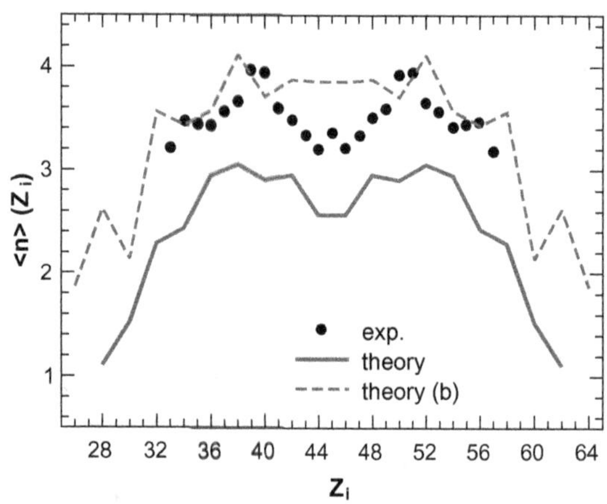

(b) ^{222}Th

Fig. 9. The calculated (lines) and experimental (symbols)[4,65] (a) TKE and (b) neutron multiplicity distributions of fission fragments for the electromagnetic-induced fission of ^{222}Th at the excitation energy $E^*_{\rm CN} = 11\,{\rm MeV}$. The lines connect the calculated points for even–even fission fragments. In the calculations of $\langle n\rangle(Z_i)$, Eq. (6) (solid lines) and Eq. (8) (dashed lines) are used.

Using other formulas,[77]

$$\langle n\rangle(Z_i) = \sum_{A_i} \int d\beta_L d\beta_H n(A_i, Z_i, \beta_i) w(A_i, Z_i, \beta_i, L = 0),$$
$$n(A_i, Z_i, \beta_i) = \sum_{i=L,H} \frac{E^*_i(A_i, Z_i, \beta_i) + U^{\rm def}_i(A_i, Z_i, \beta_i)}{B^{(i)} + 2T_i}, \quad (8)$$

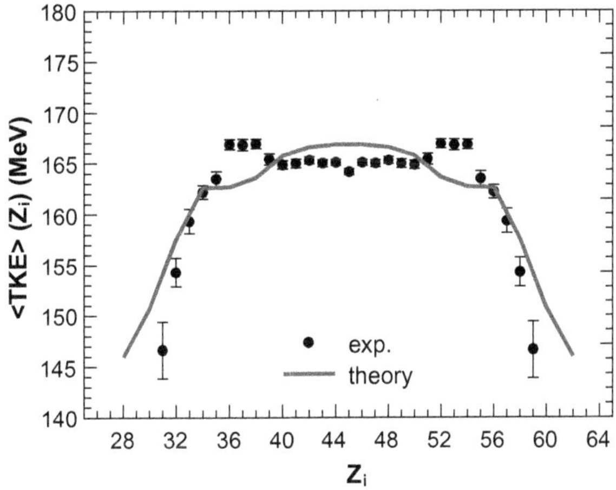

(a) ^{224}Th

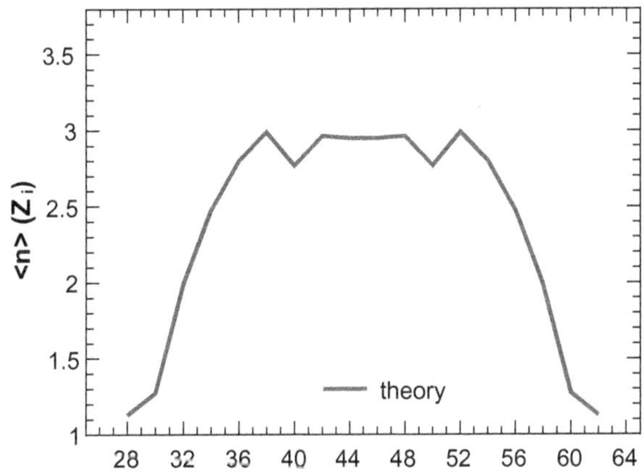

(b) ^{224}Th

Fig. 10. The same as in Fig. 9, but for the electromagnetic-induced fission of ^{224}Th at the excitation energy $E^*_{\rm CN} = 11$ MeV.

for the calculation of neutron multiplicity ($U_i^{\rm def}$, $B^{(i)}$ and $2T_i$ are the deformation energies of fragment, the average binding energy of a neutron and the average kinetic energy carried away by a neutron, respectively), we obtain better description of the absolute values of $\langle n\rangle(Z_i)$ at symmetry for ^{222,226}Th, but worse shape of the distribution $\langle n\rangle(Z_i)$. In the case of ^{230}Th (Fig. 13), the shapes of the calculated and experimental neutron multiplicity distributions match each other and the ratio between the

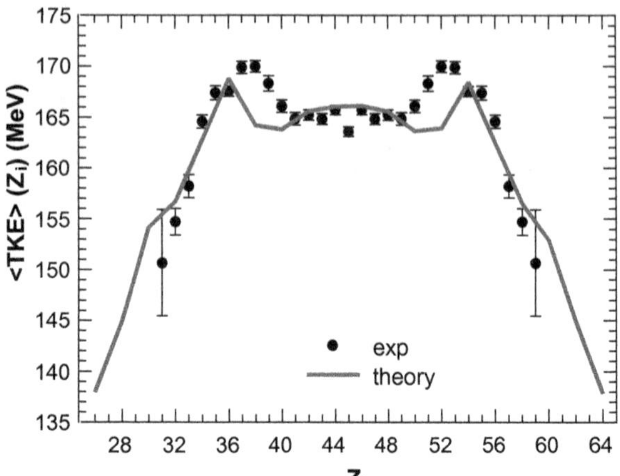

(a) ^{226}Th

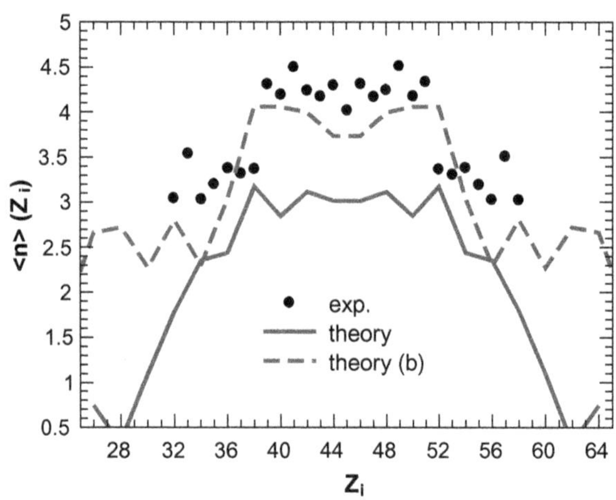

(b) ^{226}Th

Fig. 11. The same as in Fig. 9, but for the electromagnetic-induced fission of ^{226}Th at the excitation energy $E^*_{\rm CN} = 11\,$MeV.

$\langle n\rangle(Z_i)$ at symmetry and in the region $Z_i < 40$ has roughly the same value (≈ 1.7) both in experiment and in theory. However, there is a large discrepancy in the absolute values of the calculated and experimental neutron multiplicities. The measured data show an increase of $\langle n\rangle(Z_i)$ by 1 u in the region $Z_i < 40$ and almost 4 u at symmetry, as compared with the same distribution measured for the fissioning nucleus ^{222}Th. In the calculations, an increase in the neutron multiplicity is found, but not as large as in the

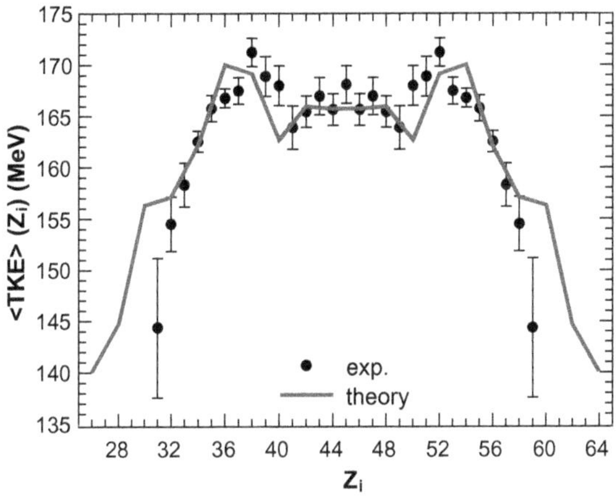

(a) ^{228}Th

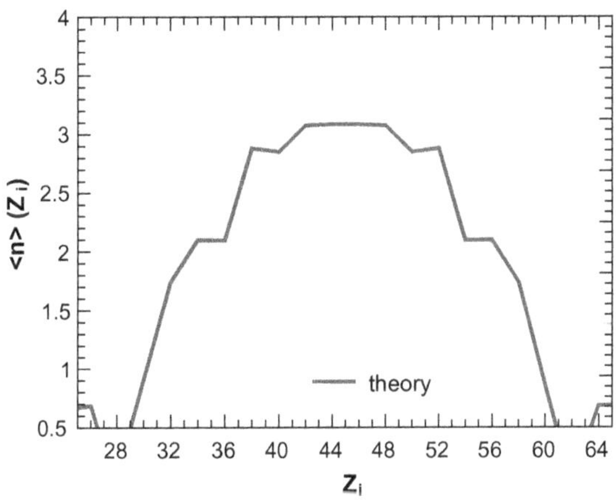

(b) ^{228}Th

Fig. 12. The same as in Fig. 9, but for the electromagnetic-induced fission of ^{228}Th at the excitation energy $E^*_{\rm CN} = 11$ MeV.

experiment. It is difficult to justify such a large increase; an increase of 3–4 u at symmetry means an increase of 25–30 MeV of the excitation energy or deformation energy. So an increase of the excitation energy alone by 25–30 MeV is extremely unrealistic, since it would lead to symmetric charge and mass distributions. According to the estimates, an increase in the deformation energy required for the neutron multiplicity at symmetry to have a value of ∼7 would mean that two fragments must have an axis ratio of more than 2.3, which is about 35% larger than that in the

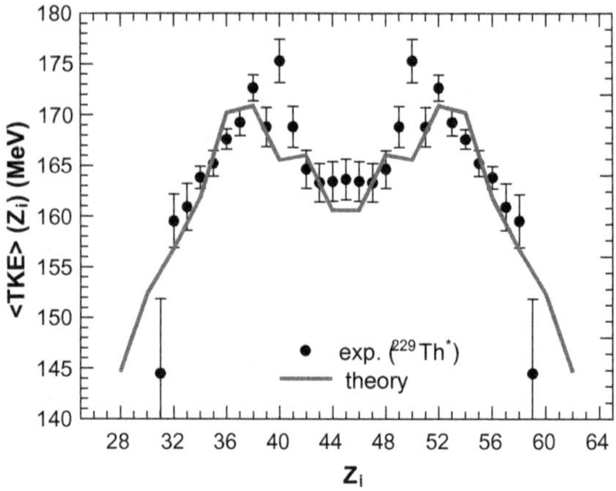

(a) ^{230}Th

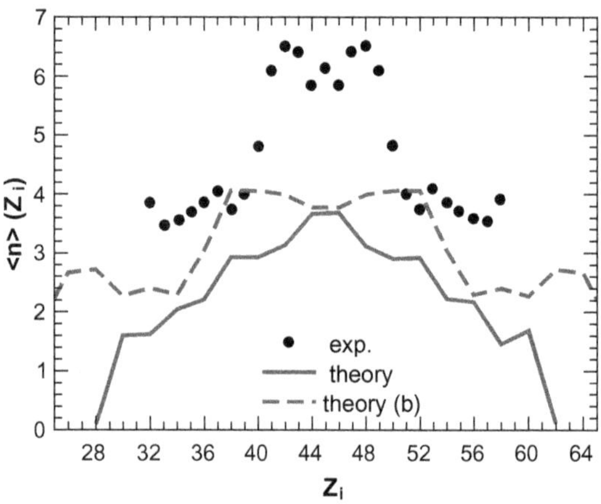

(b) ^{230}Th

Fig. 13. The same as in Fig. 9, but for the electromagnetic-induced fission of ^{230}Th at the excitation energy $E^*_{\rm CN} = 11\,\rm MeV$. There are no experimental data for ^{230}Th, and therefore, for comparison with theory, the experimental data[4] for the fissioning nucleus ^{229}Th ($E^*_{\rm CN} = 11\,\rm MeV$) are shown.

calculations. Note that since our model describes well the experimental $\langle \rm TKE \rangle$ for ^{222,226}Th, it gives realistic values of the deformation of the fragments at scission. Also, larger values of $\langle n \rangle (Z_i)$ for the fragments with $Z_i < 40$ are difficult to justify, since stronger shell effects lead to more compact configurations, so the deformation energy actually decreases and compensates for the increase of excitation energy in this region. As a result, we can conclude that that the experimental values of $\langle n \rangle$ for ^{230}Th are systematically overestimated.

As seen in Fig. 14, for ^{222,224}Th, the intrinsic excitation energy $\langle E^* \rangle(Z_i) = \langle E_L^* \rangle(Z_i) + \langle E_H^* \rangle(Z_i)$ of fragments at scission is high around symmetry and the deformation energy $\langle U^{\text{def}} \rangle(Z_i) = \langle U_L^{\text{def}} \rangle(Z_i) + \langle U_H^{\text{def}} \rangle(Z_i)$ of fragments at

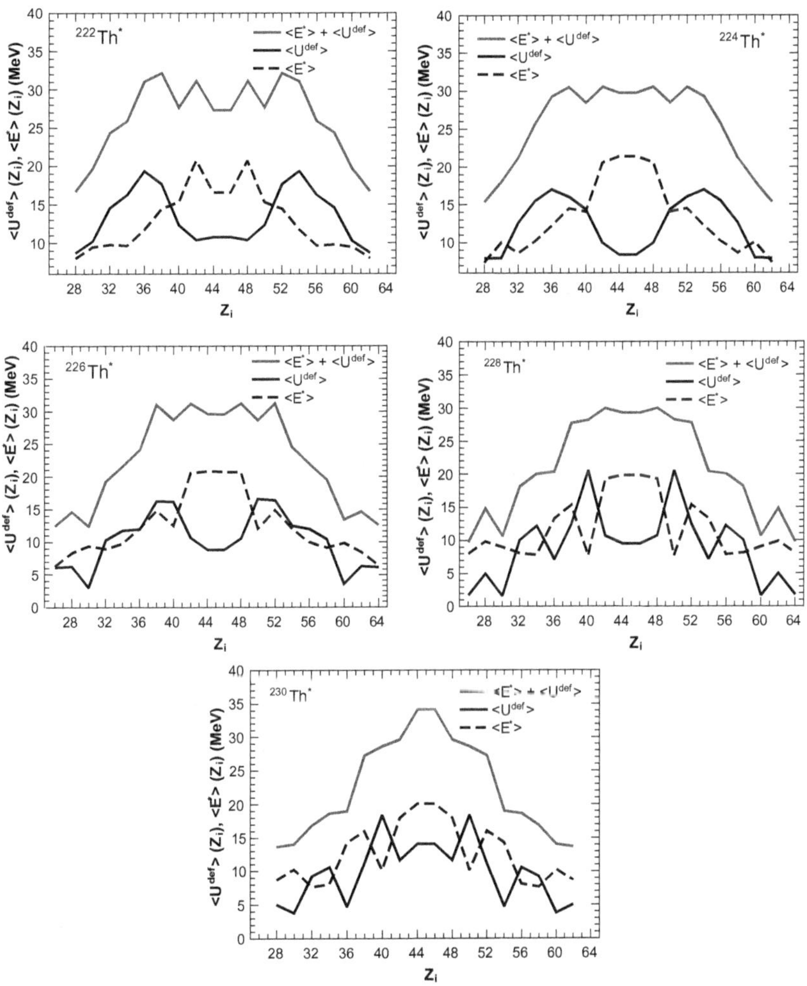

Fig. 14. The calculated total $\langle U^{\text{def}} \rangle + \langle E^* \rangle$, deformation $\langle U^{\text{def}} \rangle$ (solid lines) and excitation $\langle E^* \rangle$ (dashed lines) energies of fission fragments at scission for the electromagnetic-induced fission of ^{222,224,226,228,230}Th at the excitation energy $E^*_{\text{CN}} = 11$ MeV. The lines connect the calculated points for even–even fission fragments.

scission is low at this region, while for the asymmetric regions the opposite is true. Overall, after scission the total excitation energy of fragments is rather constant with some fluctuations in the region $35 < Z_i < 55$, and it is reflected in the rather constant value for the neutron multiplicity in this region.[65] On going from ^{224}Th to ^{230}Th, the deformation energy of fragments increases near the symmetry ($Z_L \approx 44$) and the value of $\langle U^{\rm def}\rangle + \langle E^*\rangle$ slightly increases. In the asymmetric region ($32 \leq Z_L \leq 38$), $\langle U^{\rm def}\rangle$ decreases with increasing neutron number at almost constant $\langle E^*\rangle$ which results in smaller $\langle U^{\rm def}\rangle + \langle E^*\rangle$. This leads to a change of the shape of neutron multiplicity distribution: the plateau length near symmetry decreases: $37 < Z_i < 53$ for ^{226,228}Th and $43 < Z_i < 47$ for ^{230}Th. The value of $\langle E^*\rangle \times (Z_i) + \langle U^{\rm def}\rangle(Z_i)$ at symmetry grows with isospin which leads to the increasing neutron multiplicity.

In the case of ^{230}Th, the deformation energy of asymmetric fragments at scission, which correspond to the maxima of mass yields, is smaller than that in more symmetric configurations. So, in asymmetric fission mode the fragments are stiff and less deformed because of stronger shell effects than those in the symmetric fission mode. In the case of ^{226}Th, the symmetric and asymmetric fission modes have equal weights. This results in higher overall deformation energies of the fragments at scission, compared to the ^{230}Th case. However, in Fig. 15, $\langle$TKE$\rangle$(^{230}Th)$\approx$ $\langle$TKE$\rangle$(^{226}Th)$\leq$ $\langle$TKE$\rangle$(^{228}Th) ($\langle$TKE$\rangle(Z_i)$ averaged over all Z_i) because charge/mass distribution is more asymmetric in the case of ^{230}Th. When going from ^{224}Th to ^{230}Th, the increase of deformation energies of the fragments at scission in the symmetric region and the decrease of neutron separation energies have a larger impact on $\langle n\rangle$ than the decrease of deformation energies of the fragments in the asymmetric region. As a result, $\langle n\rangle$(^{224,226,228}Th)$<$ $\langle n\rangle$(^{230}Th).

As found for a fissioning nucleus ^{222}Th, almost constant value of TKE at symmetry is associated with a fairly constant value of the neutron multiplicity. As the Th mass number increases, the role of shell corrections becomes dominant, and the TKE and neutron multiplicity distributions closely follow the changes occurring in the PES of the system. In the case of ^{226}Th, the asymmetric peaks in the charge distribution resulting from the shell corrections are closely related to the asymmetric humps found in the TKE distribution, while the symmetric peak in the charge yields is associated with lower TKE and higher neutron multiplicities at symmetry due to the LD behavior of the system. As the mass number of Th increases from 226 to 230, the strong shell effects impose a dominant asymmetric fission mode, which provides asymmetric maxima in charge yields. For the

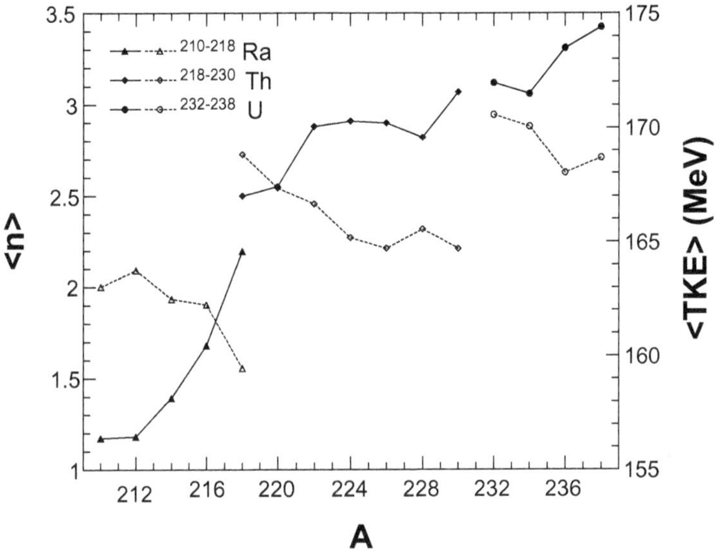

Fig. 15. The calculated average TKE (open symbols connected by dashed lines) and average number of neutrons emitted per act of fission (closed symbols connected by solid lines) as functions of mass number A of the fissioning even–even nuclei $^{210-218}$Ra, $^{218-230}$Th and $^{232-238}$U at excitation energy $E^*_{\rm CN} = 11$ MeV.

fissioning ^{230}Th, the TKE distribution also presents the shell effects and the calculated neutron multiplicities show a slight increase at symmetry. Note that the shape of the calculated distribution of neutron multiplicity is similar to the measured one, but we cannot explain the large experimental values, especially at symmetry. Nevertheless, the increase of the neutron multiplicity at symmetry with the transition from light to heavy thorium isotopes is, in our opinion, indisputable. In our opinion, the additional experimental and theoretical studies of $\langle n\rangle(Z_i)$ for ^{230}Th are desirable.

For the electromagnetic-induced fission of ^{230}U and ^{234}U at average excitation energy $E^* = 11$ MeV (Fig. 16), the asymmetric nature of the charge and TKE distributions is well reproduced. For ^{230}U, the TKE distribution at symmetry shows a rather weak dependence on Z_i, while the neutron multiplicity represents an almost constant value ($\approx$4.2) in the region $40 < Z_i < 50$ with a small hump at symmetry correlating with the minima of $\langle \mathrm{TKE}\rangle(Z_i)$ and $Y(Z_i)$. In the region $40 < Z_i < 50$, the fragments behave like LD (as for ^{230}Th). So large deformations of fragments create these features.

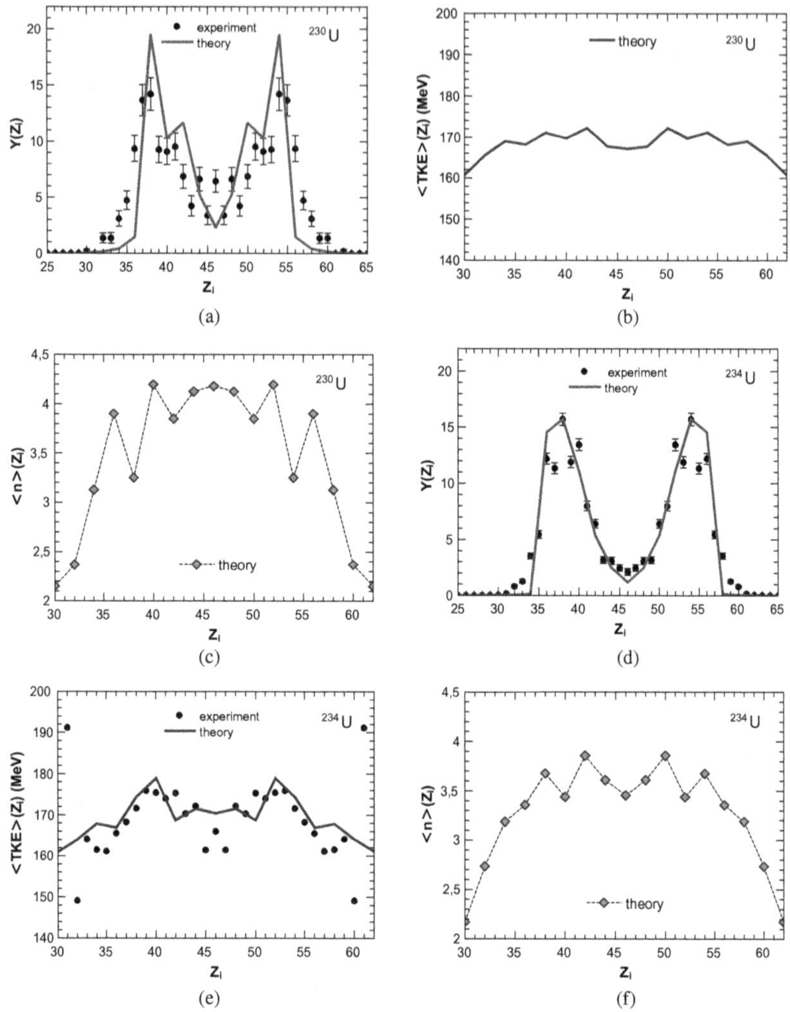

Fig. 16. The calculated (lines) and experimental (closed symbols)[4,81] (a), (d) charge, (b), (e) TKE and (c), (f) neutron multiplicity distributions of fission fragments for the electromagnetic-induced fission of ^{230,234}U at 11-MeV excitation energy and $L = 0$. The lines connect the calculated points for even–even fission fragments.

For ^{234}U, a rather weak dependence of $\langle n \rangle$ on Z_i at symmetry is again predicted. Note that the values of average neutron multiplicities at symmetry decrease with increasing mass number of uranium isotopes from 230 to 234, although for thorium isotopes the opposite trend is observed. The explanation of this behavior is as follows: As is known, neutron-deficient U isotopes provide higher charge and mass yields at symmetry than neutron-rich U isotopes with larger excitation energy at symmetry. On the other hand, the positions of the minima in the PES for symmetric fragmentations of neutron-rich uranium nuclides are located at slightly lower deformations $\beta_{L,H}$, so the deformation energies are also slightly smaller. The combined influence of these two factors generally results in slightly lower values of $\langle n \rangle$ in ^{234}U.

For the fissioning even–even nuclei $^{210-218}$Ra, $^{218-230}$Th and $^{232-238}$U, the correlation between $\langle$TKE$\rangle$ and $\langle n \rangle$ is shown in Fig. 15. There is a global trend, if $\langle$TKE$\rangle$ decreases, then $\langle n \rangle$ increases and vice versa. As seen in the case of fission of even Th isotopes, the dependencies of $\langle$TKE$\rangle$ and $\langle n \rangle$ on A are very weak around the transition from symmetric to asymmetric fission (^{226}Th). This can be used as a marker to identify the transition from symmetric to asymmetric fission, since $\langle$TKE$\rangle$ carries information about compact (noncompact) shapes in fission driven by shell (LD) effects, and $\langle n \rangle$ carries information about the deformation and intrinsic excitation energies at scission and neutron binding energies. Matching values of $\langle n \rangle$ and $\langle$TKE$\rangle$, one can conclude about symmetric–asymmetric fission transition even without knowing the mass or charge distribution. Note that weak dependencies of $\langle$TKE$\rangle$ and $\langle n \rangle$ on A are also observed in experiments.[4,64,65]

In Fig. 17, we show the calculated distributions $P_m(x)$ in the cases of electromagnetic-induced fission of nuclei ^{222,226,230}Th. As can be seen, the

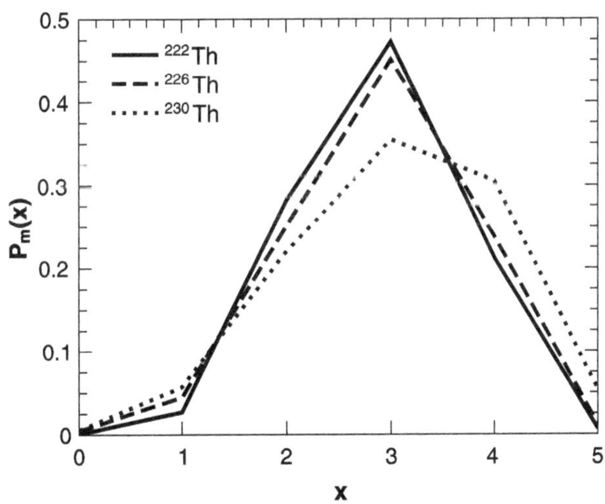

Fig. 17. The calculated (squares connected by lines) probability $P_m(x)$ for a given neutron multiplicity as a function of multiplicity, x, for the electromagnetic-induced fission of nuclei ^{222,226,230}Th ($E^*_{\rm CN} = 11$ MeV). The lines are only for an eye-guide.

distribution $P_m(x)$ becomes slightly wider with increasing mass number of Th and, correspondingly, also carries some information about the change of the shape (from the symmetric shape to the asymmetric one) of mass/charge distributions with the neutron number of Th. Note that $P_m(x \leq 1)$ for ^{230}Th is larger than those for ^{222,226}Th.

4. Isospin Dependence of Charge and Mass Distributions in Fission at Low Excitation Energies

4.1. *Even–even* $^{210-226}$*Ra and* $^{222-230}$*Th*

The experimental[4,65] and calculated charge/mass distributions resulted from the electromagnetic-induced fission of even–even $^{222-230}$Th ($E^*_{\text{CN}} = 11\,\text{MeV}$) isotopes are shown in Fig. 8. One can see that the scission-point model is suitable for describing both symmetric and asymmetric distributions of fission fragments. For ^{222}Th, the charge/mass distributions are clearly symmetric, with one relatively narrow peak around $Z/2 = 45$. For ^{224}Th, the charge/mass distributions start showing the formation of two asymmetric peaks, while for ^{226}Th, the yields around Kr–Sr and their complementary fragments are almost equal to those around Pd–Ru. As the mass number increases to $A = 228$, the symmetric peak recedes and the asymmetric mode becomes dominant. The yields of nuclei with $Z_{L,H} = 40$–50 do not go abruptly to zero; the central maximum still exists and the yields of symmetric fragments are about two times smaller than those of asymmetric ones. In the case of ^{230}Th, the charge/mass distributions are even more asymmetric.

The transition from an one-peaked to a two-peaked distribution, going through a transient three-peaked shape, can be explained by studying the driving potential

$$\langle U\rangle(Z_i) = \sum_{A_i,L} \int d\beta_L d\beta_H U(A_i, Z_i, \beta_i, R_m, L) w(A_i, Z_i, \beta_i, L) \tag{9}$$

averaged over the mass number A_i, deformations $\beta_{L,H}$ and angular momentum L. For ^{222}Th ($E^*_{\text{CN}} = 11\,\text{MeV}$), the driving potential $\langle U\rangle(Z_i)$ at $L = 0$ shows a deep minimum at $Z/2 = 45$ (Fig. 18). For the ^{224}Th nucleus, $\langle U\rangle(Z_i)$ starts displaying the formation of the second minimum around Kr–Sr, while for ^{226}Th, the second minimum becomes almost as deep as the central (symmetric) one. Even if the values of $\langle U\rangle(Z_L = 36-38)$ are slightly higher than $\langle U\rangle(Z_L = 44)$, the corresponding yields are the same. This is easily explained by the fact that in the (β_L, β_H) plane, the minima for

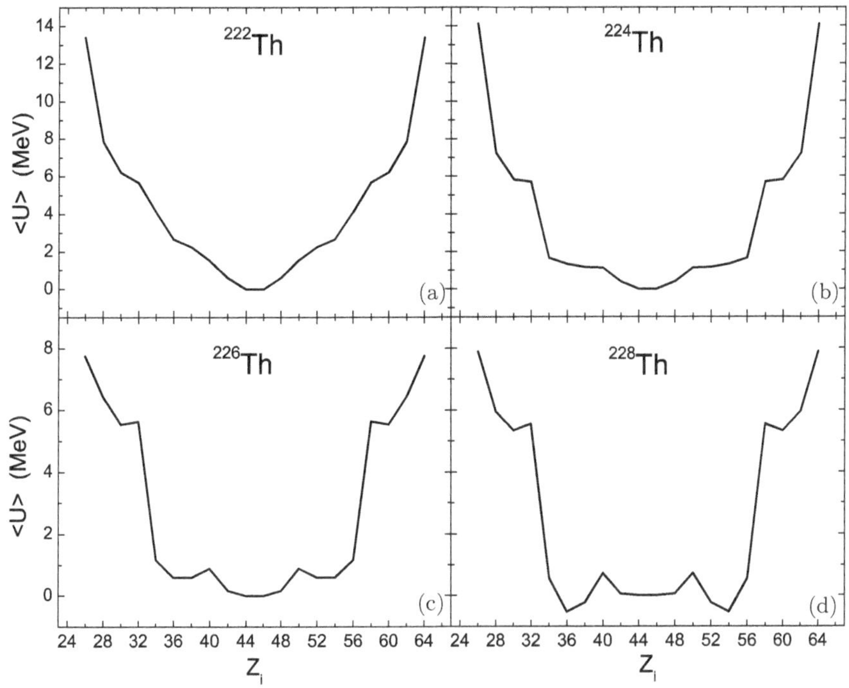

Fig. 18. The calculated driving potentials $\langle U\rangle(Z_i)$ averaged over the mass number and deformations (Eq. (9)) at 11-MeV (solid lines) excitation energy and zero spin of the initial CN for the indicated fissioning thorium isotopes. The potentials are normalized to the energy $\langle U\rangle(Z_i = 44) = 0\,$MeV. The lines connect the points corresponding to the even–even fission fragments.

Kr–Sr are wider than those for Ru–Pd. For the ^{228}Th nucleus, the asymmetric minima become deeper facilitating the asymmetric fragmentations, but the central minimum still remains.

To understand the conditions under which a mass-asymmetric minimum appears in the driving potential, one should analyze all components of the driving potential: the LD energy $\langle U^{\rm LD}\rangle(Z_i) = \langle U_L^{\rm LD}\rangle + \langle U_H^{\rm LD}\rangle$, the interaction potential $\langle V\rangle(Z_i) = \langle V^C\rangle + \langle V^N\rangle$ and the shell-correction energy $\langle \delta U^{\rm shell}\rangle(Z_i) = \langle \delta U_L^{\rm shell}\rangle + \langle \delta U_H^{\rm shell}\rangle$ (Fig. 19). The absolute value of $\langle \delta U^{\rm shell}\rangle$ is much smaller than the values of $\langle U^{\rm LD}\rangle$ and $\langle V\rangle$. The LD energy increases globally when the charge (mass) number deviates from $Z/2$ $(A/2)$. The interaction potential has the opposite global behavior. These two components depend on deformations of the DNS nuclei that create local minima in $\langle U^{\rm LD}\rangle(Z_i)$ and $\langle V\rangle(Z_i)$. Larger deformations result in smaller

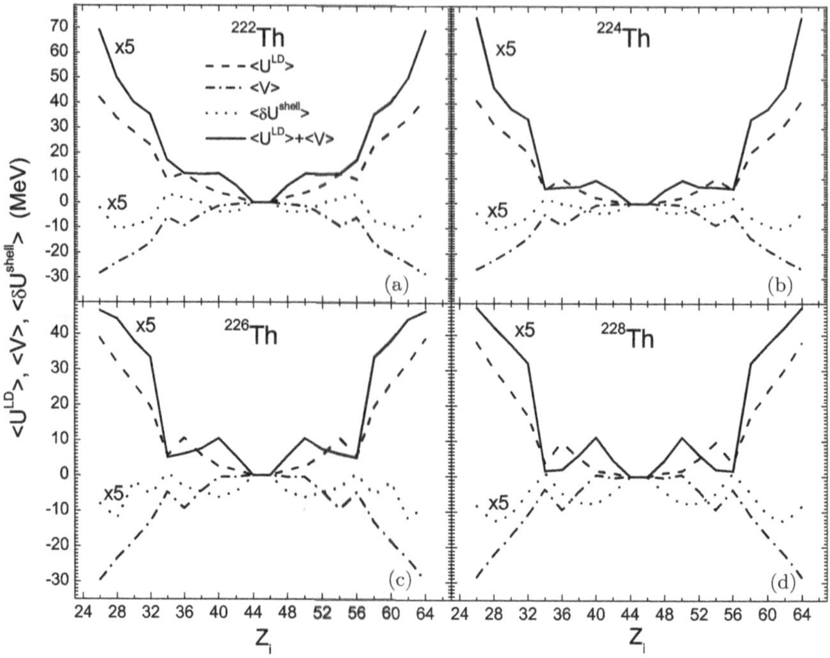

Fig. 19. The calculated components of the driving potentials $\langle U\rangle(Z_i)$ at 11-MeV (solid lines) excitation energy and zero spin of the initial CN for the indicated fissioning thorium isotopes. The LD energy $\langle U^{\mathrm{LD}}\rangle$ (dashed lines), the nucleus–nucleus interaction potential $\langle V\rangle$ (dash-dotted lines) and the shell-correction energy $\langle \delta U^{\mathrm{shell}}\rangle$ (dotted lines) are normalized to the zero energy at $Z_i = 44$. The $\langle U^{\mathrm{LD}}\rangle + \langle V\rangle$ is presented by the solid lines. The values of $\langle \delta U^{\mathrm{shell}}\rangle$ and $\langle U^{\mathrm{LD}}\rangle + \langle V\rangle$ are scaled by a factor of 5. The lines connect the points corresponding to the even–even fission fragmentations.

interaction energy and larger LD energy. The position of a minimum in the (β_L, β_H) plane depends on the shell corrections. Strong shell effects in semimagic and almost semimagic nuclei ($Z_L = 34$–38) are expressed by relatively small deformations (Fig. 20). For the nuclei with $Z_L = 40$–44, the shell effects are relatively weak because Z_L (N_L) is the midway between the proton (neutron) shell closures. This leads to almost pure LD behavior of the scission configuration and larger deformations of the fragments. Thus, the asymmetric minimum in the driving potential appears due to the interplay between the LD energy and the nucleus–nucleus interaction potential (Fig. 19). The shell effects indirectly affect (through the deformations of nuclei) the appearance of an asymmetric minimum. The direct role of the shell effects is expressed by changing the position and depth of this minimum in the driving potential. For example, for ^{226}Th, at $Z_L = 34$, 36 and 38 the

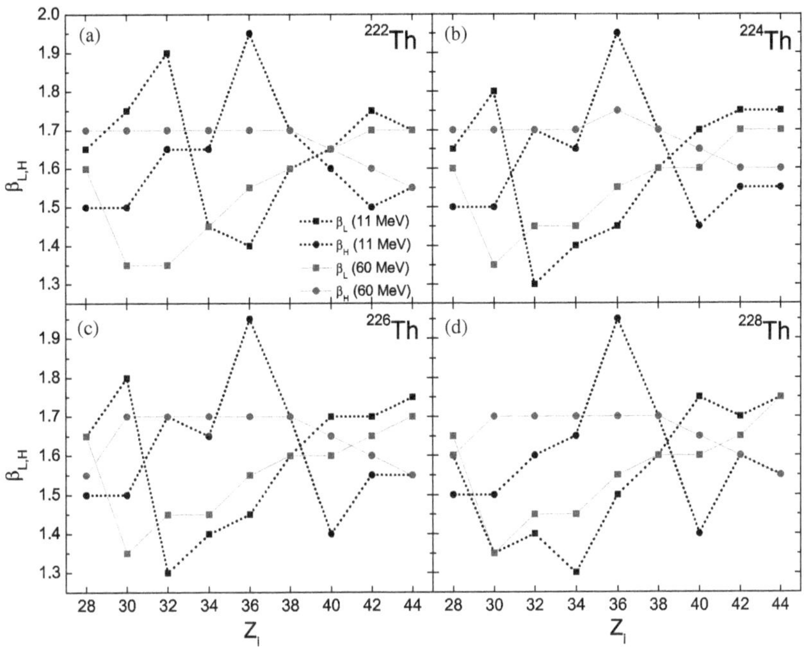

Fig. 20. The calculated average deformations $\beta_L(Z_L)$ (squares) and $\beta_H(Z_L)$ (circles) of fragments in the binary fragmentations of the indicated thorium isotopes at 11-MeV (closed symbols) and 60-MeV (open symbols) excitation energies and zero spin. The values of $A_{L,H}$ and $\beta_{L,H}$ are related to Z_L to supply the minimum of U. The lines connect the points obtained for even–even fission fragments.

DNS nuclei are very close to those with $N_L = 50$ and $N_H = 82$, which means high values of their stiffness parameter with respect to the change of deformation and that the minimum in the (β_L, β_H) plane is situated at small deformations. As a result, at $Z_L = 34$–38, $\langle V\rangle$ decreases faster than the increase of $\langle U^{\text{LD}}\rangle$ and at $Z_L < 34$ an increase of $\langle U^{\text{LD}}\rangle$ overcompensates well for a decrease of $\langle V\rangle$, and, correspondingly, an asymmetric minimum appears at $Z_L = 34$. The shell corrections cause the shift of the minimum from $Z_L = 34$ to $Z_L = 36$ and make it deeper in the driving potential (Fig. 18). It is interesting to note that only in the cases of ^{222,224}Th, the shell-correction energy $\langle\delta U^{\text{shell}}\rangle(Z_i)$ diminishes the asymmetric minimum of the LD energy plus the interaction potential in Fig. 18.

With increasing $N - Z$ of the fissioning nucleus, the LD energy (interaction potential) increases slower (faster) with charge asymmetry. This is the reason for the appearance of a pronounced asymmetric minimum in the cases of ^{226,228}Th.

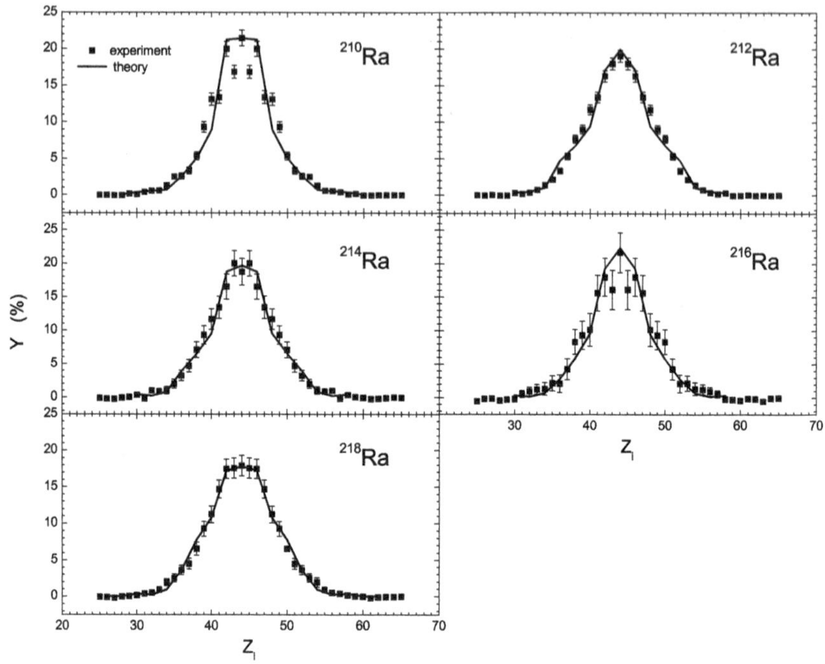

Fig. 21. The calculated charge distributions (lines) for the electromagnetic-induced fission of the indicated radium isotopes at 11-MeV excitation energy and zero spin are compared with the experimental data[4] (symbols). The lines connect the calculated points for even–even fission fragments.

The measured charge distributions resulting from the electromagnetic-induced fission of even–even $^{210-218}$Ra isotopes at 11-MeV excitation energy are clearly symmetric (Fig. 21). The evolution of the shape of the charge (mass) distribution with increasing neutron number of the fissioning nucleus Ra from $N = 132$ to $N = 138$ also displays similar features (Fig. 22) as in the case of even–even $^{222-230}$Th. The symmetric fission of ^{220,222}Ra is evident. The formation of the asymmetric peaks starts at $A = 224$. The fission of ^{226}Ra generates an asymmetric distribution. A small symmetric contribution still exists, although not so evident as in the fission of ^{228,230}Th.

4.2. *Even–even $^{242-256}$Cf, $^{250-264}$Fm, $^{250-264}$No and $^{262-266}$Rf*

In Fig. 23, the calculated mass distributions of fission fragments of even–even nuclei ^{242,244,246,248,250,252}Cf are compared with the available experimental data.[102–105] In the case of ^{242,244}Cf, the mass distributions

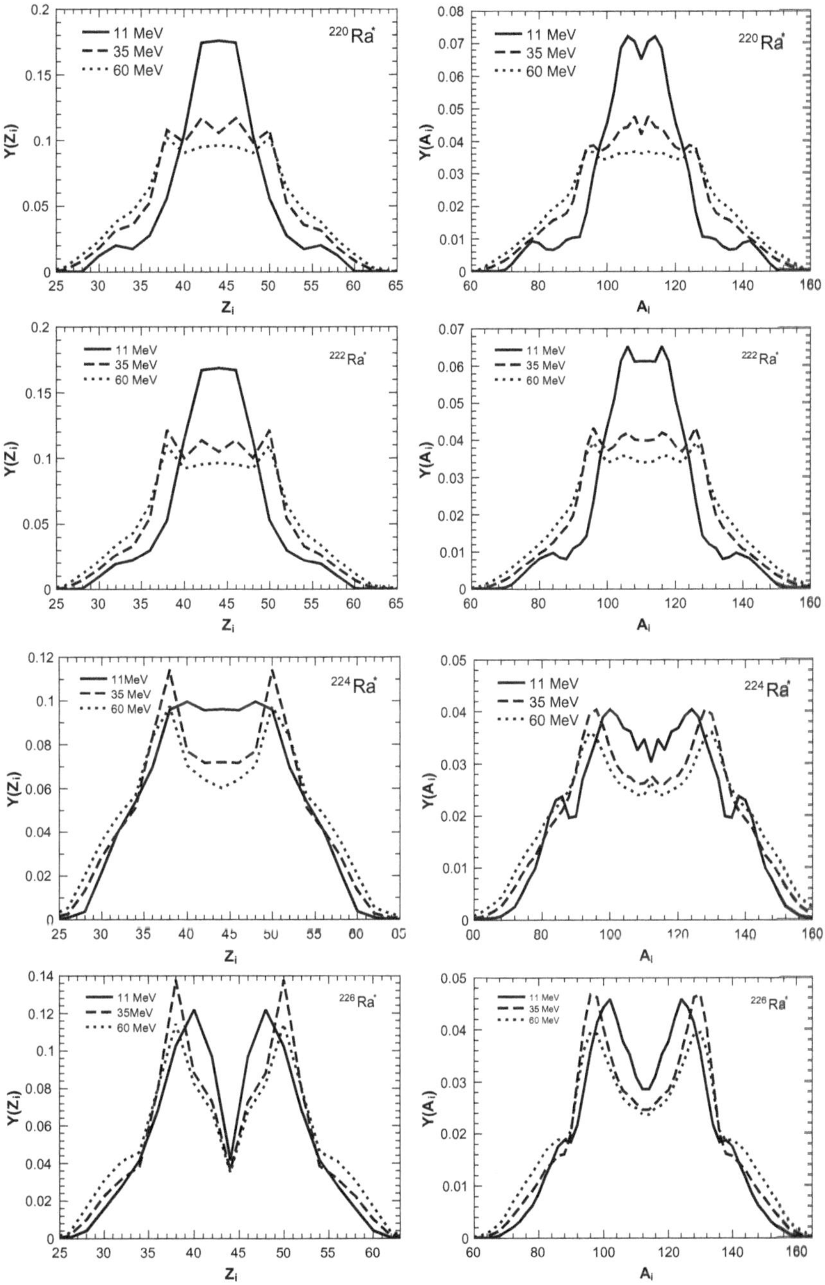

Fig. 22. The same as in Fig. 8 but for the Ra isotopes.

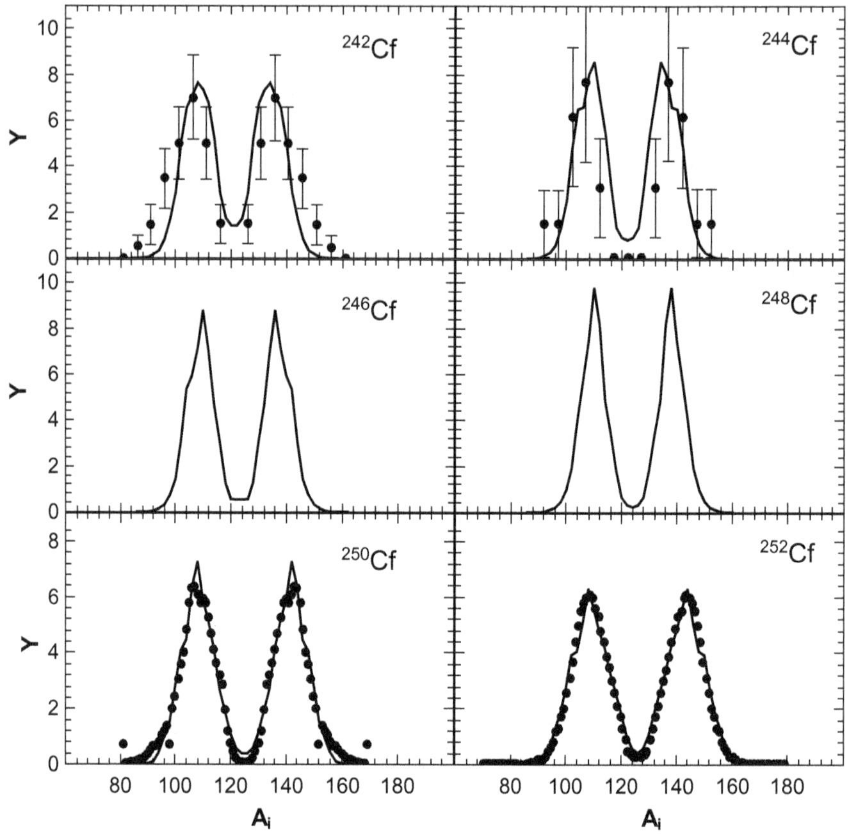

Fig. 23. The calculated mass distributions (lines) are compared with the experimental data (symbols) for the electron-capture delayed fission of the ^{242,244}Cf daughter nuclei[102,103] and for the spontaneous fission of ^{250,252}Cf.[104,105] The calculations are performed for the even–even mass fragmentations and at $L = 0$.

have been measured from the electron-capture delayed fission (ECDF) of ^{242}Es,[102,103] while the mass distributions for the fission ^{250,252}Cf result from spontaneous or thermal-neutron-induced fission, respectively.[104,105] The calculations agree better with the latter cases, where the mass numbers of fragments are measured more precisely. In the case of the ECDF, the data have large error bars and the mass yields are averaged over five mass units. This can hide some of the finer features of the distributions. We attribute a small discrepancy between the theory and the ECDF experimental data to this fact. The positions of the maxima from the ECDF data are also rather unclear, as the maximum of the experimental yield in the fission of ^{242}Cf is around $A_L \sim 103$ but $A_L \sim 106$–108 for the fission of ^{250,252}Cf. By comparison, the position of the asymmetric maximum does not change with the neutron number of the fissioning Th, Ac and U nuclei.[4] Nevertheless, the asymmetric nature of the mass distributions of fragments is evident and conserved in the fission of Cf isotopes. It is worthwhile to mention a small increase of yields near symmetry with decreasing neutron number of the fissioning nucleus. This can be explained by the reduction of B_{qf} near the symmetry that promotes high yields. The same behavior is observed, for example, in electromagnetic-induced fission of neutron-deficient U isotopes.[4,71]

The calculated charge and mass distributions of ^{250,252,254,256}Cf, ^{250,252,254,256}Fm and ^{250,252,254,256}No are presented in Figs. 24 and 25. For the fissioning nuclei ^{254,256}Cf (symbols), the experimental data[106,107] correspond to the pre-neutron emission mass yields. The data[16,106] for the nuclei ^{254,256}Fm are corrected by the post-scission neutron emission. One can observe the asymmetric shape of the mass and charge distributions for the Cf isotopes with $A = 250$, 252, 254 and 256 both in the calculated results and in the experimental data. In the fission of ^{254,256}Fm, the measured mass distributions present two asymmetric peaks which are again backed by theoretical results. In fact, the asymmetric mass and charge distributions are predicted for the lighter ^{250,252}Fm isotopes as well (Figs. 24 and 25). In the fission of ^{252,254}No (Figs. 24 and 25), the yields show a clear double-peaked distribution in good agreement with the experimental data.[17,111] For the nucleus with $A = 250$, the distribution starts showing the formation of symmetric peak (^{122,124}Sn+^{128,126}Te), while for ^{256}No the mass yields display asymmetric and symmetric peaks.

The transition from two-peaked to one-peaked mass distribution can be explained by studying different components of the average driving potential $\langle U\rangle(A_i)$ (averaged over the charge number Z_i and deformations $\beta_{L,H}$), namely the LD energy $\langle U^{\rm LD}\rangle(A_i) = \langle U_L^{\rm LD}\rangle + \langle U_H^{\rm LD}\rangle$, the nucleus–nucleus interaction potential $\langle V\rangle(A_i) = \langle V^C\rangle + \langle V^N\rangle$ and the shell-correction energy $\langle \delta U^{\rm shell}\rangle(A_i) = \langle \delta U_L^{\rm shell}\rangle + \langle \delta U_H^{\rm shell}\rangle$ (Fig. 26). The strong shell effects in double magic, semimagic and almost magic nuclei result in the relatively small deformations (close to $\beta_{L,H} \sim 1$). The shell effects indirectly affect (through the deformations of nuclei) the appearance of minima in $\langle U^{\rm LD}\rangle$ and $\langle V\rangle$, and, correspondingly, in $\langle U^{\rm LD}\rangle + \langle V\rangle$ (Fig. 26). The direct role of the shell effects is expressed by their ability to enhance or suppress the formations of the minima in $\langle U^{\rm LD}\rangle + \langle V\rangle$ and, correspondingly, in $\langle U\rangle$. For heavy actinides (in contrast to the pre-actinides and light actinides), the

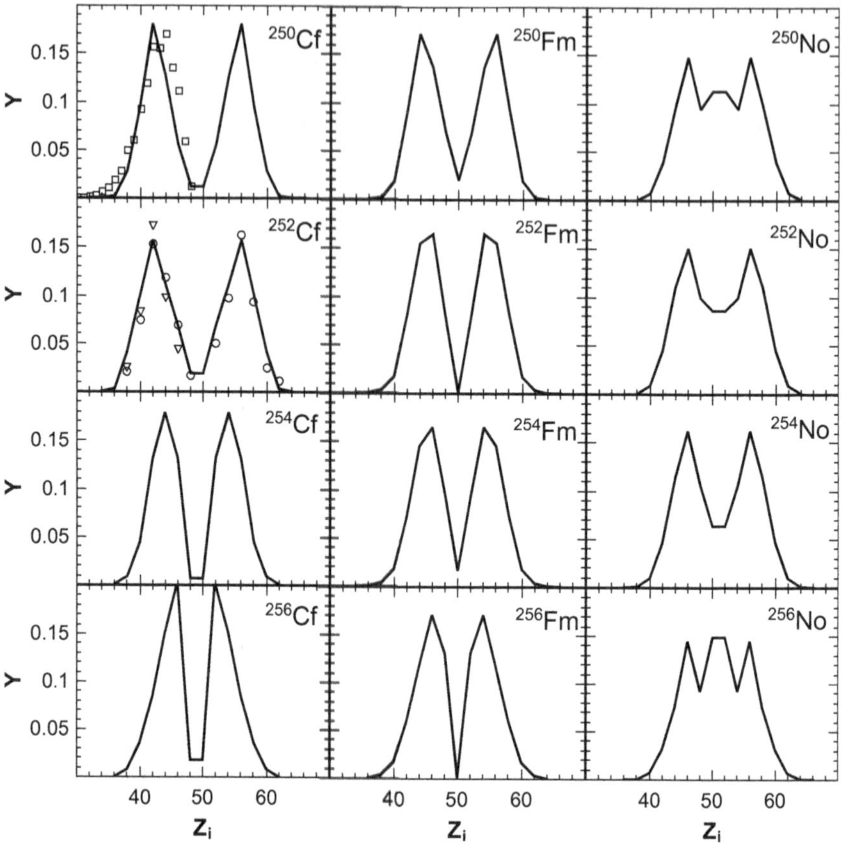

Fig. 24. The calculated charge distributions (lines) resulting from the spontaneous fission of ^{252,254,256}Cf, ^{250,252,254,256}Fm and ^{250,252,254,256}No and the thermal-neutron-induced fission of ^{249}Cf ($E^*_{\text{CN}} = 6.2$ MeV). The calculations are performed for the even–even mass fragmentations and at $L = 0$. The distributions are normalized to unity. The experimental data (symbols) are from Refs. 108–110.

microscopic shell corrections are quite strong due to low excitation energy of the system and large occurrence of magic nuclei, especially in symmetric fragmentations, with $Z_{L,H} \sim 50$ and $N_{L,H} \sim 82$. For example, for ^{250}No (Fig. 26), the macroscopic driving potential $\langle U^{\text{LD}} \rangle + \langle V \rangle$ and the total driving potential $\langle U \rangle$ show two asymmetric minima plus the central one. The asymmetric ones are, however, enhanced by the presence of the shell effects. The combined effect results in a deeper minimum for the asymmetric fragmentations, and in turn their mass yields are higher than those at

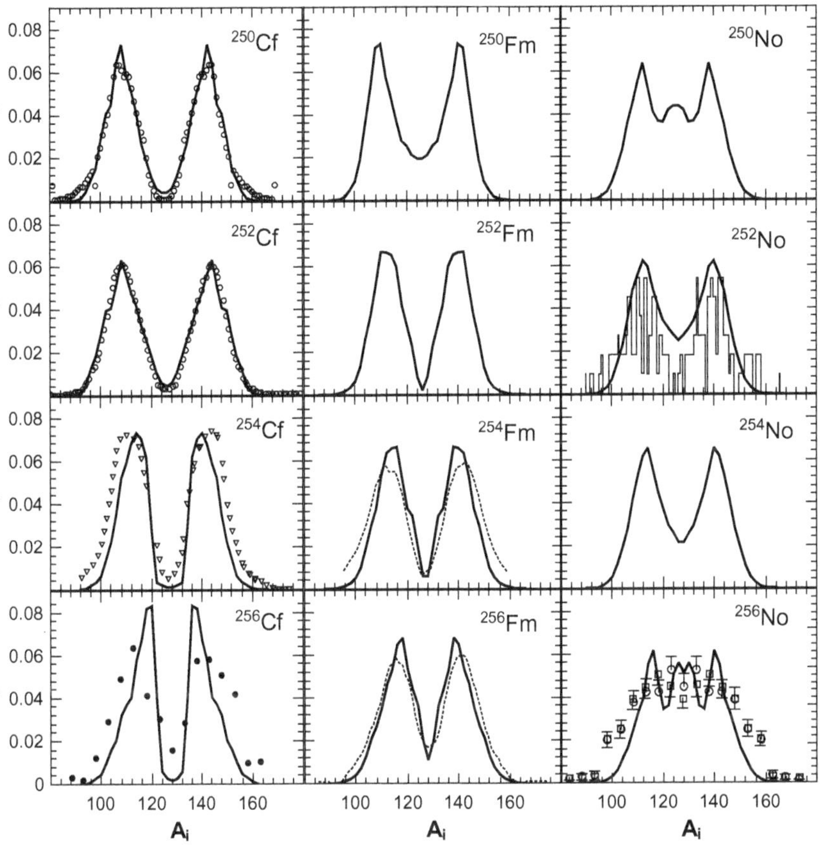

Fig. 25. The calculated mass distributions resulting from the spontaneous fission of ^{250,252,254,256}Cf, ^{250,252,254,256}Fm and ^{250,252,254,256}No isotopes. The calculations are performed for the even–even mass fragmentations and at $L = 0$. The distributions are normalized to unity. The experimental data for the spontaneous fission of Cf (symbols), Fm (dotted lines) and No (symbols and histogram) nuclei are from Refs. 16, 17, 20, 104–107 and 111.

$A_i \sim A/2$. In the case of fissioning nuclei ^{252,254}No, the shell effects work against the macroscopic energy $\langle U^{\rm LD}\rangle + \langle V\rangle$ to avoid the formation of the symmetric minimum (Fig. 26). The net result is the two-peaked mass distributions observed for these nuclei. With increasing mass number of the CN further to $A = 256$, the microscopic effects suppress the asymmetric minima to a large degree, in the favor of the formation of a shallow symmetric minimum. This explains the reduction of the asymmetric character of the

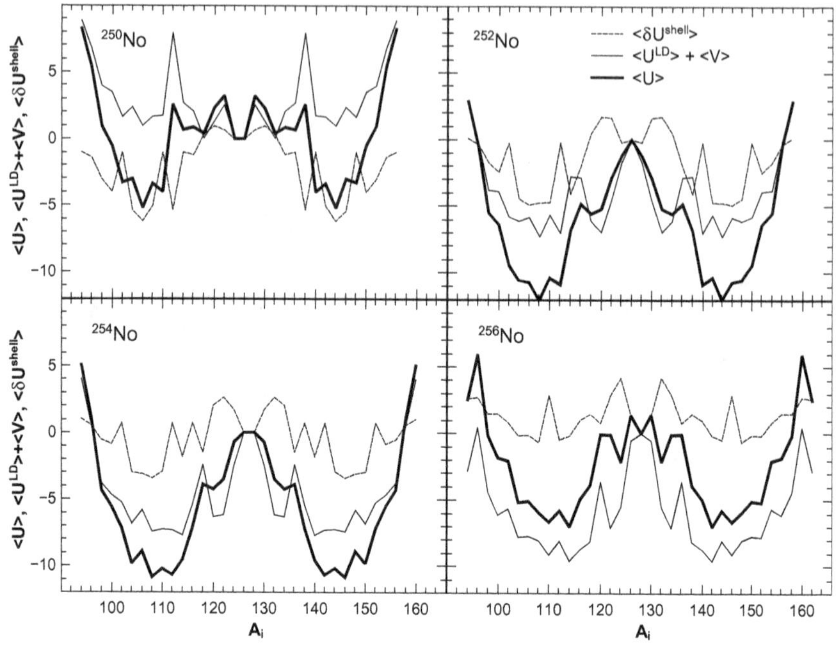

Fig. 26. The calculated average driving potential $\langle U \rangle$ (thick solid lines), shell corrections $\langle \delta U^{\mathrm{shell}} \rangle$ (dashed lines) and the sum of the macroscopic energies (the LD energy plus the nucleus–nucleus interaction potential energy) $\langle U^{\mathrm{LD}} \rangle + \langle V \rangle$ (thin solid lines) as functions of the fragment mass number A_i for the spontaneous fission of ^{250,252,254,256}No isotopes. The calculations are performed for the even–even fragmentations. All the components are normalized to the corresponding energies of the symmetric fragmentation.

mass distribution and the enhancement of the symmetric yields observed both in the calculations and in experiment.

For ^{256}No, even though the symmetric minimum is higher by about 5 MeV than $\langle U \rangle (A_L = 114)$ (Fig. 26), the corresponding yields are almost the same (Figs. 24 and 25). This is easily explained by the facts that many almost symmetric configurations with ^{126,128,130}Sn and ^{126,128}Cd and basically one asymmetric configuration with ^{114}Pd contribute to the mass yields, and the potential minimum on the (β_L, β_H) surface for the symmetric configuration is five–six times wider than that for the asymmetric configuration. One can see from Eq. (4) that by integrating over a larger area of the PES, the yields can recover the energy difference and be comparable at $A_L = 126{-}128$ and $A_L = 114$.

In the case of low-energy fission of Ra and Th isotopes, the transition from a single-peaked to a two-peaked charge distribution is mainly due to the indirect (through the deformations of nuclei) influence of the shell effects which form the asymmetric minimum by facilitating a large number of asymmetric fragmentations and by establishing the minimum position, where the light fragment has the neutron number close to $N_L = 50$, or the heavy fragment has the neutron or charge numbers close to $N_H = 82$ or $Z_H = 50$. For heavy actinides, the shell effects directly and indirectly modify the symmetric/asymmetric minimum in the driving potential $\langle U \rangle$. Thus, one can conclude that the shell effects are more important for the fissioning heavy actinides than for the fissioning pre-actinides and light actinides.

In Figs. 27–32, the calculated charge and mass distributions resulting from the spontaneous and induced fission of ^{254,256,258,260,264}Fm, ^{258,260,262,264}No and ^{262,264,266}Rf isotopes are presented. The measured mass distributions[112–117] are well reproduced for the spontaneous fission of ^{262}No, ^{262}Rf, and for the thermal-neutron-induced fission of ^{255,257}Fm. In the fission of ^{262}No and ^{262}Rf, the measured[116,117] and calculated mass distributions are symmetric with the large widths. In contrast to the experimental data,[20] the calculated mass distribution for ^{258}Fm is rather wide and has asymmetric bumps, even though the symmetric yields are extremely enhanced, and is strikingly similar to the experimental one for ^{257}Fm (Fig. 27).[18] For ^{258}No (Fig. 29), the symmetric mass yields are larger than the asymmetric mass yields, which is in good agreement with the experimental data.[20]

The calculated charge distributions of spontaneous fission of ^{254,256,258,260,264}Fm have the asymmetric shape (Fig. 28). The tendency to a more symmetric charge splitting of fermium isotopes with increasing neutron number is clearly seen. In the case of spontaneous fission of ^{258}Fm (Fig. 28), the charge distribution has two peaks divided by a small minimum for the symmetric Sn + Sn charge split. This is explained by the fact that a high yield of Cd ($Z_L = 48$) or Te ($Z_H = 52$) mainly originates from three mass fragmentations ^{122}Cd+^{136}Te, ^{124}Cd+^{134}Te and ^{126}Cd+^{132}Te which have almost the same potential energies U. Yields of the symmetric charge split mostly originate from the ^{126}Sn+^{132}Sn configuration, for which the potential energy U is larger by about 2 MeV than that for the Cd + Te configurations. There are two more mass fragmentations ^{124}Sn+^{134}Sn and ^{128}Sn+^{130}Sn which contribute to the symmetric yields. Since their potential energies are higher by 4 MeV and 3 MeV, respectively, than those for the Cd + Te configurations, their contributions to the symmetric yields are rather small.

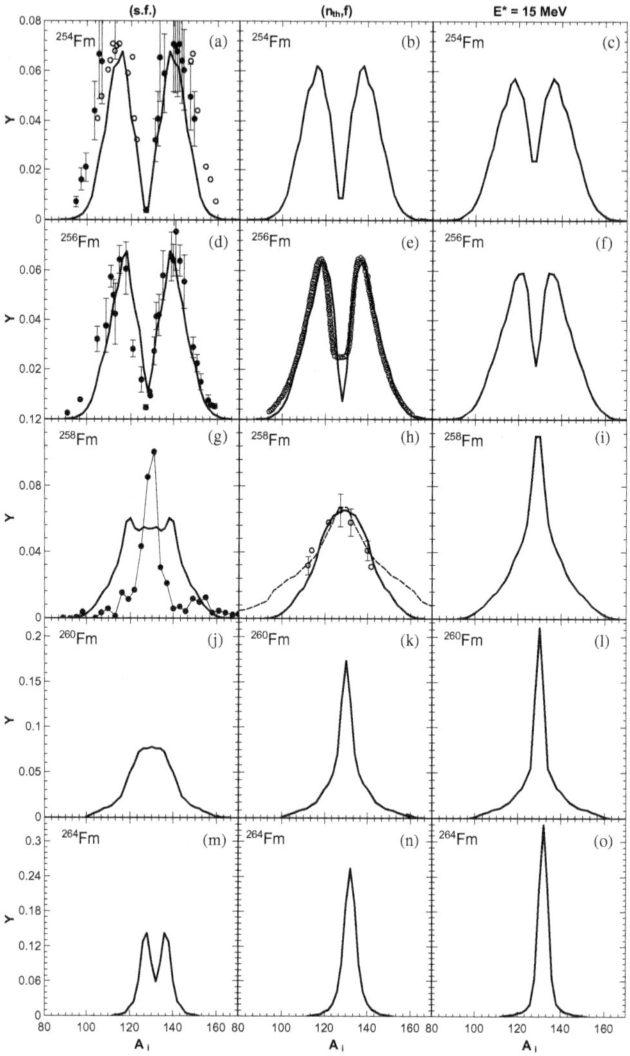

Fig. 27. The calculated mass distributions (solid lines) resulting from the spontaneous ($E^*_{\rm CN} = E^* = 0\,{\rm MeV}$, $L = 0$, the first column) and induced ($E^*_{\rm CN} = E^* = 15\,{\rm MeV}$, $L = 0$, the third column) fission of the indicated nuclei ^{254,256,258,260,264}Fm and the thermal-neutron-induced ($E^*_{\rm CN} = E^* \sim 6.3\,{\rm MeV}$, $L = 0$, the second column) fission of nuclei ^{253,255,257,259,263}Fm. The calculations are performed for even–even mass and charge fragmentations. The distributions are normalized to unity. The open symbols represent experimental data of Refs. 20 and 112–115. The spontaneous fission data for ^{258}Fm are multiplied by a factor 0.8. In panel (h), the dashed line represents the provisional mass deduced in Ref. 115, and the open square represents radiochemically deduced secondary yield, while the open circles are the derived primary distributions.[113]

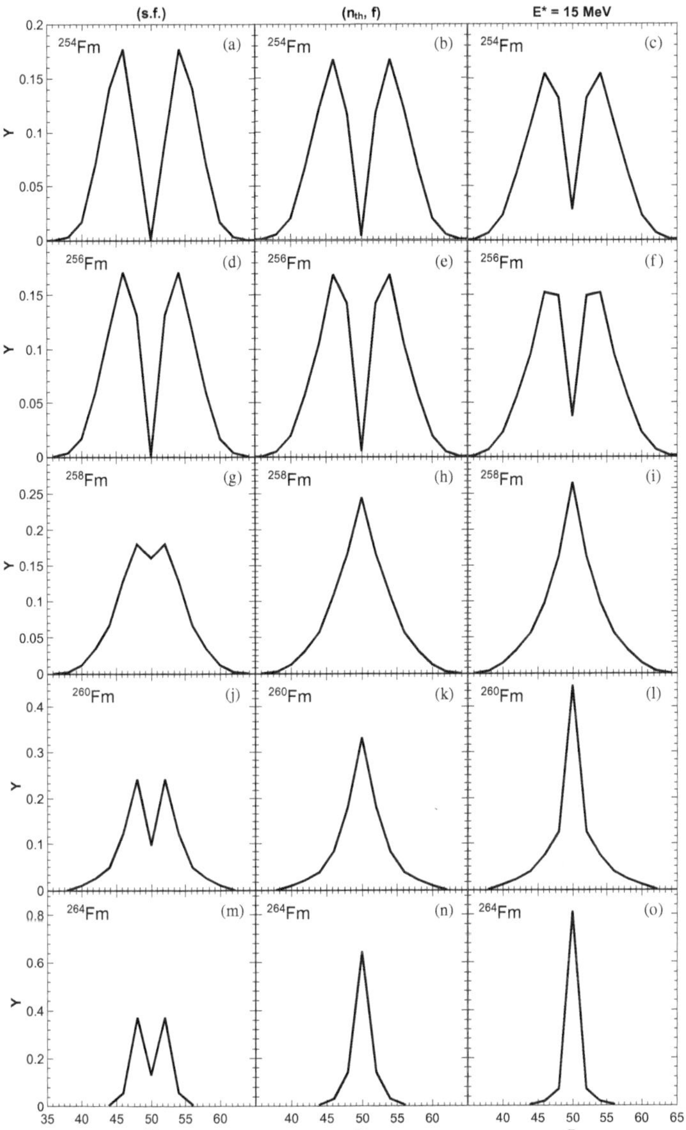

Fig. 28. The calculated charge distributions (solid lines) resulting from the spontaneous ($E^*_{\rm CN} = E^* = 0\,{\rm MeV}$, $L = 0$, the first column) and induced ($E^*_{\rm CN} = E^* = 15\,{\rm MeV}$, $L = 0$, the third column) fission of the indicated nuclei ^{254,256,258,260,264}Fm and the thermal-neutron-induced ($E^*_{\rm CN} = E^* \sim 6.3\,{\rm MeV}$, $L = 0$, the second column) fission of the nuclei ^{253,255,257,259,263}Fm. The calculations are performed for even–even mass and charge fragmentations. The distributions are normalized to unity.

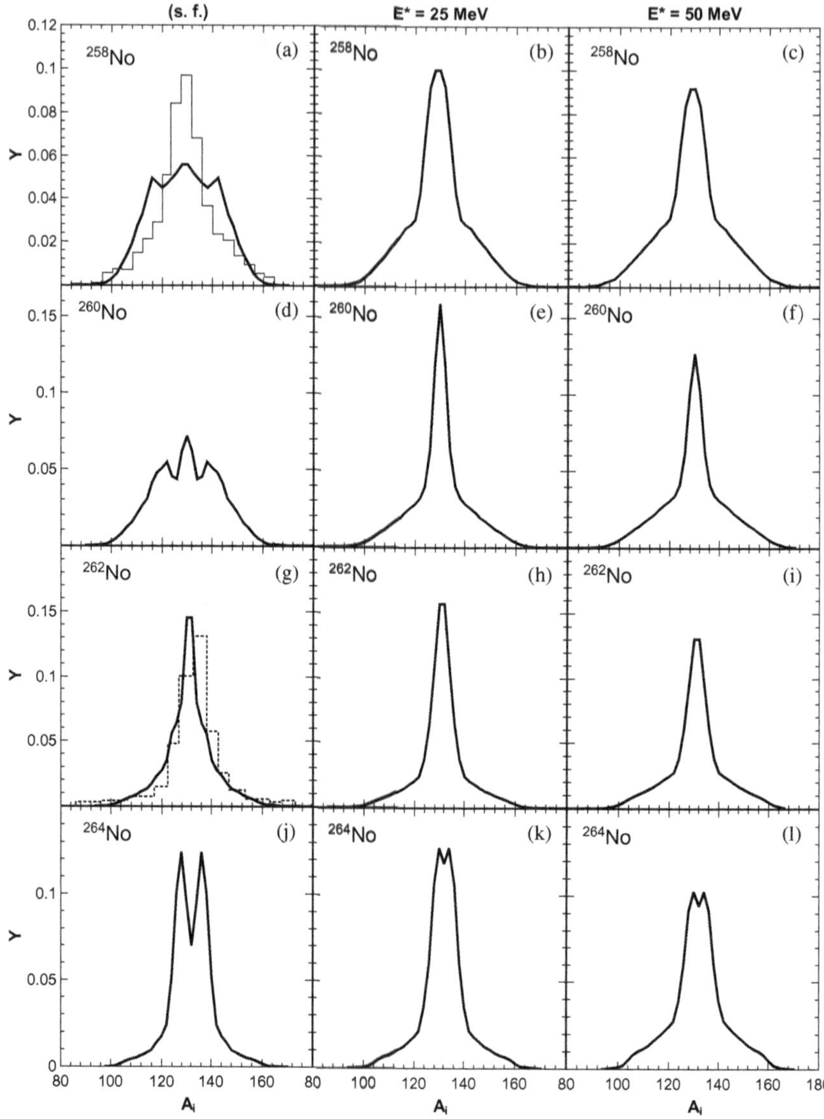

Fig. 29. The calculated mass distributions (solid lines) resulting from the fission of the indicated nuclei ^{260,262,264}No at excitation energies $E^*_{\rm CN} = E^* = 0$ MeV (the first column), $E^*_{\rm CN} = E^* = 25\,$MeV (the second column) and $E^*_{\rm CN} = E^* = 50\,$MeV (the third column). The calculations are performed at $L = 0$ and for even–even mass and charge fragmentations. The distributions are normalized to unity. In panels (a) and (g), the histograms represent the experimental data of Refs. 20 and 116, respectively.

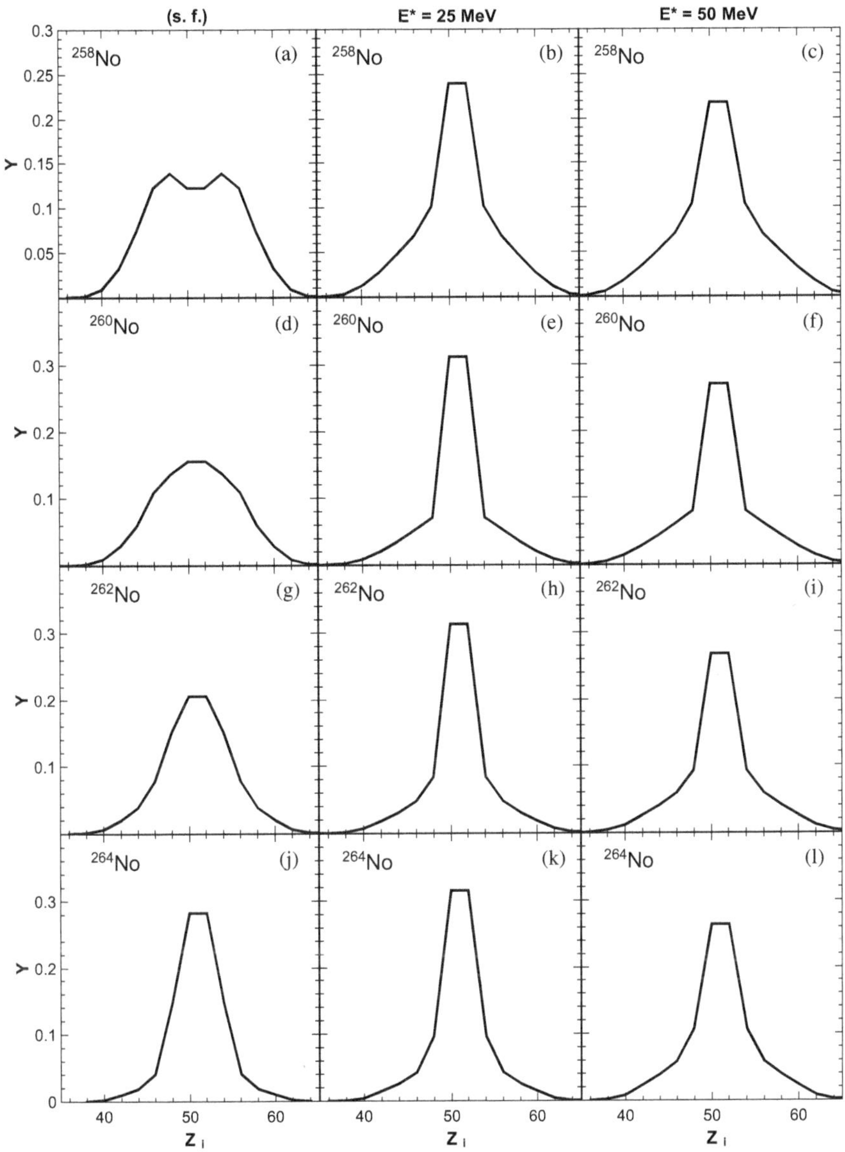

Fig. 30. The calculated charge distributions (solid lines) resulting from the fission of the indicated nuclei ^{260,262,264}No at excitation energies $E^*_{CN} = E^* = 0\,$MeV (the first column), $E^*_{CN} = E^* = 25\,$MeV (the second column) and $E^*_{CN} = E^* = 50\,$MeV (the third column). The calculations are performed at $L = 0$ and for even–even mass and charge fragmentations. The distributions are normalized to unity.

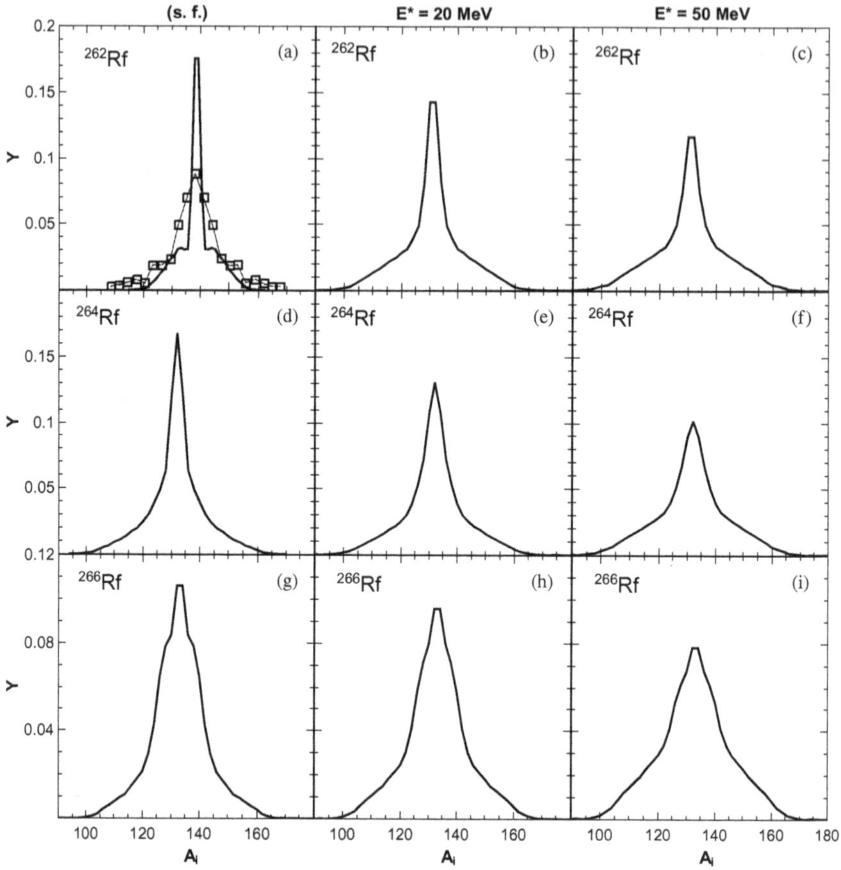

Fig. 31. The same as in Fig. 29 but for the indicated nuclei ^{262,264,266}Rf. In panel (a), the symbols connected by line represent the experimental data of Ref. 117.

5. Unexpected Asymmetry of Charge and Mass Distributions in Fission at High Excitation Energies

A comparison between the experimental and calculated mass or charge distributions resulting from the reactions $^{235}\mathrm{U}(n,\ f)$ at $E_n = 6\,\mathrm{MeV}$ and $14\,\mathrm{MeV}$ ($E^*_{\mathrm{CN}} = 12.5\,\mathrm{MeV}$ and $20.55\,\mathrm{MeV}$), $^{238}\mathrm{U}(n,\ f)$ at $E_n = 32.8, 45.3$ and $59.9\,\mathrm{MeV}$ ($E^*_{\mathrm{CN}} = 37.6$, 50.1 and $64.7\,\mathrm{MeV}$), $^{240}\mathrm{U}^*$ and $^{244}\mathrm{Pu}^*$ at $E^*_{\mathrm{CN}} = 15$, 25, 35, 45 and $55\,\mathrm{MeV}$, $^{244}\mathrm{Cm}^*$ at $E^*_{\mathrm{CN}} = 23\,\mathrm{MeV}$ and $^{250}\mathrm{Cf}^*$ at $E^*_{\mathrm{CN}} = 46\,\mathrm{MeV}$ are shown in Figs. 33–37. As can be seen, the calculated distributions are in good agreement with the experimental data.[21,25,29,118] The influence of excitation energy on the peak-to-valley ratio in the yields is well described.

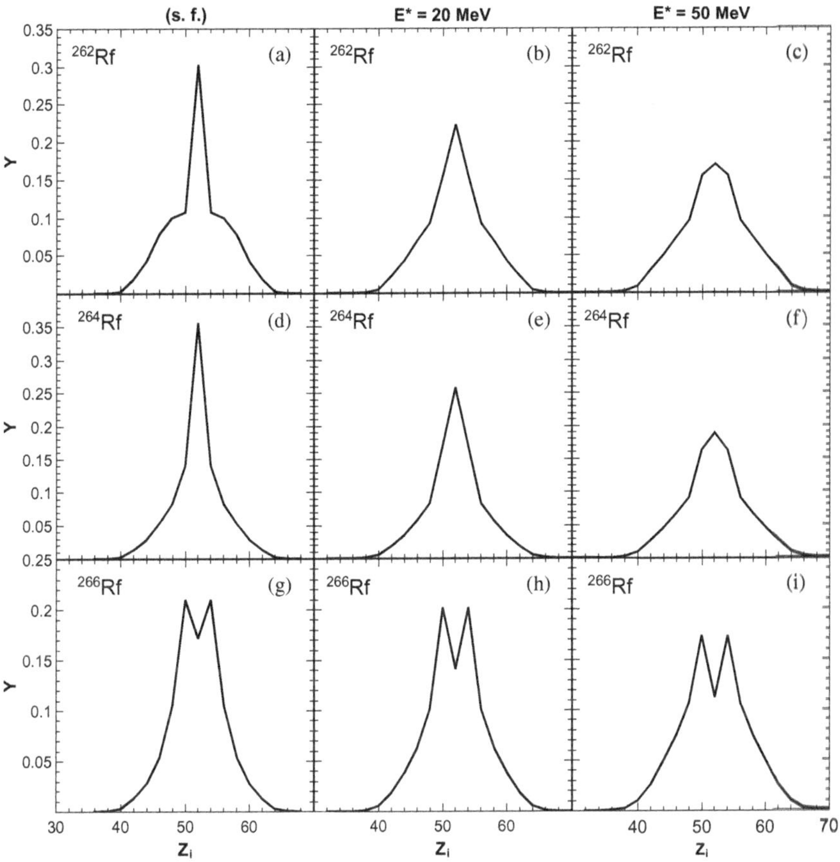

Fig. 32. The same as in Fig. 30 but for the indicated nuclei ^{262,264,266}Rf.

In addition, the scission-point model allows us to describe well and to predict the isotopic distributions of fission fragments $Z = 36, 38, 40$ and 42 in the ^{235}U $(n,\ f)$ reaction at various incident neutron energies (Figs. 38 and 39).[69] In Fig. 39, the only notable change in the yields with energy is the shift of position of the maximum toward lower mass (neutron) numbers. This is easily explained by the increase of neutron evaporation from the fragments with excitation energy.

In Figs. 33–37, the results indicate that the combined energy-dependent structures on the PES govern the yields of the fragments. The excitation energy reduces the shell effects, which causes a widening of the minima in the deformation space and migration of the minima in the PES (see Fig. 40).

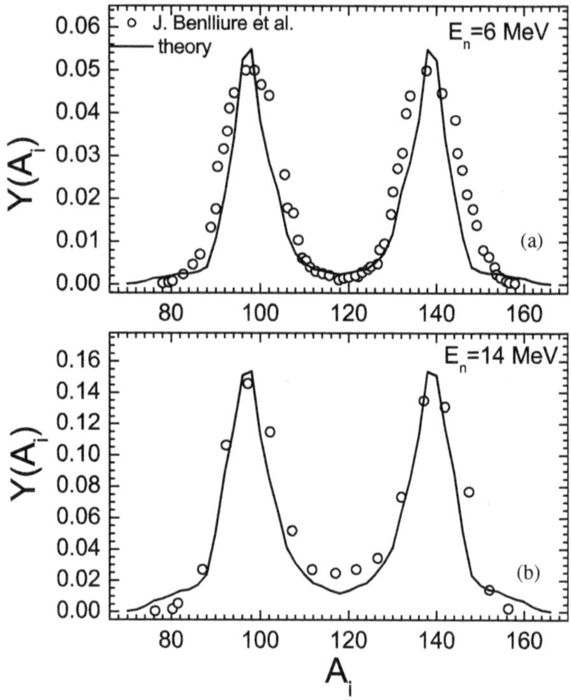

Fig. 33. The calculated mass distributions (solid lines) in the ^{235}U(n, f) reaction at neutron incident energies (a) 6 MeV and (b) 14 MeV and $L = 0$ without employing the multi-chance fission assumption. The experimental data (symbols) are from Ref. 118.

As a result, the mass and charge distributions are changed. However, the influence of excitation energy is not so dramatic. The calculated mass/charge distributions have a pronounced asymmetric shape even at incident neutron energies $E^*_{\mathrm{CN}} \geq 50$ MeV, which need experimental verification.

5.1. *Even–even* $^{220-226}$*Ra and* $^{222-230}$*Th*

The predicted charge/mass distributions of the fissioning even nuclei $^{222-230}$Th at excitation energies 35 MeV and 60 MeV are represented in Fig. 8. The charge distributions of ^{226,228}Th at higher excitation energies emphasize the asymmetric aspect of the distribution, making the symmetric peak smaller and narrower, and the asymmetric maxima wider and taller. This is easily explained by the fact that with higher excitation energy the previously inaccessible asymmetric configurations are involved (Fig. 2). For the energy range considered, the central peak does not disappear.

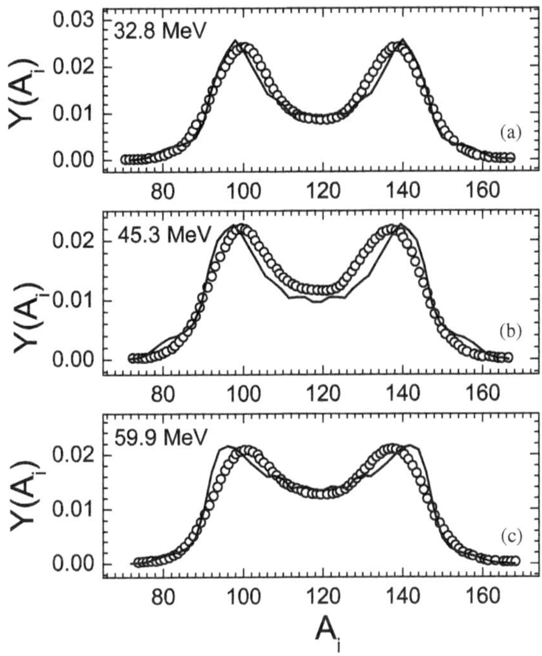

Fig. 34. The calculated fragment mass distributions (solid lines) for the $^{238}U(n,f)$ reaction at incident neutron energies (a) 32.8, (b) 45.3 and (c) 59.9 MeV and $L = 0$ without employing the multi-chance fission assumption. The lines connect the calculated yields of even–even fission fragments. The experimental data (symbols) are from Ref. 21.

For ^{222,224,226,228}Th, the average driving potentials (Figs. 41 and 42) match the yields in Fig. 8 at 60-MeV excitation energy. As can be seen, at 60-MeV excitation energy $\langle U\rangle(Z_i)$ has asymmetric minima resulting in larger yields of the asymmetric mode. The excitation energy reduces the shell effects, which causes a widening of the minima in the deformation space (β_L, β_H) and migration of the minima in the PES (Fig. 2). In the case of fissioning Th isotopes for Se–Kr charge fragmentations, the minima widen with excitation energy faster than for symmetric fragmentations. However, the influence of the excitation energy is not so spectacular. The deformations $\beta_L(Z_L)$ and $\beta_H(Z_L)$ of the fragments corresponding to the asymmetric minimum of the driving potential do not change strongly, especially for the configurations with semimagic or close-to-semimagic nuclei Se–Kr ($N_L = 50$ and $N_H = 82$; Fig. 20). So the deformations of the light nuclei with $Z_L =$ 34–36 remain to be smaller than the deformations of the DNS nuclei with $Z_L = 38$–44. As a result, $\langle V\rangle(Z_i)$ decreases faster than the increase of

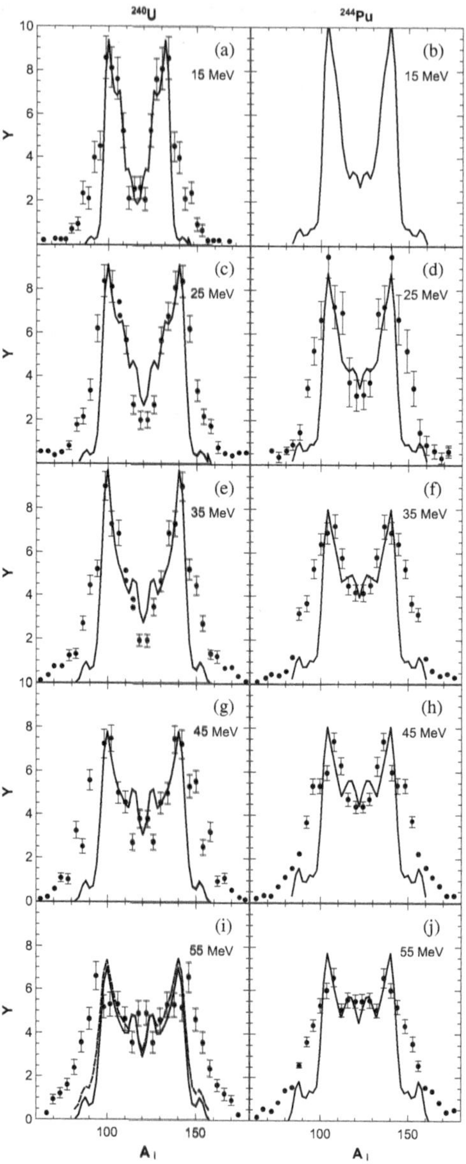

Fig. 35. The calculated mass distributions (solid lines) for the fission of $^{240}U^*$ (the first column) and $^{244}Pu^*$ (the second column) at the indicated excitation energies and $L = 0$ without employing the multi-chance fission assumption. In plot (i), the calculated mass yields with (dashed line) and without (solid line) employing the multi-chance fission assumption are presented. The experimental data (symbols) are from Ref. 25. Note that the experimental mass resolution is 6.5 amu.

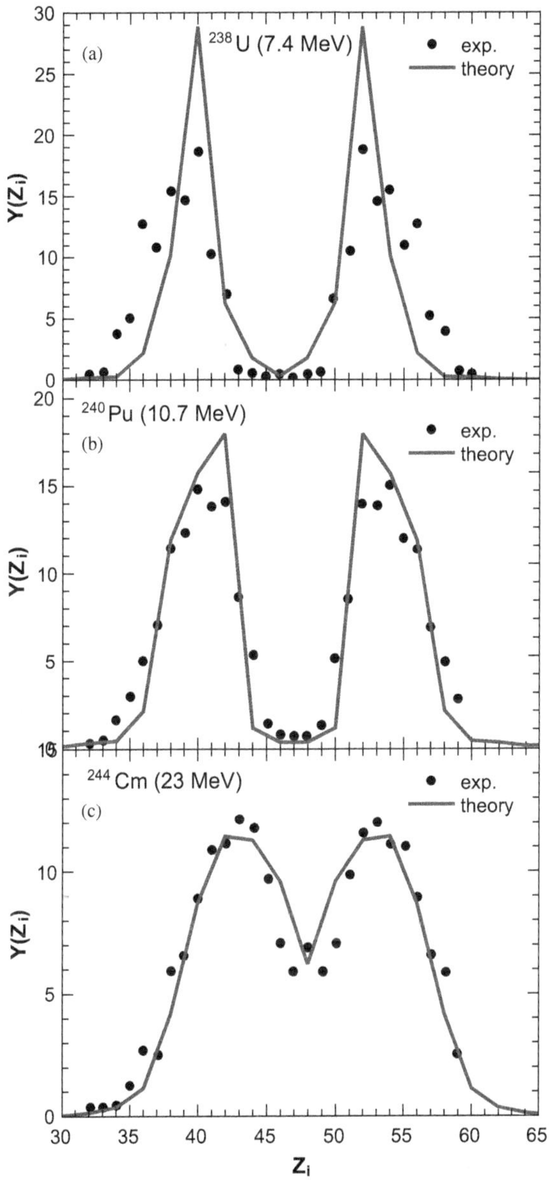

Fig. 36. Calculated charge distributions (solid lines) resulting from the fission of (a) ^{238}U ($E^*_{\rm CN} = 7.4\,$MeV), (b) ^{240}Pu ($E^*_{\rm CN} = 10.7\,$MeV) and (c) ^{244}Cm ($E^*_{\rm CN} = 23\,$MeV) at $L = 0$ compared with the experimental data (symbols).[29]

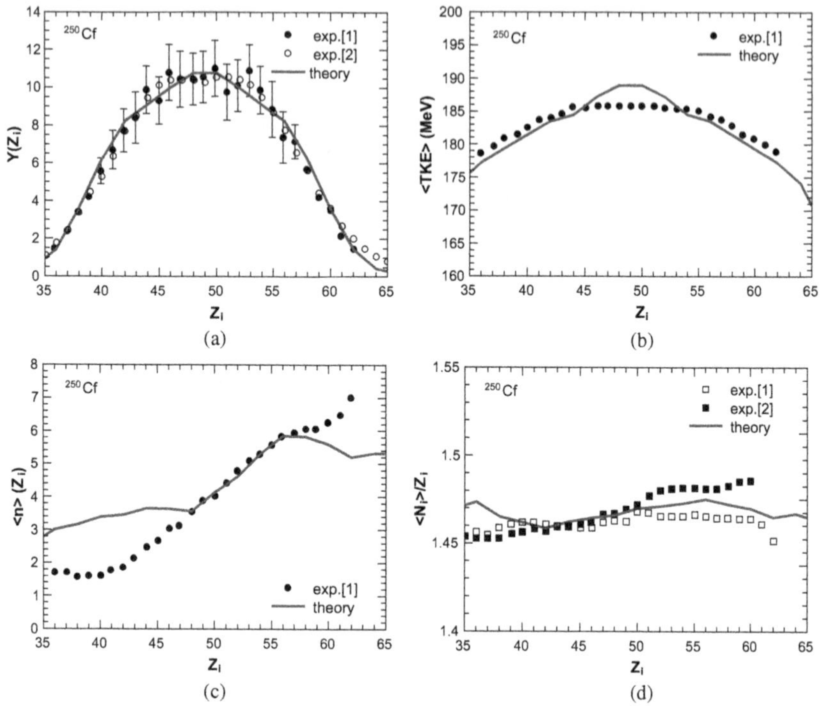

Fig. 37. Calculated (lines) and experimental (symbols) charge and TKE distributions (panels (a) and (b)) as functions of the charge number of one of the fragments resulting from the fission of ^{250}Cf at an excitation energy of 46 MeV. Panel (c) shows the average number of neutrons emitted from one fragment and panel (d) shows the neutron excess ratio $\langle N_i\rangle/Z_i$, both as functions of the charge number of the light fragment. Solid lines and symbols are theoretical calculations and experimental data of Refs. 29 and 58, respectively.

$\langle U^{\rm LD}\rangle(Z_i)$ from $Z_L = 38$ to $Z_L = 34$ and at $Z_L < 34$ their behavior is opposite and a minimum at $Z = 34$ is produced (Fig. 42). So the asymmetric minimum in $\langle U^{\rm LD}\rangle(Z_i) + \langle V\rangle(Z_i)$ and, correspondingly, in the driving potential is produced due to the deformation effect. With increasing excitation energy of the fissioning nucleus, the LD energy $\langle U^{\rm LD}\rangle(Z_i)$ (or the LD surface energy) increases slower with charge asymmetry. This leads to a pronounced asymmetric minimum in the cases of ^{222,224}Th.

What is the main condition under which a charge-asymmetric minimum does not appear in the driving potential? From the study, we have the following: With increasing charge asymmetry of the scission configuration, the increase of the LD energy or the LD surface energy should be faster than the decrease of the nucleus–nucleus interaction potential.

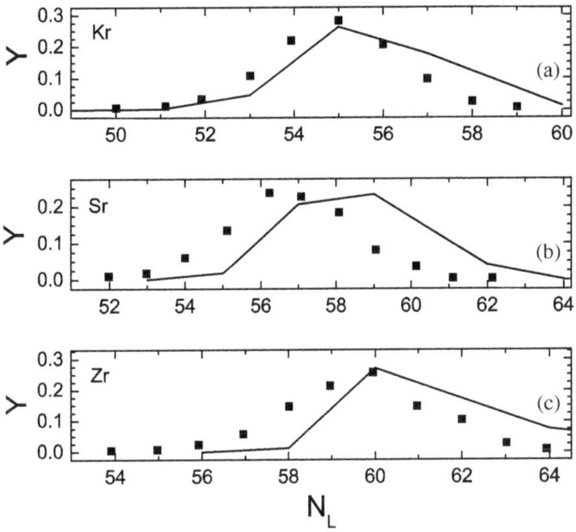

Fig. 38. The calculated mass yields (solid lines) of the (a) Kr, (b) Sr and (c) Zr isotopes as functions of neutron number N_L in the fission of ^{235}U by thermal neutron ($L = 0$). The neutron emissions from the primary fragments are taken into consideration. The experimental data (symbols) are from Ref. 86.

The charge and mass distributions of the fissioning nuclei ^{220,222,224,226}Ra at 35-MeV and 60-MeV excitation energies are shown in Fig. 22. In all cases, the asymmetric component is strongly enhanced because with increasing excitation energy previously inaccessible asymmetric configurations are populated. For the fissioning nucleus ^{226}Ra, the symmetric component is completely suppressed.

For electromagnetic-induced fission of even $^{220-226}$Ra and $^{222-228}$Th isotopes and neutron-induced fission of U or Pu isotopes, the behavior of charge distributions with increasing energy is different. In the case of neutron-induced fission of ^{238}U, the yields of symmetric fragments monotonically increase with neutron energy,[21] but the asymmetric shape remains even at neutron incident energies of 60 MeV. The symmetric shape of the charge distributions of Th fission fragments is not conserved with increasing energy, and the yields around $Z/2$ decrease in contrast to the enhancing yields of asymmetric fission fragments.

5.2. *Even–even* $^{248-256}$*Cf*

In Figs. 43–49, the predicted charge and mass distributions of fragments resulting from the fission of the excited even–even isotopes $^{248-256}$Cf are

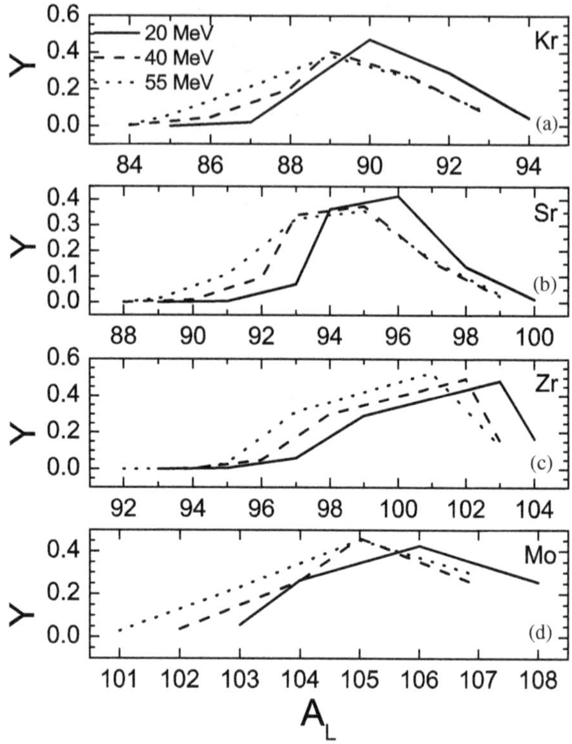

Fig. 39. The calculated isotopic trends for the nuclei (a) Kr, (b) Sr, (c) Zr and (d) Mo in the ^{235}U(n,f) reaction at different indicated neutron energies ($L = 0$).

presented. Some general noteworthy aspects are observed. For the isotopes of Cf with increasing excitation energy, almost symmetric components (Cd + Sn) of the mass and charge distributions are enhanced with respect to the asymmetric components Mo + Ba and Ru + Xe. The fast evolution of the symmetric yields is found with increasing excitation energy from 0 MeV to 15–25 MeV. The saturation of the symmetric components is reached at about $E^*_{\text{CN}} = 55$ MeV. At the excitation energy of 45 MeV, the asymmetric shape of the distributions is preserved, with the same being more pronounced in the mass distributions. The charge distribution in the fission of ^{248}Cf shows the formation of a small almost symmetric peak with charge split Cd + Sn. At the highest excitation energies of 55 MeV and 65 MeV, the fission of ^{252,254,256}Cf has the asymmetric mass yields, with the same being less pronounced with the reduction of neutron number in the fissioning nucleus. For the fission of ^{250}Cf at $E^*_{\text{CN}} = 55$ MeV and 65 MeV, the mass

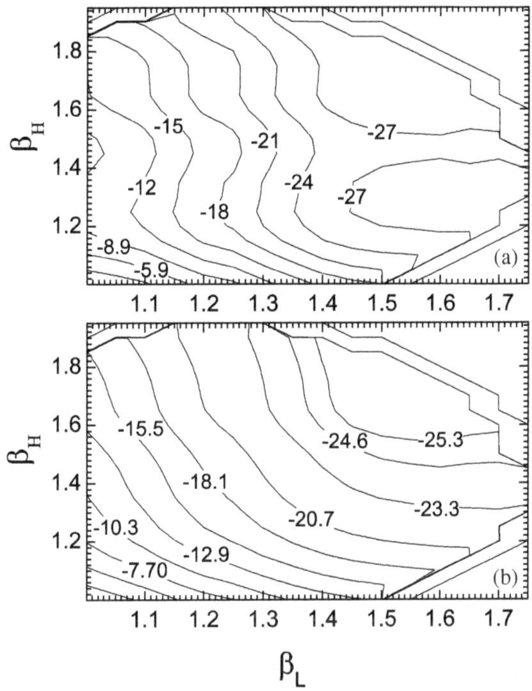

Fig. 40. The calculated PESs (in MeV) for the DNS with $(A_L, Z_L) = (96, 36)$ and $(A_H, Z_H) = (140, 56)$ at (a) $E^*_{\rm CN} = 6.5$-MeV and (b) $E^*_{\rm CN} = 61.5$-MeV excitation energies and zero spin. The value of the potential energy is normalized to the energy of the DNS with the spherical nuclei $(\beta_L = \beta_H = 1)$. The deformation β_i $(i = L, H)$ is given as the ratio between the major and minor semiaxes of the nucleus i. In the nuclear shape parametrization used, the value of the quadrupole deformation parameter is approximately $\beta_i - 1$.

distribution has the quasi-Gaussian shape with a large plateau on the top that indicates that the weight of the asymmetric fission mode is almost comparable with that of the symmetric fission mode. In the case of fission of ^{250}Cf at $E^*_{\rm CN} = 46$ MeV (Fig. 37), the measured secondary charge and mass distributions demonstrate the conservation of the asymmetric components.[29] The experimental width of the charge distribution extends at FWHM over 16 charge numbers. Note that in the case of fissioning $^{204-208}$Rn, $^{210-218}$Ra and ^{222}Th with the symmetric nature of fission, the width of the measured distribution is 8 u at FWHM.

The charge distribution resulting from the fission of ^{248}Cf at excitation energy in the range between 25 MeV and 35 MeV displays both almost symmetric Cd + Sn and asymmetric Ru + Xe modes with relatively equal

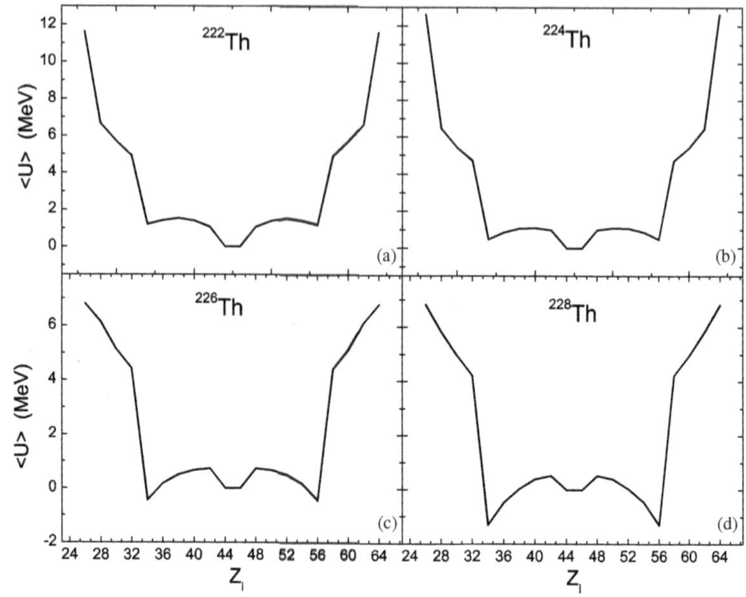

Fig. 41. The same as in Fig. 18 but for 60-MeV excitation energy and zero spin of the initial CN.

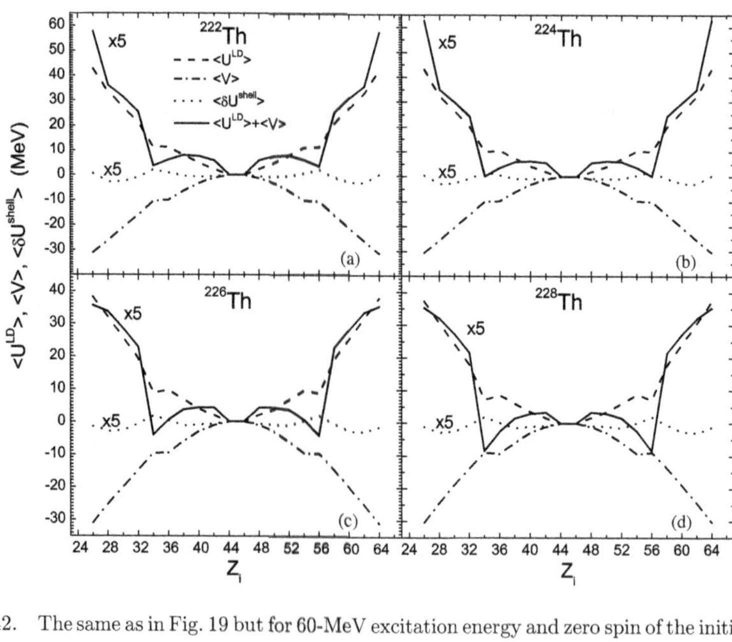

Fig. 42. The same as in Fig. 19 but for 60-MeV excitation energy and zero spin of the initial CN.

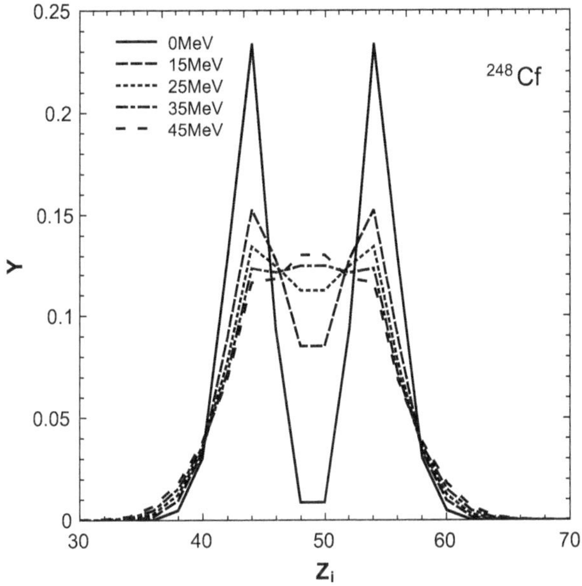

(a)

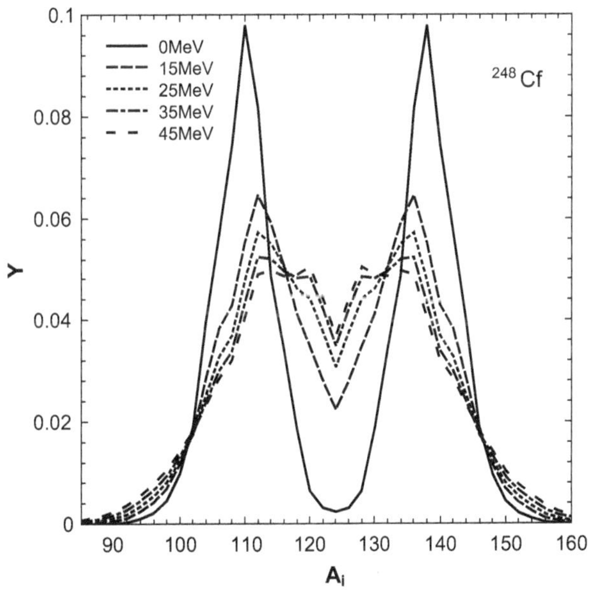

(b)

Fig. 43. The calculated (a) charge and (b) mass distributions of fragments resulting from the fission of ^{248}Cf at the indicated excitation energies. The calculations are performed for the even–even mass and charge fragmentations.

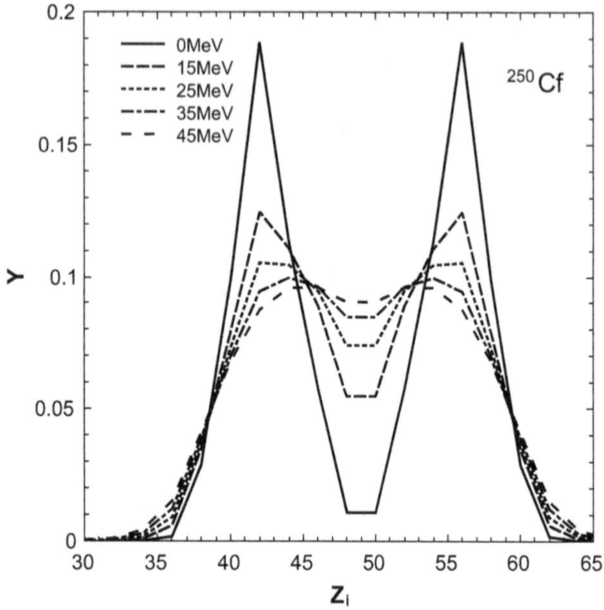

(a)

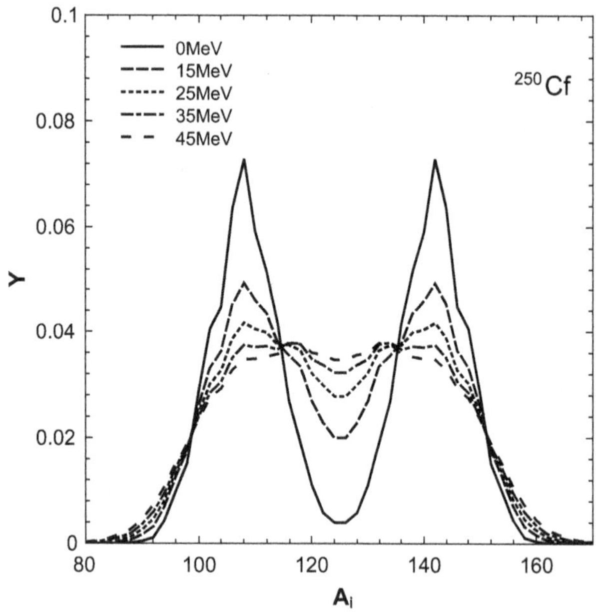

(b)

Fig. 44. The same as in Fig. 43, but for the fission of ^{250}Cf.

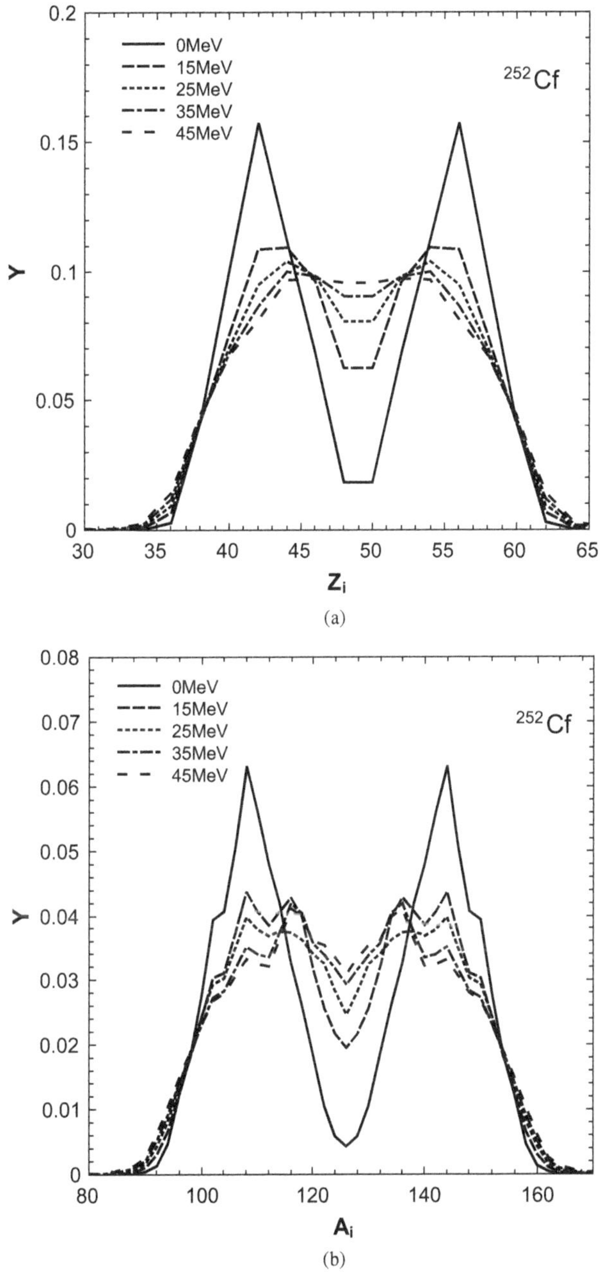

Fig. 45. The same as in Fig. 43, but for the fission of ^{252}Cf.

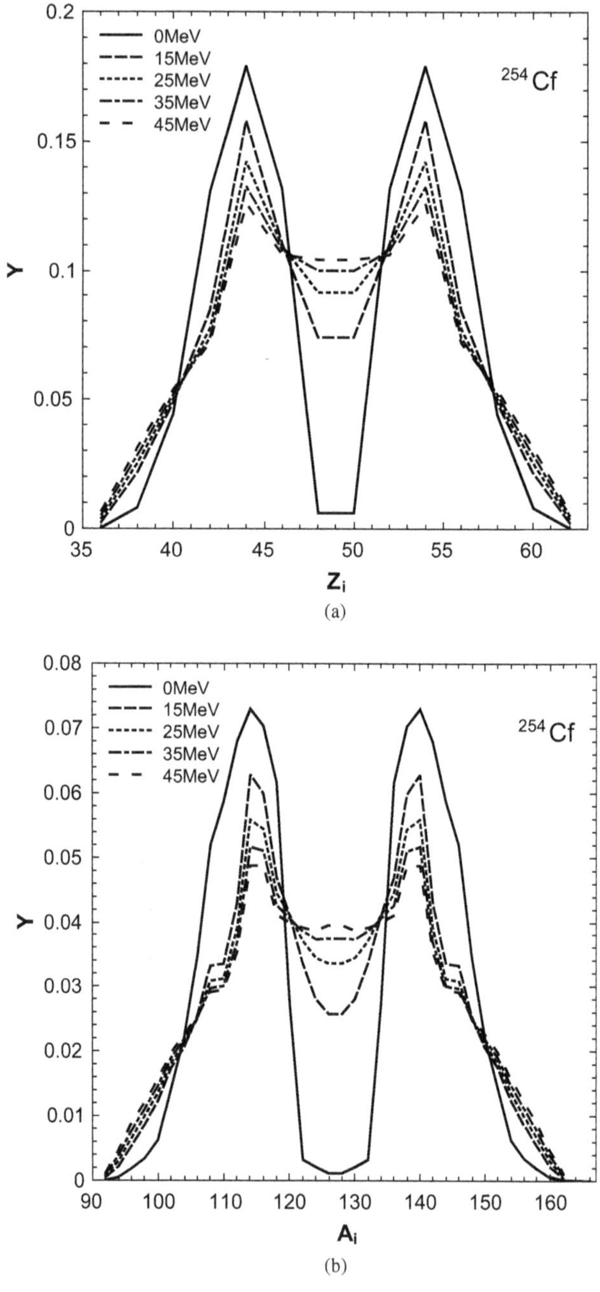

(a)

(b)

Fig. 46. The same as in Fig. 43, but for the fission of ^{254}Cf.

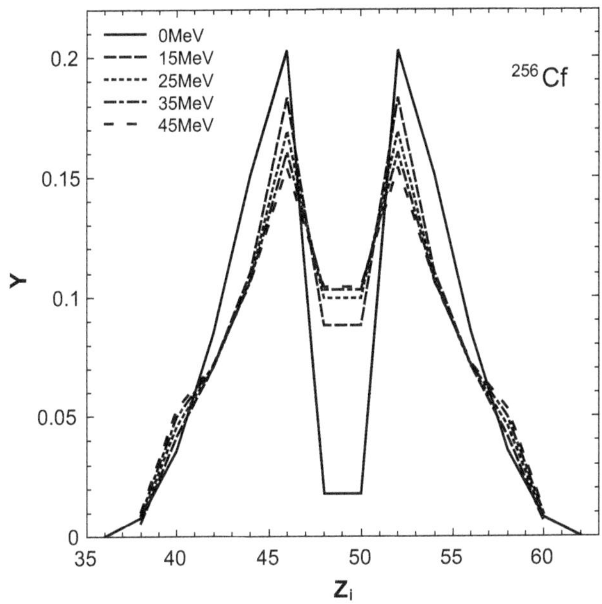

(a)

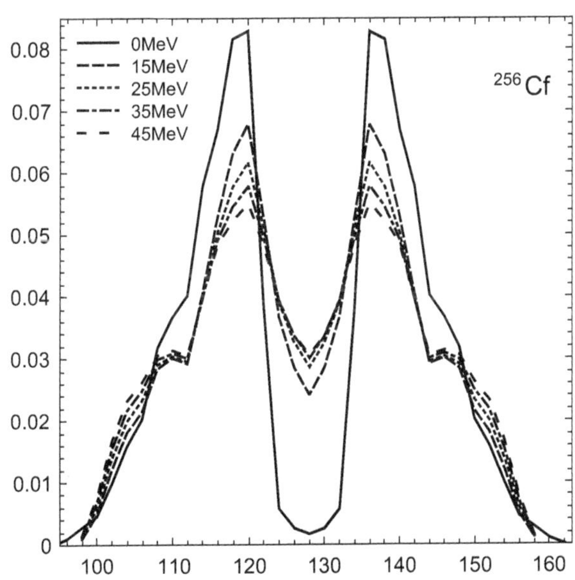

(b)

Fig. 47. The same as in Fig. 43, but for the fission of ^{256}Cf.

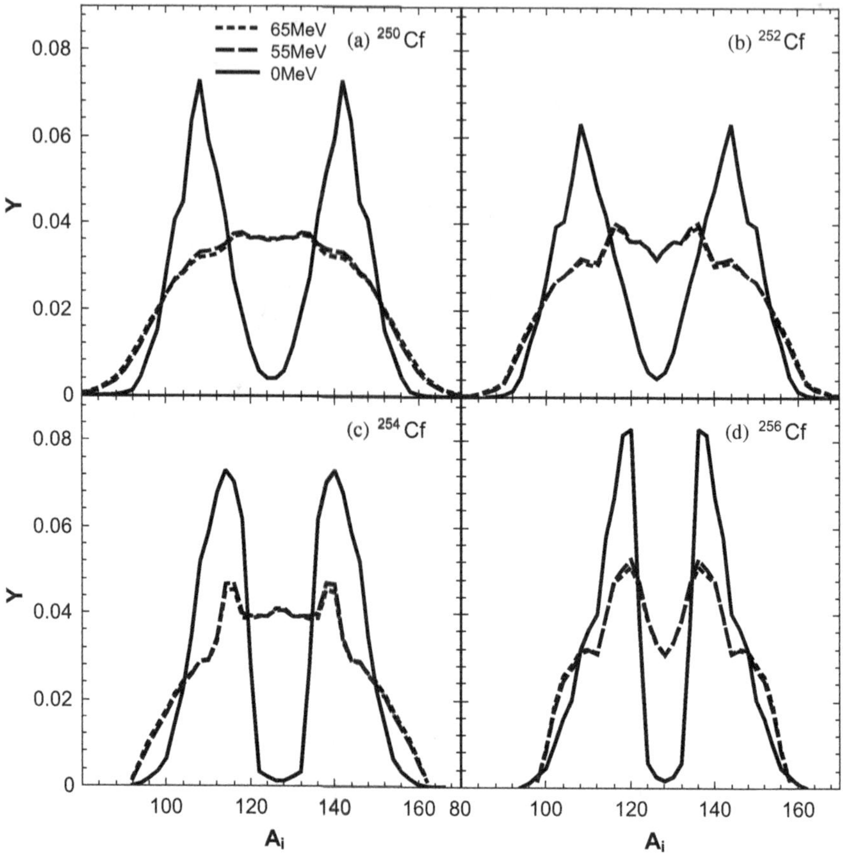

Fig. 48. The calculated mass distributions of fragments resulting from the fission of (a) ^{250}Cf, (b) ^{252}Cf, (c) ^{254}Cf and (d) ^{256}Cf at the indicated excitation energies. The calculations are performed for the even–even mass and charge fragmentations.

weights (Fig. 43). Thus, the transition from two-peaked to four-peaked charge distribution occurs with increasing excitation energy. This behavior is not found for other californium nuclei, even at higher excitation energies.

The fast evolution of the symmetric component observed with increasing excitation energy from 0 MeV to 15–25 MeV can be attributed to the fast shell melting of the magic (Sn) and near-magic (Cd) nuclei. At low excitation energies, the stiffness of the nuclei provide a spherical shape for the two nuclei of almost symmetric configuration Cd + Sn, which leads to a larger value of the interaction potential, and, correspondingly, to a larger value of

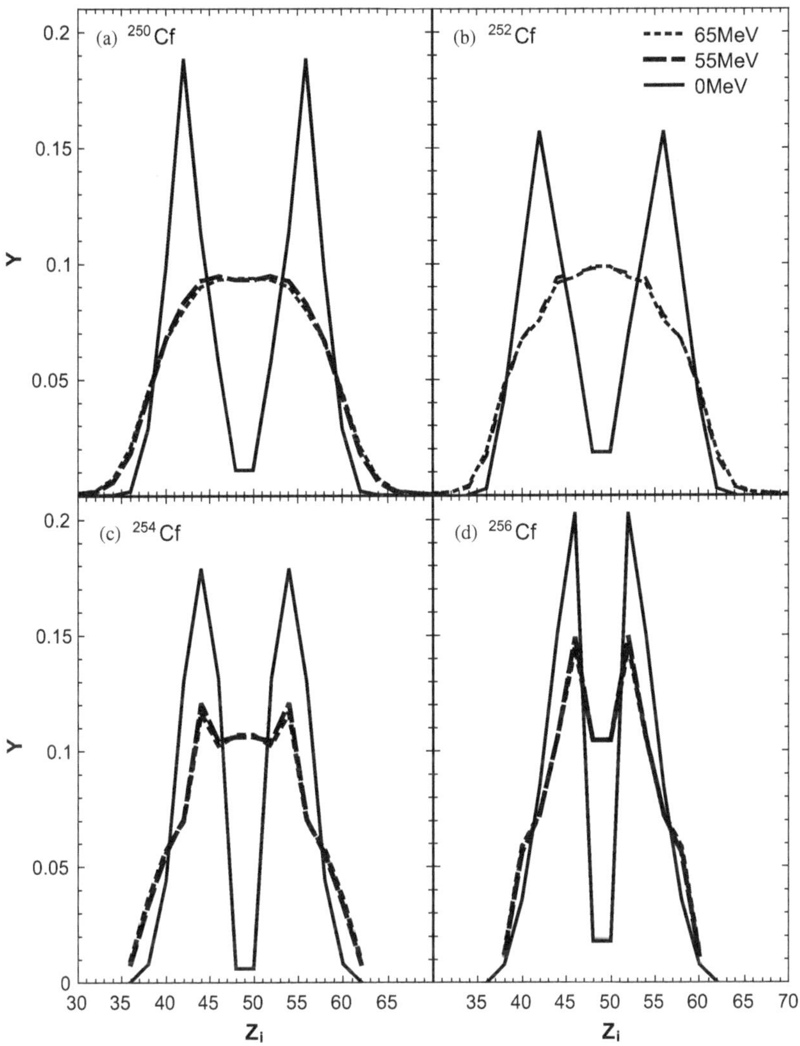

Fig. 49. The calculated charge distributions of fragments resulting from the fission of (a) ^{250}Cf, (b) ^{252}Cf, (c) ^{254}Cf and (d) ^{256}Cf at the indicated excitation energies. The calculations are performed for the even–even mass and charge fragmentations.

the total potential energy U, compared to the more asymmetric Mo + Ba or Ru + Xe configurations. As the excitation energy is increased the shell effects wash away and the lowering nuclear stiffness provides a "softer" symmetric configuration, in which the value of U decreases (the decrease in the Coulomb term of the LD potential and the decrease of the interaction potential are not compensated by the increase of the surface energies of the two nuclei), leading to higher yields. In this way, one can understand the shift to more symmetric mass and charge distributions with increasing $E^*_{\rm CN}$.

As can be seen in Figs. 43, 48 and 49, there is a difference between the shapes of mass and charge yields in the fission of ^{248}Cf at $E^*_{\rm CN} = 45$ MeV and ^{252}Cf at $E^*_{\rm CN} = 55$–65 MeV. For the fission of ^{248,252}Cf, the pronounced asymmetric mass distributions of fission fragments coexist with the charge distributions with almost symmetric peak at Cd + Sn.

The saturation of the symmetric mass and charge components with increasing excitation energy (Figs. 48 and 49) can be understood in the following way. For each mass and charge fragmentation, the configurations with the highest yields correspond to local minimum on the PES (β_L, β_H). This minimum on the PES results from the competition between the macroscopic interaction and LD energies and the microscopic shell corrections at scission. The strong shells also affect the macroscopic parts of the potential energy by keeping the minimum energy at small deformations (β_L, β_H). With increasing excitation energy, the shell effects are washed out and the stiffness k_i of the nuclear surface decreases. At the same time, the temperature-dependent terms of the macroscopic part of U begin to act stronger on the PES. The combined effect is the enlargement of the minimum on the PES and its shift toward larger deformations. At large $E^*_{\rm CN}$, the shell effects are completely damped, the surface stiffness becomes minimal and the minima on the PES reach their maximum widths and final positions dictated by the macroscopic part of U. At this point the yields reach the maximal values, and further increase of excitation energy leads only to the population of more asymmetric accessible (by the total energy conservation) configurations. Note that the width of the minimum is mainly governed by the stiffness parameters of two nuclei — a large k_i value provides a narrow and deep minimum, while a small k_i value leads to a wide and shallow minimum. In Fig. 50, the example of the increase of the width of the PES minimum with increasing excitation energy is presented for the ^{108}Mo+^{142}Ba fragmentation of the ^{250}Cf nucleus at $E^*_{\rm CN} = 0$, 15, 25, 35 and 45 MeV. This fragmentation corresponds to the maxima of the charge and mass distributions in the spontaneous and thermal-neutron-induced fission of ^{250}Cf. At the given $E^*_{\rm CN}$, the scission configurations with the potential energies in the range between the minimum value of $U = U_{\rm min}(E^*_{\rm CN})$ and $U = U_{\rm min}(E^*_{\rm CN}) + 2$ MeV are enclosed in contour lines. The inspection of Eq. (3) reveals that these configurations provide the highest yields. As can be seen, the minimum on the

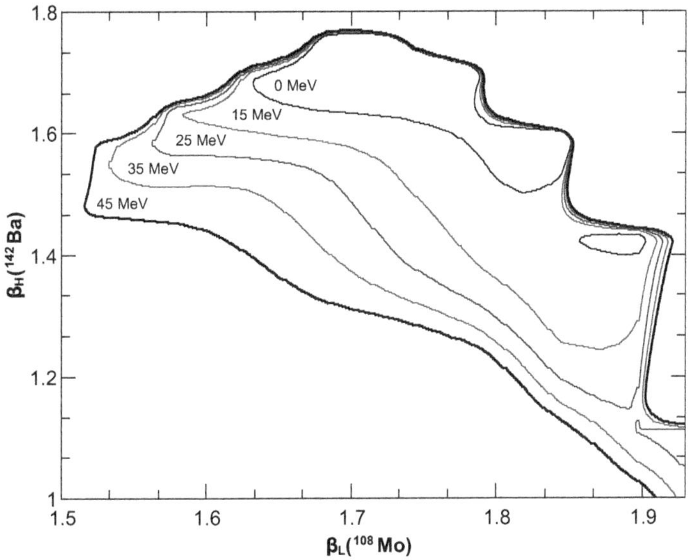

Fig. 50. The scission configurations with the potential energies between $U_{\rm min}$ and $U_{\rm min} + 2\,\rm MeV$ are enclosed in contour lines on the plane (β_L, β_H) for the fragmentation ^{250}Cf$\to$^{108}Mo+^{142}Ba at the indicated $E^*_{\rm CN}$.

PES becomes wider and shallower with increasing $E^*_{\rm CN}$. There is saturation of the expansion of this minimum at $E^*_{\rm CN} > 35\,\rm MeV$.

In Table 2, the calculated average charge $\langle Z_H\rangle$, neutron $\langle N_H\rangle$ and mass $\langle A_H\rangle$ numbers of the heavy fragments resulting from the fission of cold and excited even–even isotopes $^{248-256}$Cf are presented. The values of $\langle Z_H\rangle$ and $\langle N_H\rangle$ weakly increase with $E^*_{\rm CN}$ up to $\sim$25 MeV and are almost unchangeable at higher energies. The asymmetric fission is characterized by the average proton $\langle Z_H\rangle = 54$ and neutron $\langle N_H\rangle = 80-86$ numbers in the heavy fragments. Such behavior is a reflection of the increase of the symmetric yields compared to the asymmetric ones. It is worth to note that stabilization of $\langle Z_H\rangle$ and $\langle N_H\rangle$ is related to the saturation of the widths of the PES minima with $E^*_{\rm CN}$. So, we do not expect that with increasing excitation energy the distributions are only peaked around $(Z/2,\ A/2)$, but reach a maximum "height" at a certain value of $E^*_{\rm CN}$ and then become wider with increasing $E^*_{\rm CN}$. In Table 2, the peak-to-valley ratios of the charge and mass distributions decrease toward the values $\simeq$1$-$1.5, pointing out again a saturation of these distributions.

Table 2. The calculated average charge $\langle Z_H \rangle$, neutron $\langle N_H \rangle$ and mass $\langle A_H \rangle$ numbers of the heavy fission fragments and peak-to-valley ratios $P/V(Z)$ and $P/V(A)$ of the charge and mass distributions, respectively, in the fission of even–even nuclei $^{248-256}$Cf at different excitation energies $E^*_{\rm CN}$ (Figs. 43–47).

Nucleus	$E^*_{\rm CN}$ (MeV)	$\langle Z_H \rangle$	$\langle N_H \rangle$	$\langle A_H \rangle$	$P/V(A)$	$P/V(Z)$
^{248}Cf	0	54.4	83.6	138.0	42.5	27.1
	15	53.6	79.7	133.3	1.8	2.9
	25	53.4	78.4	131.8	1.2	1.9
	35	53.4	78.9	133.3	1.0	1.5
	45	53.4	78.8	133.2	0.9	1.4
^{250}Cf	0	55.6	84.1	139.7	17.4	18.4
	15	54.9	81.8	136.7	2.3	2.5
	25	54.7	81.2	135.9	1.4	1.5
	35	54.6	80.4	135.0	1.2	1.2
	45	54.6	80.2	134.8	1.1	1.0
^{252}Cf	0	55.5	87.1	142.6	8.6	14.7
	15	54.8	83.9	138.7	1.7	2.2
	25	54.6	83.3	137.9	1.3	1.5
	35	54.6	82.4	137.0	1.1	1.4
	45	54.6	81.6	136.2	1.0	1.3
^{254}Cf	0	54.9	86.4	141.3	29.2	63.2
	15	54.0	85.7	139.7	2.1	2.4
	25	53.9	85.7	139.5	1.6	1.7
	35	53.9	85.7	139.5	1.3	1.4
	45	53.9	85.7	139.5	1.2	1.2
^{256}Cf	0	55.5	85.5	141.0	11.4	48.3
	15	53.3	83.1	136.4	2.0	2.8
	25	53.3	82.7	136.0	1.7	2.2
	35	53.3	82.7	136.0	1.6	1.9
	45	53.3	82.7	136.0	1.5	1.8

5.3. *Even–even $^{254-264}$Fm, $^{258-264}$No and $^{262-266}$Rf*

As can be seen in Figs. 27 and 28, in the case of spontaneous fission of ^{260}Fm with the symmetric mass distribution, the charge distribution is asymmetric with the most probable charge fragmentation Cd + Te. With increasing excitation energy, the charge distribution becomes symmetric (Figs. 28 and 51). For the thermal-neutron-induced fission of ^{257}Fm (Figs. 27 and 28), the experimental and calculated mass distributions are symmetric whereas the predicted charge distribution is asymmetric. However, at $E^*_{\rm CN} \geq$ 15 MeV, the difference between these distributions disappears. For the induced fission of ^{254}Fm at $E^*_{\rm CN} = 35$ MeV and 50 MeV (Figs. 51 and 52), the shape of mass yields becomes symmetric in contrast to the shape of the charge distribution. This unexpected difference between the shapes of mass and charge yields is worth to be studied experimentally.

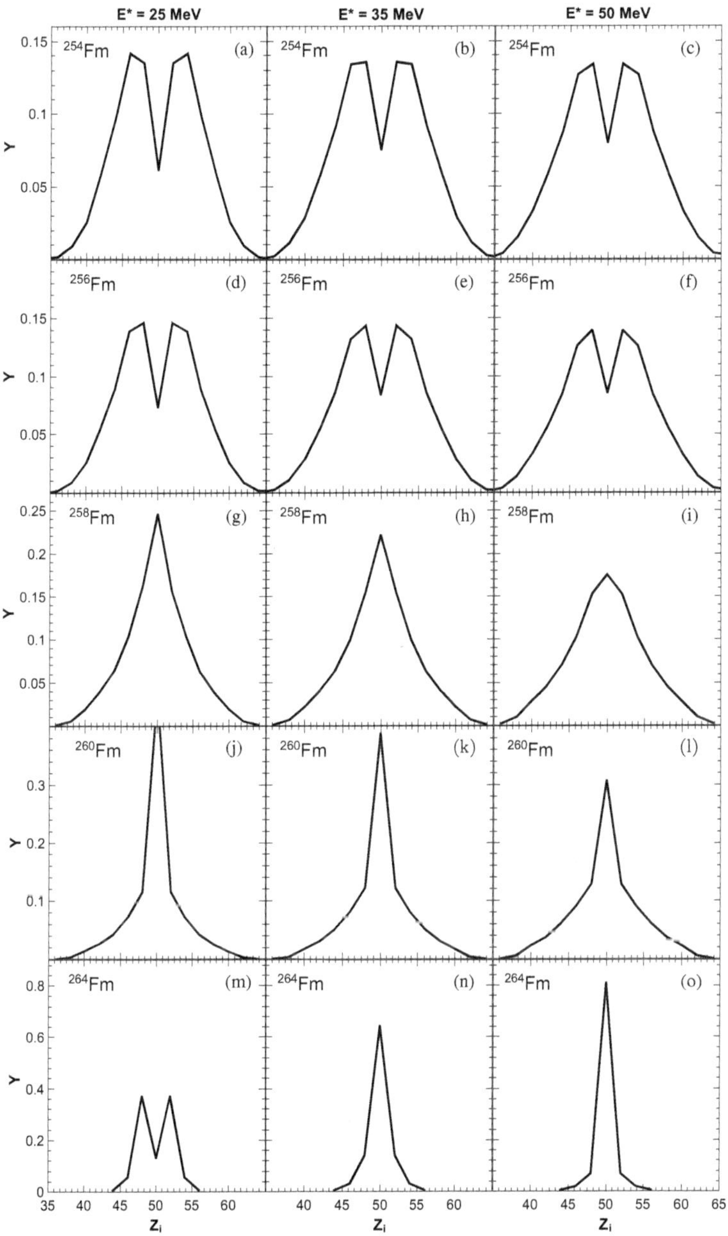

Fig. 51. The same as in Fig. 28 but for $E^*_{\rm CN} = E^* = 25$-MeV (the first column), $E^*_{\rm CN} = E^* = 35$-MeV (the second column) and $E^*_{\rm CN} = E^* = 50$-MeV (the third column) excitation energies. The calculations are performed at $L = 0$.

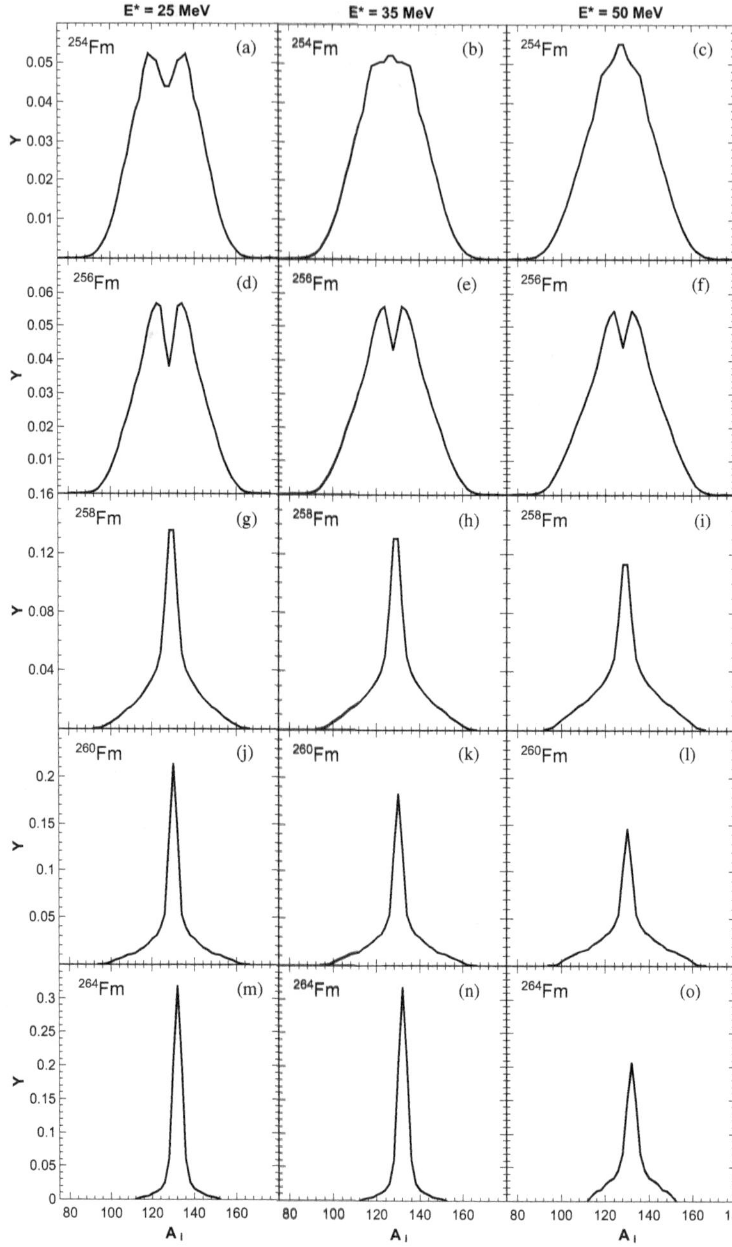

Fig. 52. The same as in Fig. 27 but for $E^*_{\rm CN} = E^* = 25$-MeV (the first column), $E^*_{\rm CN} = E^* = 35$-MeV (the second column) and $E^*_{\rm CN} = E^* = 50$-MeV (the third column) excitation energies. The calculations are performed at $L = 0$.

As can be seen in Fig. 53, the interplay between the LD surface energy and the nucleus–nucleus interaction potential is the main reason for the appearance or disappearance of the asymmetric fission mode. This interplay depends on excitation energy and isospin of the fissioning nucleus. The shell effects affect directly and indirectly (through the deformations of nuclei) the appearance of asymmetric potential minima. For example, for the

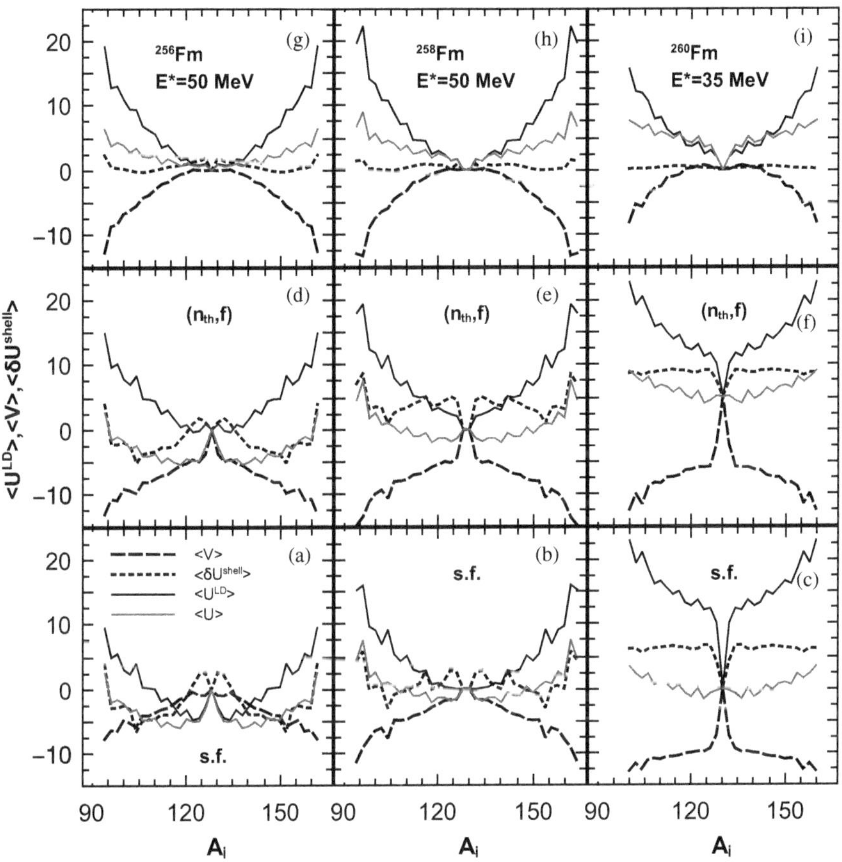

Fig. 53. (Color online) For the indicated spontaneous (s.f.) of ^{256,258,260}Fm, thermal-neutron-induced fission of ^{255,257,259}Fm and induced fission of ^{256,258,260}Fm at $E^*_{\rm CN} = 50$ MeV and $L = 0$, the calculated average driving potentials $\langle U \rangle$ (red line), the average macroscopic LD energies $\langle U^{\rm LD} \rangle$ (solid line) and interaction energies $\langle V \rangle$ (dashed line) and the average shell corrections $\langle \delta U^{\rm shell} \rangle$ (dotted line) as functions of the mass number A_i of one of the fragments. In the case of ^{256,258}Fm, the values of the shell corrections $\langle \delta U^{\rm shell} \rangle$ are multiplied by a factor 5 for better visibility. In panel (c), the values of $\langle V \rangle$ are multiplied by a factor 0.5. The "origins" of all components of potential energy are taken as those of the symmetric fragmentation.

spontaneous fission of ^{258}Fm (Fig. 53), the driving potential $\langle U\rangle(A_i)$ shows two relatively wide asymmetric minima separated by a small local maximum around $A_L \sim A/2$. The shell corrections here play a major role, as they establish the positions of maxima in the mass yield (Fig. 27(c)) as well as its width. With increasing excitation energy, the shell effects start to be washed out but remain rather strong for almost symmetric ^{128}Sn+^{130}Sn fragmentation. This can be seen in Fig. 53(e) at $A_L \sim A/2$, where the strong shell effects fix the minima of $\langle U\rangle(A_i)$ at small values of deformations which lead to large (small) values of the interaction potential $\langle V\rangle$ (the LD energy $\langle U^{\rm LD}\rangle$) compared to the neighboring fragmentations. The damping of the shell effects in the asymmetric configurations reduces the asymmetric yields, and the nucleus is left with a dominant symmetric mode, strongly enhanced due to the shell closures for both Sn fragments. At $E^*_{\rm CN} = 50$ MeV, the shell effects are completely damped (Fig. 53(h)) and the stiffness of the nuclear surface reaches its minimal value. As a result, the mass yield becomes wider with further increasing $E^*_{\rm CN}$. Thus, the mass-yield shape is related to the PES at scission.

The transformation of the shape of the mass distribution in the fission of ^{260}Fm occurs in a similar fashion as in the fission of ^{258}Fm but much faster (Figs. 27 and 28). In the case of spontaneous fission, the distribution is relatively wide and single-peaked. For the symmetric fragmentation ^{130}Sn+^{130}Sn, the interaction and LD energies have opposite and almost canceling effects. The strong shell effect in the ^{130}Sn+^{130}Sn configuration is seen in Fig. 53(c). This also affects the macroscopic parts of the potential energy by fixing the potential minimum at small deformations (β_L, β_H). As $E^*_{\rm CN}$ increases, the shell corrections and nuclear stiffness diminish, quickly shifting the minima on the PES (for ^{130}Sn+^{130}Sn) toward larger deformations. Note that with increasing deformations of two fragments, the value of $\langle V\rangle$ (and in turn $\langle U\rangle$) considerably decreases. As a result, the mass distribution becomes narrow. The symmetric mode is already saturated at $E^*_{\rm CN} = 15$ MeV, and at larger $E^*_{\rm CN}$ the distribution only becomes wider. At high excitation energy, the minimum of $\langle U\rangle(A_i = A/2)$ is given only by the competition between the interaction and LD energies of the driving potential (Fig. 53(i)).

In the case of spontaneous fission of ^{260}No (Figs. 29 and 30), the mass distribution displays symmetric ($A_L = A_H = 130$) and asymmetric ($A_L = 122$, $A_H = 138$) peaks. For ^{260}No, the peaks of the charge distribution are at $Z_L = 50$ and $Z_H = 52$. The spontaneous fission of ^{264}No (Figs. 29 and 30) has slightly asymmetric mass distribution with peaks at $A_L = 128$ and $A_H = 136$ and charge distribution with peaks at $Z_L = 50$ and $Z_H = 52$. In the spontaneous fission of ^{266}Rf (Figs. 31 and 32), the mass yield is symmetric and the charge yield is asymmetric ($Z_L = 50$ and $Z_H = 54$). Note that in the case of spontaneous fission of ^{264}Fm (Figs. 27 and 28), the peaks of charge and mass distributions are at $Z_L = 48$, $Z_H = 52$ and $A_L = 128$, $A_H = 136$, respectively.

As a global trend for the isotopes of Fm and No, the symmetric components of the mass- and charge-yields are enhanced with increasing excitation energy (Figs. 27–30). The saturation of the symmetric components is reached at 15–30 MeV. At larger $E^*_{\rm CN}$, the distributions are getting wider. The fission of ^{260}No is a good example of this (Figs. 29 and 30). The shift to more symmetric mass and charge distributions and the saturation of the symmetric components with increasing excitation energy (Figs. 27–30) can be understood in the following way: For each mass and charge fragmentation, the configurations with the highest yields correspond to local minima on the PES (β_L, β_H). These potential minima result from the complex competition between the macroscopic interaction and LD (surface) energies and the microscopic shell corrections at scission. The strong shells also affect the macroscopic parts of the potential energy by fixing the minimum energy at small deformations (β_L, β_H). With increasing excitation energy the shell effects are washed out, and the stiffness of the nuclear surface decreases. The combined effect is the enlargement of the minima on the PES and their shift toward much larger deformations. At large $E^*_{\rm CN}$, the shell effects are completely damped, the surface stiffness becomes minimal and the minima on the PES reach their maximum widths and final position. At this point, the yields reach the maximal values, and a further increase of excitation energy leads only to the population of more asymmetric accessible configurations.

In the case of fission of ^{262,264,266}Rf (Figs. 31 and 32), the increase of excitation energy leads to the access of previously inaccessible asymmetric configurations and to the decrease of the symmetric component.

5.4. ^{180}*Hg*

The predicted mass and charge distributions for the β-delayed fission of ^{180}Tl and for the fission of ^{180}Hg at $E^*_{\rm CN} = 60$ MeV are shown in Fig. 54. Surprisingly, these distributions are asymmetric at the low and high excitation energies. This asymmetry is more surprising at low energy since a mass-symmetric split of the extremely neutron-deficient nucleus would lead to two ^{90}Zr fragments, with magic $N_i = 50$ and semimagic $Z_i = 40$.[6] One can explain the origin of asymmetry by examining the driving potentials at $E^*_{\rm CN} = 10.8$ MeV and 60 MeV (Fig. 55). As can be seen for the

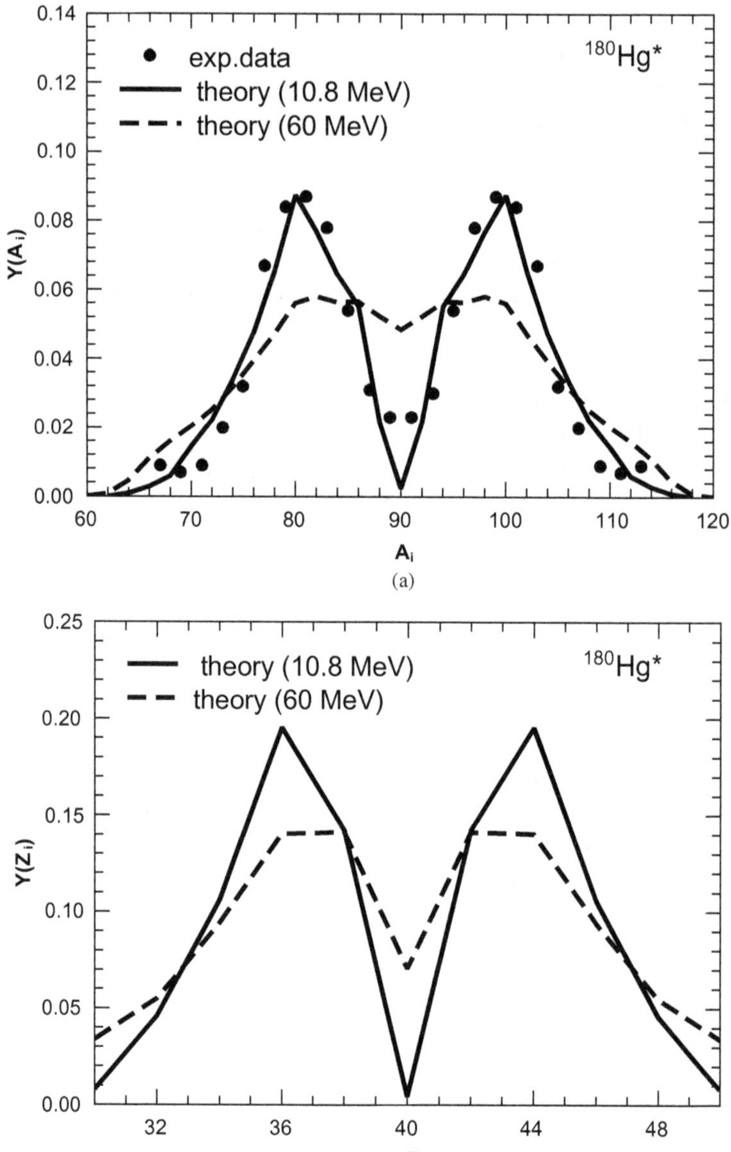

(b)

Fig. 54. The predicted (a) mass and (b) charge distributions of fission fragments for the fissioning ^{180}Hg* nucleus at 10.8-MeV and 60-MeV excitation energies and zero spin of the initial CN. The lines connect the calculated yields of even–even fission fragmentations. The experimental (symbols) mass distribution for the β-delayed fission of ^{180}Tl is from Ref. 6.

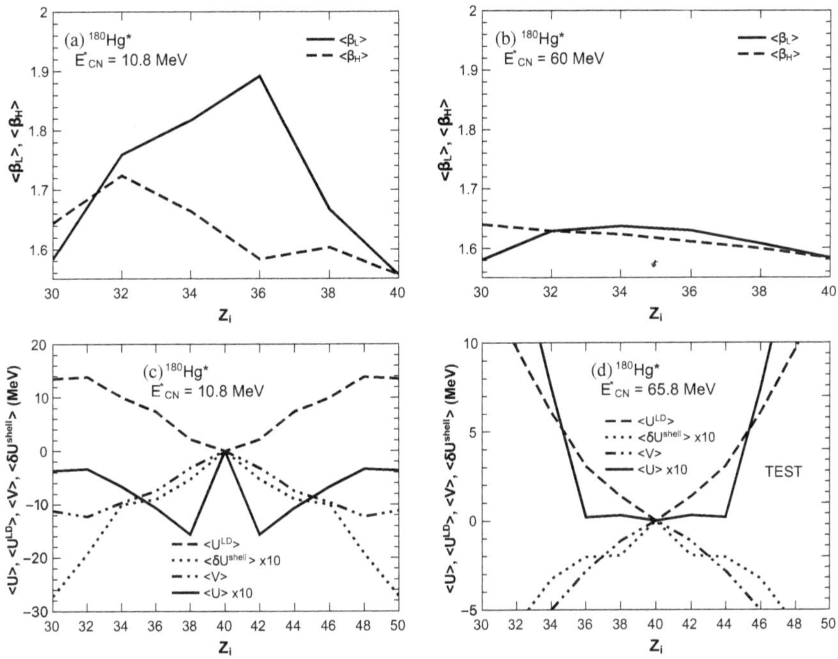

Fig. 55. The calculated deformations $\langle\beta_L\rangle(Z_L)$ (squares) and $\langle\beta_H\rangle(Z_L)$ (circles) of fragments (panels (a) and (b)), driving potentials $\langle U\rangle(Z_i)$ and components of the driving potentials (panels (c) and (d)) averaged over the mass number and deformations (Eq. (9)) at 10.8-MeV and 60-MeV excitation energies and zero spin of the initial CN for the indicated fissioning nucleus. The driving potential, the LD energy $\langle U^{\rm LD}\rangle$ (dashed lines), the nucleus–nucleus interaction potential $\langle V\rangle$ (dash-dotted lines) and the shell-correction energy $\langle\delta U^{\rm shell}\rangle$ (dotted lines) are normalized to the zero energy at $Z_i = 40$. The values of $\langle\delta U^{\rm shell}\rangle$ and $\langle U\rangle$ are scaled by a factor of 10. The lines connect the points corresponding to the even–even fission fragments.

β-delayed fission, the asymmetric configurations Sr + Mo, Kr + Ru, Se + Pd, Ge + Cd and Zr + Sn are energetically more favorable than the symmetric configurations Zr + Zr because of the interplay between the LD energy and the fragment–fragment interaction potential. The absolute values of the shell correction energies are very small compared to the macroscopic energies, and shell effects influence indirectly through the deformation of fragments (Fig. 55). The strong shell effects are expressed in relatively small deformations of the corresponding fragments. Even though $\langle U\rangle$ for the configurations Kr + Ru are higher by about ~0.7 MeV than for the symmetric divisions Sr + Mo, the large number of mass fragmentations and the large minima in the deformation plane (β_L, β_H) promote large yield in the charge split Kr + Ru, when integrating over the deformations and the mass numbers. In the case of fission of ^{180}Hg at $E^*_{CN} = 60$ MeV, the almost negligible shell effects lead to an almost pure LD behavior of the fissioning system and the symmetric fragmentations Zr + Zr are favorable in the potential energy. However, as for the β-delayed fission, the number of asymmetric fragmentations and the width of asymmetric minima are much larger, compared with the symmetric ones, and finally the charge distribution is asymmetric. The same explanation is valid in the cases of mass distributions at low and high energies.

6. Examination of Coexistence of Symmetric Mass and Asymmetric Charge Distributions of Fission Fragments

To reveal the dependence of the shape of the fission-fragment mass distribution on the excitation energy of the fissioning nucleus, the fission of ^{180}Hg* formed in the complete fusion reaction ^{36}Ar+^{144}Sm at bombarding energies $E_{\text{lab}} = 157$, 175 and 197 MeV (Fig. 56 and Table 1) is treated.[31] We assume that after the complete fusion of heavy nuclei, a statistical equilibrium is established before fission.[83] The corresponding excitation energies of ^{180}Hg* are $E^*_{CN} = 33.4$, 48 and 65.8 MeV, respectively. As can be seen, the increase of excitation energy and angular momentum in the CN leads to a more symmetric mass distribution. The model describes the experimental data rather well; however, the local symmetric minimum is present. If the experimental uncertainty $\sigma = 2.4$ u of mass resolution[31] is taken into consideration, the measured data are faithfully reproduced (Fig. 56). One can conclude that the finite experimental mass resolution of the fission fragments can have an unwanted effect, severely masking some fine structures of the mass distribution. Even at large excitation energies and/or angular momenta, the mass distribution does not exhibit smooth distributions with a plateau. Similar examples can be found in Refs. 24, 25, 31 and 72–75, where at large excitation energies the isotopes of U and Pu displayed asymmetric peaks or a wider plateau than those predicted with the LD model. So a direct or an indirect effect of the shell structure is expected even at high excitation energies and angular momenta.

The explanation of the asymmetric mass distribution at high excitation energy can be found by analyzing the relative contributions to the mass yields originating from the systems with different angular momenta L. Note that in the previous sections the calculations were performed at $L = 0$. In Fig. 57, the fission-fragment yields originating from the fissioning

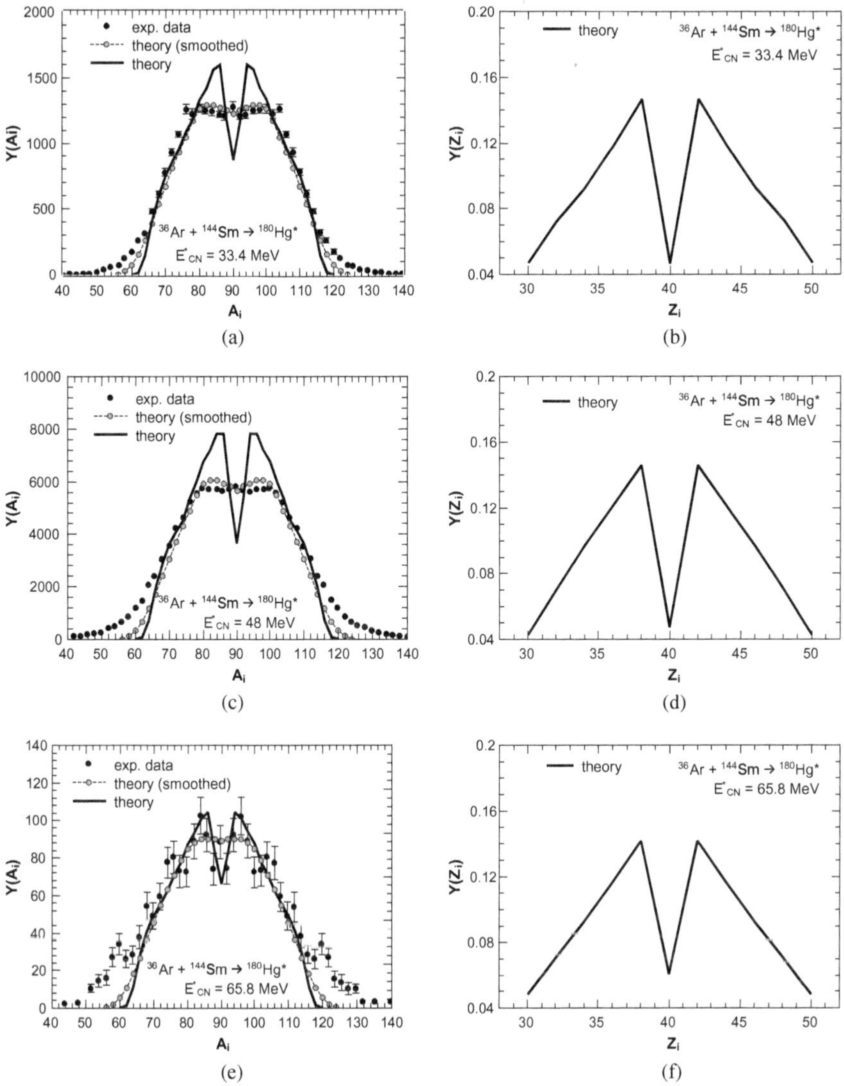

Fig. 56. Calculated mass distributions (thin solid lines) for the fission resulting from the reactions ^{36}Ar+^{144}Sm→^{180}Hg* at $E_{\rm lab} = 157, 175$ and 197 MeV compared with the experimental data (closed symbols)[31] (panels (a), (c) and (e)). The excitation energies of the CN at $L = 0$ are also given. The open symbols represent the calculated mass distributions smoothed with the Gaussian function having a width $\sigma = 2.4$ u. The thick dashed lines connecting the open symbols are for eye guidance. Panels (b), (d) and (f) present the predicted charge distributions for the same reactions.

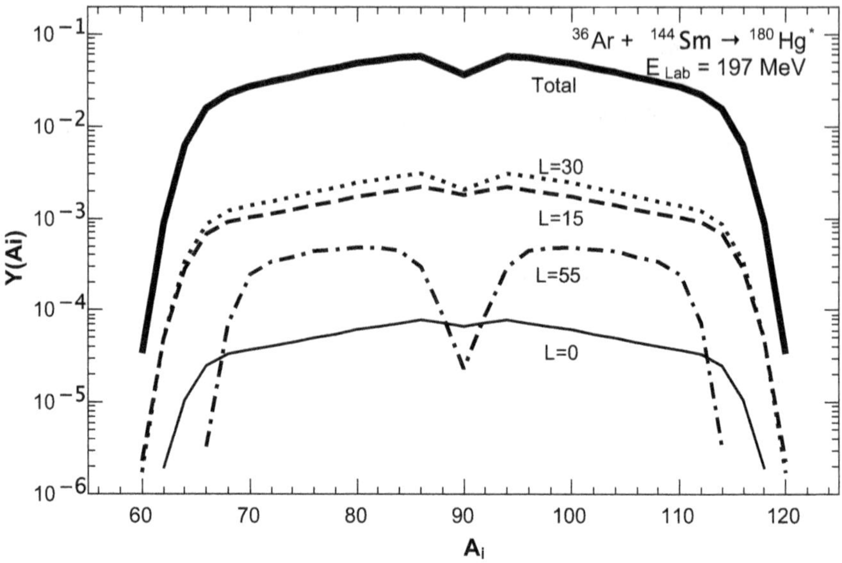

Fig. 57. The relative contributions of indicated angular momenta to the mass distribution (thick solid line) in the case of ^{36}Ar+^{144}Sm→^{180}Hg* reaction at $E_{\rm lab} = 197$ MeV ($E^*_{\rm CN}(L=0) = 65.8$ MeV). The contribution of all partial waves is shown by thin solid line.

systems with $L = 0$, 15, 30 and 55 are separately shown together with their total contribution for the ^{36}Ar+^{144}Sm→^{180}Hg* reaction at $E_{\rm lab} = 197$ MeV. At low angular momenta ($L = 0$ and 15), the yields show the asymmetric shape due to the interplay between the LD energy and the fragment–fragment interaction potential (see the last paragraph of Sec. 5). The potential energy in the mass asymmetry coordinate is rather smooth due to the large excitation energy. The largest contributions originate from $L = 10$–40. As the angular momentum increases, the rotational energy increases leaving the system with less excitation energy, resulting in stronger shell effects, and therefore in a more pronounced asymmetric structure of mass yields. Also, larger L values (smaller excitation energies) lead to smaller relative yields and fewer mass-asymmetric configurations available (Fig. 57).

The effect of excitation energy and angular momentum on the average potential energy $\langle U\rangle(A_i)$ as a function of the fragment mass number A_i is seen in Fig. 58. The average values of fission-fragment deformations are also shown in Fig. 58 for different values of L. For $L = 0$ and $L = 40$, the fragment deformations are almost independent of the mass number and are equal in two fragments. This is due to the large excitation energy presented

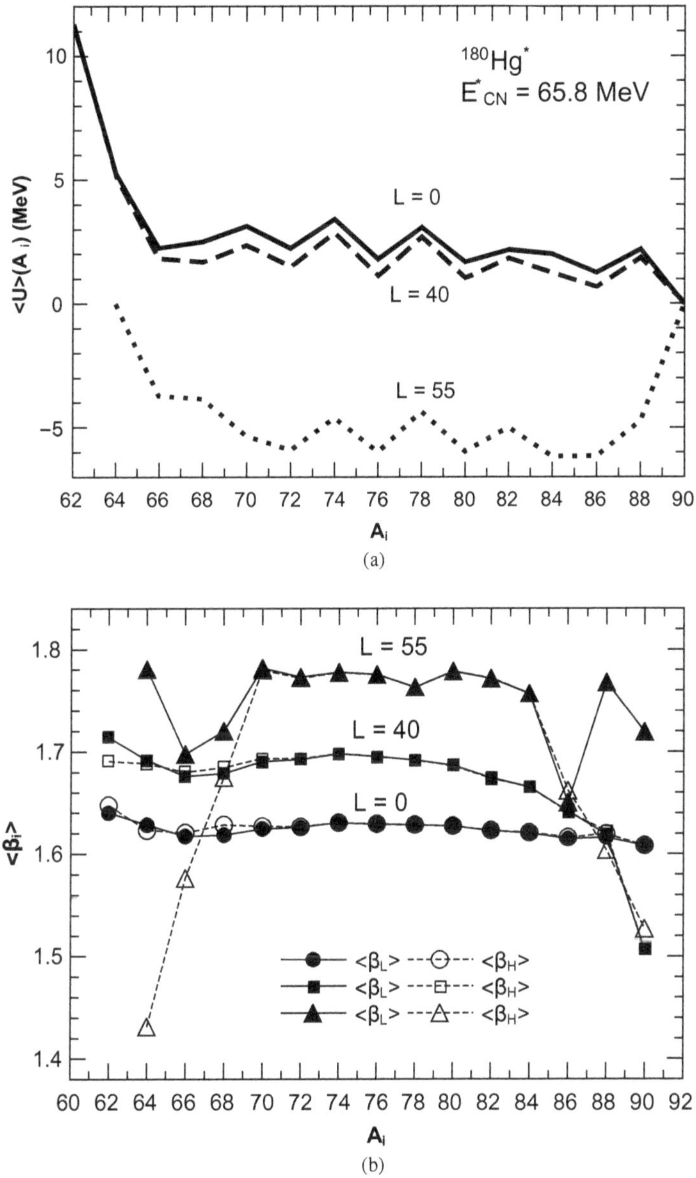

Fig. 58. For the fissioning nucleus ^{180}Hg*, the calculated average potential energy $\langle U\rangle(A_i)$ as a function of mass number of one of the fragments at the indicated angular momenta L. The driving potential is normalized such that $\langle U\rangle(A/2) = 0$ (panel (a)). The calculated average deformations β_H and β_L of the two fragments as functions of their mass numbers at the indicated values of L (panel (b)).

in the system. As the angular momentum increases, more elongated configurations become energetically favorable due to the centrifugal part of the potential. At large values of L, large fragment elongation is favorable, but the diminishing excitation energy leads to small deformations of magic nuclei. This can be seen in Fig. 58 at $L = 55$ for $A_H = 116$ (Sn) and $A_L = 90$ (Zr). At these mass numbers, the centrifugal part of the potential becomes large, leading to the maxima seen in Fig. 58 and, correspondingly, to the minima in the mass yield seen in Fig. 57 at $L = 55$. So, the mass distribution is the sum of low L, almost structureless, distributions and high L distributions which reflect the microscopic shell effects. Note also that as the value of the angular momentum increases, the excitation energy decreases, and at the highest L values not all configurations are energetically available.

In Fig. 59, the mass distributions resulting from the ^{40}Ca+^{142}Nd→^{182}Hg* reaction are compared with the available experimental data[32] at excitation energies $E^*_{\rm CN} = 33$, 58 and 75 MeV of CN (Table 1). As can be seen, the experimental data[32] are rather well reproduced. One can compare the results in Fig. 59 with those for ^{180}Hg* (Fig. 56). If the excitation energies of ^{180,182} Hg* are approximately the same, the additional two neutrons in ^{182}Hg* make the symmetric fragments to have $N_L = N_H = 50$ neutron shells nonclosed as in the case of ^{180}Hg*. This makes the ratio between the asymmetric maxima and the symmetric minimum in the ^{40}Ca+^{142}Nd→^{182}Hg* reaction, Fig. 59, smaller than the one in Fig. 56. For example, one can compare the mass distribution depicted in Fig. 59, where $E^*_{\rm CN}(L = 0) = 58$ MeV, with the one in Fig. 56, with $E^*_{\rm CN}(L = 0) = 65.8$ MeV. Even though the excitation energy is higher in the case of ^{180}Hg*, the peak-to-valley ratio is larger. Note that the values of the maximum angular momentum for these two systems are quite similar (Table 1).

The calculated mass distributions from the fissioning ^{190}Hg* formed in the ^{36}Ar+^{154}Sm→^{190}Hg* reaction are shown in Fig. 60 and compared with the experimental data at bombarding energies $E_{\rm lab} = 157$, 165 and 175 MeV or the corresponding excitation energies $E^*_{\rm CN} = 56$, 62.4 and 70.5 MeV[31] (Table 1). With increasing excitation energy, neither the experimental data nor the theoretical results show significant changes in this excitation energy range. This was pointed out in Ref. 31. However, the mass distribution exhibits a much smaller minimum at symmetry than those in the fission of ^{180,182}Hg*. The explanation of this becomes evident from the comparison of the average potential energy U as a function of the mass number of a fragment at different angular momenta L (Figs. 58 and 61).

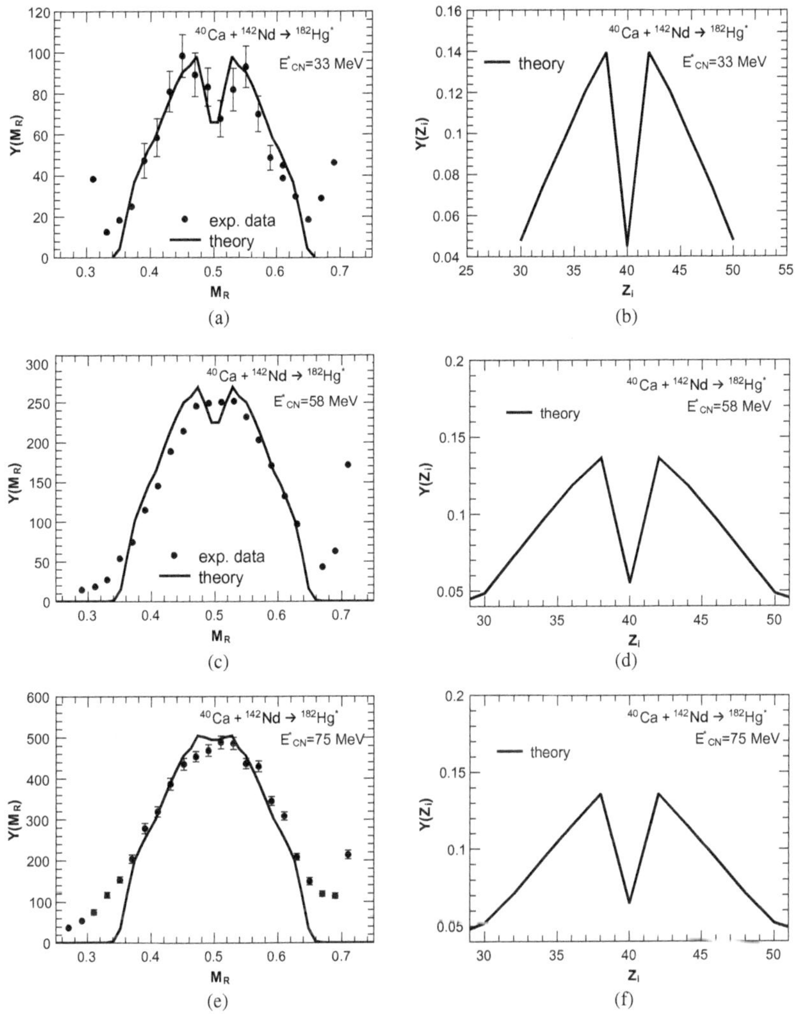

Fig. 59. The same as in Fig. 56 but for the ^{40}Ca+^{142}Nd→^{182}Hg* reaction at $E^*_{\rm CN}(L = 0) = 33$, 58 and 75 MeV (panels (a), (c) and (e), respectively). Here, $M_R = A_i/(A_L + A_H)$ is the fragment mass ratio. In panels (b), (d) and (f), the predicted charge distributions are shown. The experimental data are taken from Ref. 32.

In Figs. 56, 59 and 60, the charge distributions in the fission of ^{180,182,190}Hg* are predicted. They have similar features that are an asymmetric double-peaked nature and a large peak-to-valley ratio at low excitation energies. As the excitation energy and angular momentum of the CN increase, this ratio decreases. While the charge distributions in these

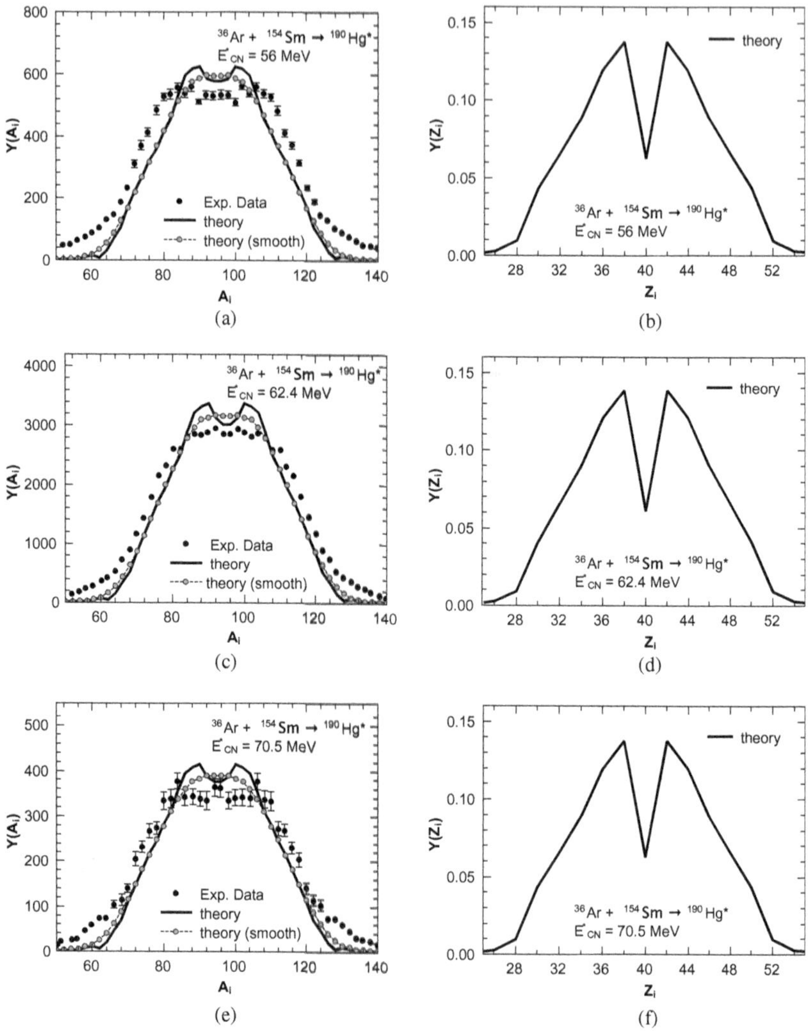

Fig. 60. The same as in Fig. 56 but for the ^{36}Ar+^{154}Sm→^{190}Hg* reaction at $E^*_{\rm CN}(L=0) = 56$, 62.4 and 70.5 MeV (panels (a), (c) and (e), respectively). In panels (b), (d) and (f), the predicted charge distributions are shown. The experimental data are taken from Ref. 31.

reactions are quite similar, the mass distributions bear more discrepancies. This important observation means that the average N_i/Z_i ratio of the fragments is not constant and depends on the excitation energy and angular momentum of the fissioning nucleus.[29] The fission of ^{182}Hg* at $E^*_{\rm CN} = 75$ MeV (Fig. 59) is quite interesting: It exhibits almost symmetric mass

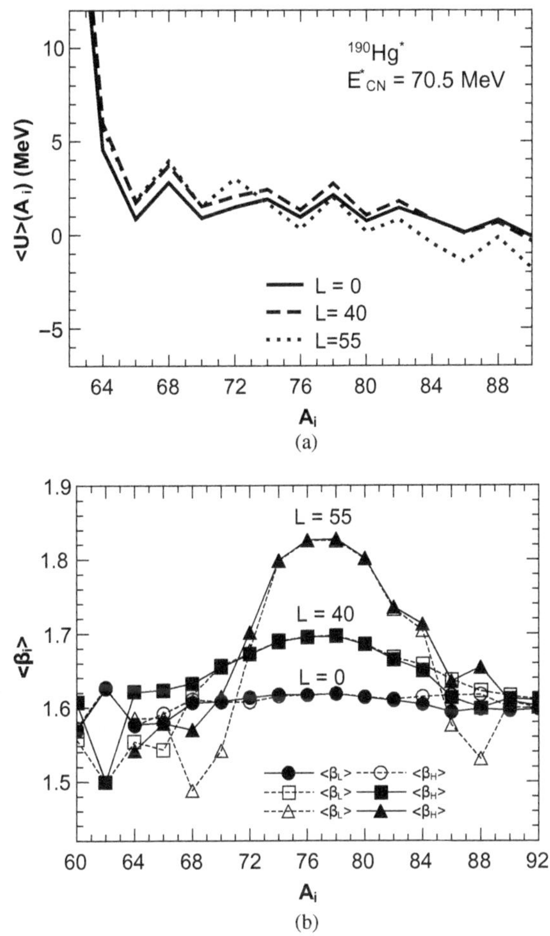

Fig. 61. (a) The calculated average total potential energy $\langle U\rangle(A_i)$ as a function of the mass number of one of the fragments at the indicated angular momenta L. The driving potential is normalized such that $\langle U\rangle(A/2) = 0$. (b) The calculated average deformations β_H and β_L of the two fragments as functions of their mass numbers at the indicated values of L.

distributions and clearly asymmetric charge distributions. This effect can be easily understood on the basis of the fact that Sr and Mo nuclei appear in several mass fragmentations with rather large yields. The difference between the mass and charge yields is reflected in different dependencies of the average potential energy on mass and charge numbers of fragments (Figs. 61 and 62). While $U(A_i)$ (Fig. 61) has a local minimum at $A_L = 90$ and all L

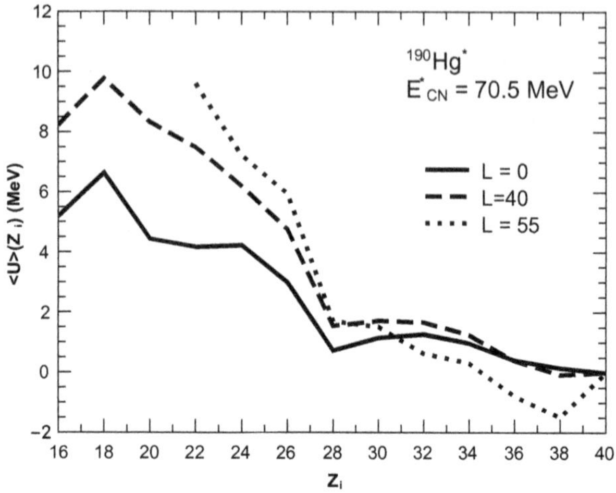

(a)

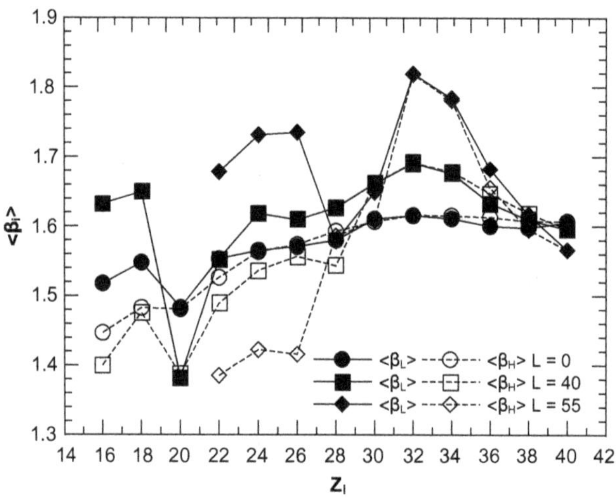

(b)

Fig. 62. (a) The calculated average total potential energy $\langle U\rangle(Z_i)$ as a function of the charge number of one of the fragments at the indicated angular momenta L. The driving potential is normalized such that $\langle U\rangle(Z/2) = 0$. (b) The calculated average deformations β_H and β_L of the two fragments as functions of their charge numbers at the indicated values of L.

considered, the dependence of the average potential energy $U(Z_i)$ does not exhibit a minimum at $Z_{L,H} = 40$. Moreover, at $L = 55$, $U(Z_i)$ has a minimum at $Z_L = 38$ and a maximum at $Z_{L,H} = 40$ (Fig. 62).

In Fig. 63, the mass and charge yields in the α-induced fission of ^{194}Pt (the fissioning nucleus is ^{198}Hg* (Table 1)) at $E^*_{\rm CN} = 49$ MeV are presented

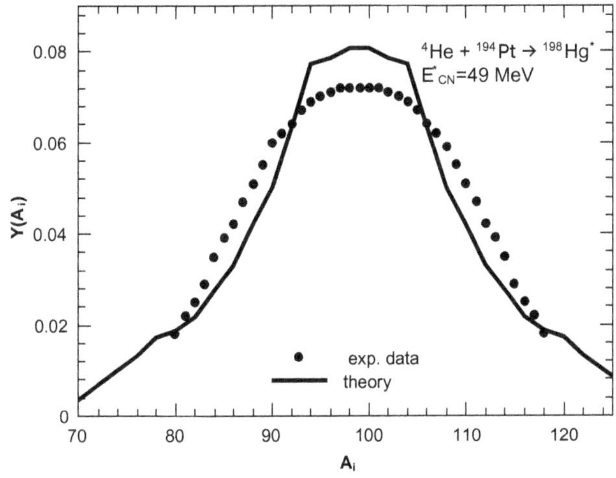

(a)

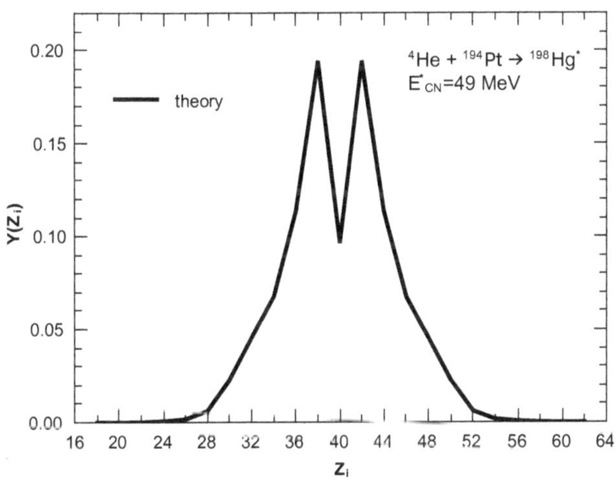

(b)

Fig. 63. (a) The experimental (symbols) and calculated (line) mass distributions for the $\alpha+^{194}\mathrm{Pt}\rightarrow^{198}\mathrm{Hg}^*$ reaction. (b) The predicted charge distribution for the same reaction. The excitation energy of CN at $L = 0$ is $E^*_{\mathrm{CN}}(L = 0) = 49\,\mathrm{MeV}$. The experimental data are taken from Ref. 119.

and compared with the experimental data.[119] The $^{198}\mathrm{Hg}^*$ nucleus shows mass-symmetric fission which differs from that of $^{180,182,190}\mathrm{Hg}^*$. The predicted charge distribution, however, is asymmetric, with a peak-to-valley ratio ~ 2. However, this ratio is smaller than those in the fission of $^{180,182,190}\mathrm{Hg}^*$.

In the ^{48}Ca+^{154}Sm→^{202}Pb* reaction at $E^*_{\rm CN} = 49$, 57 and 95 MeV (Fig. 64 and Table 1), the experimental data[120] are well reproduced. The fission of ^{202}Pb* clearly exhibits symmetric mass and charge distributions at all excitation energies. In the experiment at the lowest energy, the asymmetric shoulders are observed. However, the latter are related to the

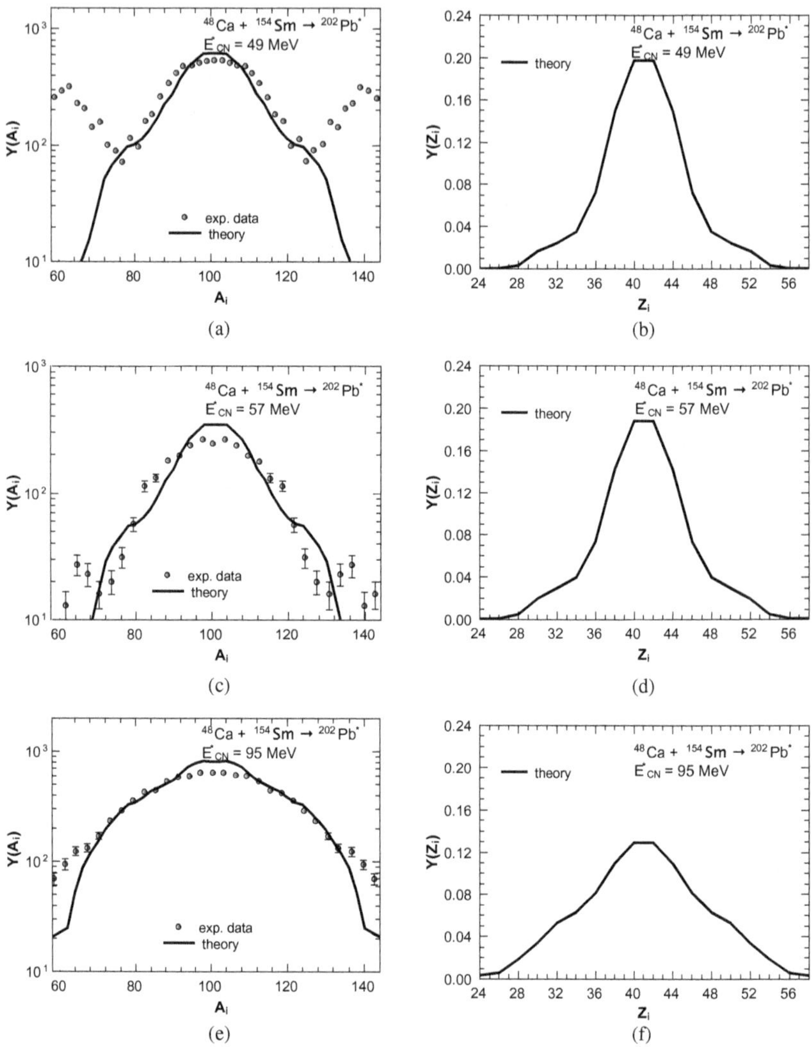

Fig. 64. (a), (c), (e) The same as in Fig. 56 but for the ^{48}Ca+^{154}Sm→^{202}Pb* reaction at the indicated excitation energies $E^*_{\rm CN}(L = 0) = 49$, 57 and 95 MeV. (b), (d), (f) The predicted charge distributions are shown. The experimental data are taken from Ref. 120.

dynamical quasi-fission process because the fission data mix with quasi-fission data.[33–35,120–126] Note that in the experiment, as well as in the theory, the distributions become wider with increasing excitation, as more mass-asymmetric configurations are reached. This effect can be experimentally investigated.

7. Excitation Energy Dependence of the Fission-Fragment Neutron Excess Ratio

The PES is very sensitive to the mass-to-charge ratio of fission fragments. As shown in Refs. 14, 29, 30, 58 and 66, the measured neutron excess ratios defined as the average neutron number per proton, i.e., $\langle N_i \rangle / Z_i$ ($\langle N_i \rangle$ is the average neutron number of the primary fission fragment with the charge number Z_i), are not constant and governed by the shell effects. Ratios $\langle N_i \rangle / Z_i$ of primary fission fragments resulting from the fission of ^{240}Pu at $E^* = 6.54$, 10 and 20 MeV and $E^* = 7.5$, 10.5, 12.5 and 16.5 MeV are analyzed within the dynamical models based on the macroscopic–microscopic[127] and self-consistent microscopic[128] approaches, respectively.

Employing an improved scission-point model,[69–78] one can analyze $\langle N_{L,H} \rangle / Z_{L,H}$ in the primary fission fragments (prior to the neutron evaporation stage). In Fig. 65, the theoretical neutron excess ratios $\langle N_i \rangle / Z_i$ as a function of the fission-fragment charge number Z_i are compared with the experimental data[29] for the fission of nuclei ^{238}U*, ^{240}Pu* and ^{244}Cm*. For the same fissioning nuclei, the theoretical and experimental[29,87,92–94] charge and mass distributions of fission fragments and the average number of neutrons emitted from the fission fragment with the mass number A_i are shown in Figs. 3, 5 and 36. Since the fragments are more deformed at scission than in their ground states, they are relaxed to the ground states after the DNS decays and the energies of deformations are transformed into the fragment intrinsic excitation energies. The position of the maximum of the calculated mass distribution of light fragments in Fig. 3 is shifted toward the experimental points if neutron emission from the primary fragments is taken into account according to Fig. 5. The charge distributions in Fig. 36 are not affected by the neutron emission.

In Fig. 65, the calculated neutron excess ratio $\langle N_i \rangle / Z_i$ for primary fragments strongly depends on Z_i, exhibiting a structure which relates to the shell structure of fragments. In the case of the fissioning ^{238}U at excitation energy $E^*_{\rm CN} = 7.4$ MeV (Fig. 65(a)), the minimum located at $Z_L = 32$ is related to the closed shell $N_L = 50$, as the minimum of the PES

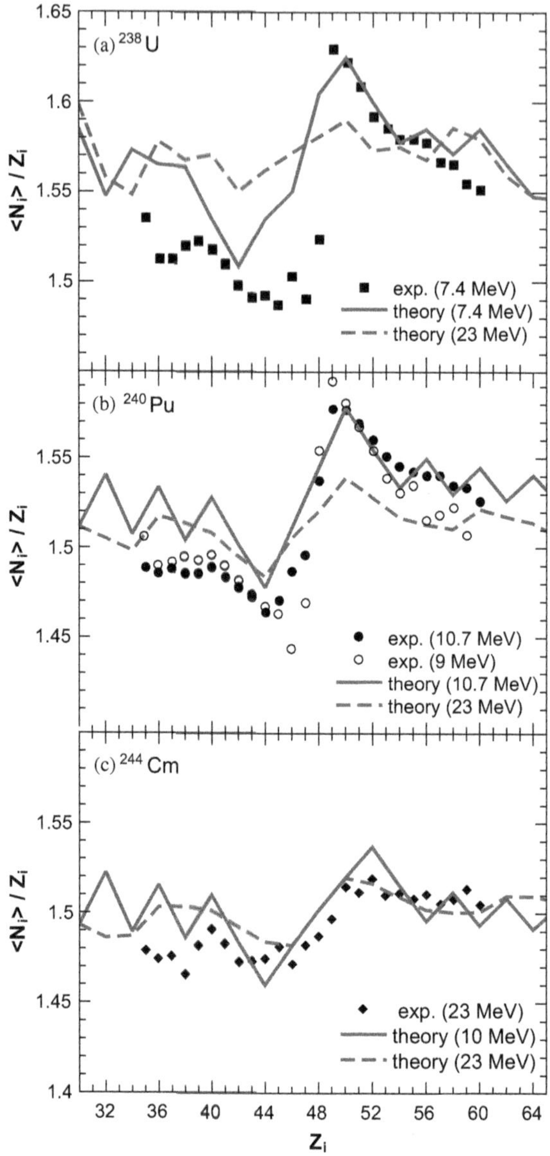

Fig. 65. Calculated $\langle N_i \rangle / Z_i$ ratios for the primary fission fragments as functions of Z_i (lines) for ^{238}U (a) at the excitation energies 7.4 MeV (solid line) and 23 MeV (dashed line), ^{240}Pu (b) at the excitation energies 10.7 MeV (solid line) and 23 MeV (dashed line) and ^{244}Cm (c) at the excitation energies 10 MeV (solid line) and 23 MeV (dashed line). The available experimental data are taken from Ref. 29 (closed symbols) and Ref. 58 (open symbols). The excitation energy $E^*_{\rm CN}$ of the fissioning nucleus is presented in parentheses.

in the charge-mass coordinates comes from the configuration ^{82}Ge+^{156}Nd. This minimum is related to the maximum of $\langle N_i\rangle/Z_i$ at $Z_H = 60$. At $Z_L = 38$, the neutron shell $N_H = 82$ starts to influence. In this case, the most likely configuration is ^{98}Sr+^{140}Xe; however, the configurations ^{102}Sr+^{136}Xe and ^{100}Sr+^{138}Xe with $N_H = 82$ and $N_H = 84$, respectively, exhibit minima which are very close in energy to the minimum corresponding to the ^{98}Sr+^{140}Xe fragmentation. The same can be noted for the fragmentations Zr + Te: The minimum of the PES is supplied by the fragmentations ^{102}Zr+^{136}Te, ^{104}Zr+^{134}Te and ^{106}Zr+^{132}Te with close potential energies (within ~0.3 MeV). As the atomic number of a light fragment increases, a heavy fragment becomes double magic. Thus, the ^{106}Mo+^{132}Sn configuration becomes likely. The shell effects in the heavy fragment hinder a large neutron excess in the light fragment, resulting in the saw-tooth structure in Fig. 65. Note that the closed shell $N_H = 82$ is responsible for the decrease of the theoretical $\langle N_i\rangle/Z_i$ in the intervals $Z_H = 50$–54 and $Z_L = 38$–42. In this case, the magic neutron number $N_H = 82$ in the heavy fragment tries to keep the neutron number of the light fragment.

In the case of the fissioning ^{240}Pu at $E^*_{\rm CN} = 10.7$ MeV (Fig. 65(b)), the average neutron excess of primary fragments has a similar structure as in the case of ^{238}U ($E^*_{\rm CN} = 7.4$ MeV). The maximum obtained at $Z_L = 32$ and the minimum observed at $Z_L = 34$ are explained by the magic shell $N_L = 50$ of the light fragment in the fragmentations ^{82}Ge+^{158}Sm and ^{84}Se+^{156}Nd. The effect of the $N_H = 82$ shell is seen in the configurations ^{102}Zr+^{138}Xe and ^{104}Zr+^{136}Xe, which exhibit minima on the PES with close energies. The influence of this neutron shell is also seen in the ^{106}Mo+^{134}Te configuration. In the same manner as in the ^{238}U case, the combined effect of both $Z_H = 50$ and $N_H = 82$ shells creates the saw-tooth feature shown in Fig. 65. In the minimum of $\langle N_i\rangle/Z_i$ at $Z_L = 44$, the most probable DNSs are ^{108}Ru+^{132}Sn and ^{110}Ru+^{130}Sn. Note that the structure in $\langle N_i\rangle/Z_i$ (Fig. 65) appears in calculations for primary fragments. Therefore, the neutron emission from the fission fragments is not the reason for this structure. From Figs. 5 and 65(b), it follows that the neutron emission would deepen the minimum of $\langle N_i\rangle/Z_i$ at $Z_i = 44$, i.e., highlight the saw-tooth structure.

In the case of the fissioning ^{244}Cm, the 23-MeV excitation energy is much higher than those discussed above, so the shell effects are damped and, as a result, the neutron excess ratio $\langle N_i\rangle/Z_i$ does not exhibit such a strong dependence on the charge number of the fission fragment. In Figs. 65(a) and 65(b), the predicted neutron excess ratios of the fission fragments in the cases of the fissioning nuclei ^{238}U and ^{240}Pu at the initial excitation energy of

23 MeV are also presented. As can be seen, the saw-tooth character of $\langle N_i \rangle / Z_i$ is less pronounced.

Figures 37(a) and 37(b) show the calculated charge and TKE distributions for the fission of ^{250}Cf at an excitation energy of 46 MeV. In Fig. 37(c), the average number of neutrons emitted by a fragment with charge number Z_i is compared with experimental data of Ref. 58. One striking feature of Fig. 37(c) compared to Fig. 5 is the fact that in the case of high-energy fission of ^{250}Cf, the $\langle n \rangle$ distribution does not have the saw-tooth shape of Pu, but rather a steady increase in neutron multiplicity as the heavier fragment receives more and more excitation energy in proportion to its mass number. In Fig. 37(d), the neutron excess $\langle N_i \rangle / Z_i$ of one of the fragments is presented. In this case, the values are fairly constant, which agrees with the experimental data.[29,58] The explanation for this is the same as discussed earlier, namely, the "melting" of shell effects at high excitation energy.

Our conclusions about the neutron excess in the fission fragments are consistent with the conclusions of Refs. 127 and 128, namely that with increasing excitation energy, fragments tend to have the same neutron-to-proton ratios as in a fissioning compound nucleus. However, it is worth noting that in our case, the transition to a constant $\langle N_i \rangle / Z_i$ is slower with increasing excitation energy, than in the cases presented in Refs. 127 and 128.

8. Suggestion for Examination of a Role of Multi-Chance Fission

The multi-chance fission process is fission after consecutive neutron evaporations. In this process, the fissioning nuclei with a smaller number of neutrons have lower excitation energy and relatively stronger shell effects than in the original CN. After neutron evaporation the excitation energy of the nucleus is reduced by the binding and kinetic energies of the emitted neutron.

To describe the yields of fission of actinides ($^{236-240}$U, $^{239-242}$Np, $^{241-244}$Pu and ^{250}Cf) at high excitation energies ($E^*_{\rm CN} > 20$ MeV) with the models of Refs. 25, 129 and 130, one should take the multi-chance fission into account. In Refs. 25, 129 and 130, the mass distribution of fission fragments originates from the first-chance–last-chance fission (i.e., up to sixth-chance fission of ^{240}U at $E^*_{\rm CN} = 40$–50 MeV). With the multi-chance fission the calculated distribution of fission fragments dramatically changes with respect to that obtained in the first-chance fission. It is obvious that the first-chance fission in the models mentioned leads to a symmetric shape of the mass (charge) distribution due to the LD feature of the fissioning nucleus at higher excitation energy. In Refs. 25, 129 and 130, the fission after evaporation of several neutrons shows the asymmetric mass distribution that is typical for low-energy fission of the actinides with charge numbers $Z \geq 92$. Thus, the asymmetric component of the mass (charge) distribution is mainly attributed to the late-chance fission.

With small changes of the weights in the multi-chance fission, the ratio between the symmetric and asymmetric yields can strongly change.[25,129–131] The estimates of the percentages in the multi-chance fission are difficult because of the absence of reliable information about the competition between particle and gamma emission and fission as a function of excitation energy and angular momentum. This leads to large uncertainties in the calculations and to the dependence of the results on the parameters used. However, the asymmetry of the mass (charge) distribution at high excitation energies can only be caused by the multi-chance fission in the models of Refs. 25, 129 and 130.

The models of Refs. 25 and 130 are dynamical models for the calculation of fission observables. Exploiting the expected strongly damped character of nuclear dynamics, the authors of Ref. 130 treated the nuclear shape evolution in analogy to the Brownian motion on the five-dimensional fission PES. This novel approach requires the only new global parameter, namely the critical neck size at which the mass split is frozen in. In the model of Ref. 25, the evolution of a nuclear shape, defined by three collective coordinates, is traced from the CN state to the scission point by solving the Langevin equations, and the fission-fragment mass distributions are calculated with the Monte Carlo method. The model of Ref. 129 is a semiempirical model with 50 parameters adjusted by experimental data. Five fission channels are considered. The mass and charge distributions are calculated assuming a statistical population of states in the fission valleys at freeze-out. The freeze-out time is different for various collective coordinates.

The improved scission-point model[69–78] describes the yields for the fission of ^{235}U$(n,\ f)$ at $E_n = 6$ MeV and 14 MeV ($E^*_{\rm CN} = 12.5$ MeV and 20.55 MeV; Fig. 33), ^{238}U$(n,\ f)$ at $E_n = 32.8$, 45.3 and 59.9 MeV ($E^*_{\rm CN} = 37.6$, 50.1 and 64.7 MeV; Fig. 34) and ^{240}U* and ^{244}Pu* at $E^*_{\rm CN} = 15$, 25, 35, 45 and 55 MeV (Fig. 35) without employing the multi-chance fission assumption. In the model, the fission at the first chance seems to be the most important and the contribution of the multi-chance fission does not change the shape of the mass (charge) distribution (see Fig. 35). This is supported by the experimental data[36] in which the number of pre-scission neutrons does not exceed 2 at excitation energies $\sim$50 MeV of actinides considered.

The role of the multi-chance fission is usually estimated with the statistical model considering the competition between particle (mainly neutrons) evaporation and fission. The results depend on the parameters and assumptions used. Using the experimental values of the neutron separation energy,[132] the calculated fission barrier heights,[133] the level density parameters $a_n = A/10\,\mathrm{MeV}^{-1}$ and $a_f = 1.05a_n$ for neutron emission and fission, respectively, and the pairing gap $\Delta = 12/\sqrt{A}$ MeV, one can calculate the ratio of the neutron emission width to the total decay width $\Gamma_n/\Gamma_{\mathrm{tot}}$. The height of the fission barrier consists of the LD[134] and shell components. The shell component is damped with excitation energy. The damping coefficient is set as $E_d = 20$ MeV. The parameters listed are similar to those used in the previous works.[135] With these parameters the fission fractions are about 34%, 25% and 16% at the first, second and third steps, respectively, for ^{240}U at $E^*_{\mathrm{CN}} = 50$ MeV. For ^{226}Th at $E^*_{\mathrm{CN}} = 50$ MeV, the fission fractions have similar values: 37%, 23% and 17%. Thus, about 75% of fission events occur during emission of two pre-scission neutrons. Taking into account this fission fractions, the distributions in Fig. 2 remain almost unchanged. The calculated fission fraction at the second step for ^{238}Pu at $E^*_{\mathrm{CN}} = 30$ MeV is similar to that presented in Ref. 136. Thus, the calculated results seem to be reliable.

In Ref. 37, the method of pure experimental (model-independent) verification of the multi-chance fission assumption and two kinds of models is suggested. As is known, the experimental[4] charge distributions in the electromagnetic-induced fission of even–even isotopes ^{218,220,222}Th ($E^*_{\mathrm{CN}} = E_\gamma = 11$ MeV) are clearly symmetric, with one relatively narrow peak. For ^{224}Th, the distribution starts showing the formation of two asymmetric peaks, while for ^{226}Th the yields around Kr–Sr and their complementary fragments are equal to those around Pd–Ru. As the mass number of Th increases up to $A = 228$, the symmetric peak recedes and the asymmetric mode becomes dominant.[4] Thus, in the low-energy fission, the transition from an one-peaked to a two-peaked charge distribution occurs with increasing neutron number through the transient three-peaked shape in which the symmetric and asymmetric components of mass (charge) distribution have almost the same weights. The symmetric mode tends to increase strongly with decreasing neutron number.

In the transfer reaction ^{40}Ca+^{226}Ra→^{228}Th*+^{38}Ar ($Q_{tr} = -0.1$ MeV) or ^{32}S+^{226}Ra→^{228}Th*+^{30}Si ($Q_{tr} = -1.6$ MeV), one can produce the nucleus ^{228} Th* at excitation energies E^*_{CN} from 0 MeV to 60 MeV. If the multi-chance fission is important, the contributions to the fission yields from the neutron-deficient Th isotopes formed after the emission of several neutrons should increase with E^*_{CN}. The fission of $^{218-224}$Th shows the symmetric mass (charge) distributions at low (see the experimental data) and high energies. As a result, the shape of the charge (mass) distribution of the fission fragments of ^{228}Th* will be predominantly asymmetric at low excitation energies, with a gradual increase and final domination of the symmetric mode toward higher excitation energies. For example, the charge distribution of fission fragments originating from the primary nucleus ^{228}Th* at $E^*_{CN} = 50$ MeV (six-chance fission is possible) should be symmetric based on the model of Refs. 129 and 130. In contrast to that, an opposite behavior will be obtained if the first-chance fission assumption used in the improved scission-point model[69–78] is correct (see Fig. 8).

9. How to Understand the Anomaly in Charge Yield of Mo/Sn Fragments in Fission Reaction ^{238}U(n, f) at Low Excitation Energies

For the first time, the fast-neutron-induced fission of ^{238}U at energy just above the fission threshold has been recently studied with a novel experimental technique which involves the coupling of the high-efficiency γ-ray spectrometer (MINIBALL) to the inverse kinematics neutron source (LICORNE) to extract charge yields of fission fragments via the γ–γ coincidence spectroscopy.[137] The measured charge yields for a given charge number Z have been compared to the JEFF3.1.1[138] and ENDF.B7v1[136] evaluations or models and the statistical fission model, GEF.[129] As determined, there is a reasonable agreement (with accuracy 20%) for many nuclei. However, the significant discrepancies of up to 600% have been observed for isotopes of Mo and Sn. The authors of Ref. 137 have concluded that since the used models significantly overestimate the standard I fission mode, the spherical shell effects in the nascent fission fragments are less important for low-energy fast-neutron-induced fission than for the thermal-neutron-induced fission.

One can employ the improved version of the statistical scission-point fission model[139] to study the anomaly in the charge yield of Mo/Sn fragments in fission reaction ^{238}U(n, f). The experimental[137] and

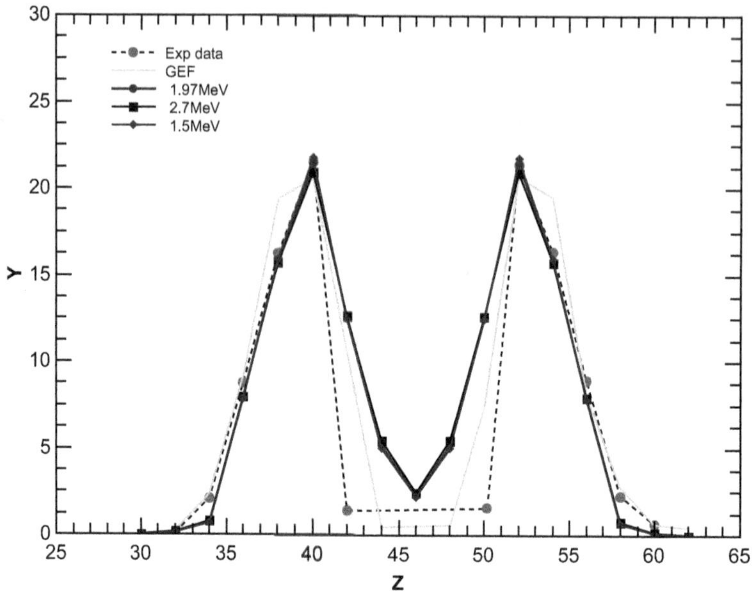

Fig. 66. The calculated[139] charge distributions (lines) for neutron-induced fission ^{238}U(n, f) at the indicated neutron energies $E_n = 1.5$, 1.97 and 2.7 MeV are compared with the experimental data[137] (symbols) and the results of GEF model.[129] The lines connect the calculated yields of even-Z fission fragments.

calculated[139] charge distributions resulted from the neutron-induced fission of ^{238}U at neutron energies $E_n = 1.5$, 1.97 and 2.7 MeV are shown in Fig. 66. A bombarding neutron spectrum has a mean energy of 1.72 MeV and a spread at a half maximum is approximately 1 MeV.[137] As can be seen, calculations within the scission-point model[139] are almost insensitive to the reasonable variation of the neutron energy. The charge distribution of the fission fragments is well described, except for Mo/Sn pairs. For this charge fragmentation, the ratio of theoretical and experimental yields is about one order of magnitude. So, there is a huge discrepancy between experiment and the scission-point model for Mo/Sn fragmentations. Our results are consistent with the ones of the models JEFF3.1.1,[138] ENDF.B7v1[136] and GEF.[129] One can conclude that the possible explanation of this anomaly in charge yield is not related to the spherical shell effects in the nascent fission fragments because in the scission-point model the shell effects are important for low-energy neutron-induced fission as well as for the thermal-neutron-induced fission. Because the yields for detectable even–even isotopes were deduced by measuring the intensities of all observable transitions feeding the lowest 2^+ state, coincident with the corresponding $2^+ \to 0^+$ transition to the ground state 0^+, one can suggest that the population of the states with spin I larger than 2 is suppressed for the scission-point configurations Mo + Sn with the spherical stiff heavy nuclei Sn. The explanation of this fact is as follows. At the scission point, a fissioning system is like a DNS whose intrinsic degrees of freedom are in thermal equilibrium. However, the shape of the DNS is not equilibrated around the orientation with minimum potential energy. The angular vibrations generate the rotational energy and spin of the fragments of binary fissions. As shown in Refs. 140–146, the bending vibrations are the main source by which the fragments gain spin of about 2–8[110,147] at the scission point, while the fissioning nucleus has initially zero or almost zero spin. The primary fission fragments have the average angular momenta of about $\langle I \rangle = 2\text{--}8$.[110,147] In the bending mode, the rolling of nuclear surfaces occurs which is suppressed by larger stiffness of the potential when one of the DNS nuclei is spherical. Because the excitation energies of the DNS configurations with magic Sn are smaller than for other neighboring charge splits, the population of the excited states of bending mode is less probable (the temperature is much smaller than the frequency of zero-point bending vibrations). Moreover, the scission-point DNS has almost nonrotating spherical nucleus and the bending mode would need larger rotational energy because of larger orbital momentum of the whole DNS compensating the spin of the rotating deformed light fragment Mo. Therefore, the bending mode becomes energetically unfavorable. As a result, the probability of population of states with spin I larger than 2 is small. This simply explains why the theory and experiment differ. One of the assumptions of the γ–γ-coincidence technique to measure yields is that the γ-ray multiplicity does not vary much with the charge split.

Applying the model of Ref. 145, one can estimate the average spin of the light deformed fragment ^{107}Mo if the DNS heavy fragment ^{132}Sn is spherical at scission: $\langle I(\text{Mo})\rangle = 2.6$ and $\langle I(\text{Sn})\rangle = 0$. The corresponding spin distribution for ^{107}Mo is

$$P(I) = \gamma_0^2(2I+1)e^{-\left(I+\frac{1}{2}\right)^2\gamma_0^2}, \tag{10}$$

where $\gamma_0^2 = \frac{\hbar}{\Im_b\omega_b} = 0.08$ ($\Im_b$ and ω_b are the moment of inertia and frequency of the bending mode). The total gamma-multiplicity is about $\langle I(\text{Mo})\rangle/2 \approx 1$. For comparison, the absolute value of the average spins of the fragments ^{101}Zr/^{138}Te is about $\langle I \rangle = 4.6$ ($\gamma_0^2 = 0.03$). Here, the total gamma-multiplicity is about 5. In the other DNS with the deformed nuclei, the bending mode generates almost the same values of the fragment spin like in the ^{101}Zr+^{138}Te configuration. So, the Mo/Sn pairs are populated on average with much less spin than the other fragment pairs. Note that in Ref. 148 it was also shown experimentally that in the spontaneous fission of ^{252}Cf the population of spin in the configuration with magic Sn is less probable than in the other neighboring charge splits.

10. Conclusions

As shown above, the improved scission-point model of the fission is able to consistently and reliably describe charge, mass, TKE and neutron multiplicity distributions of fission fragments of nuclei in the range of even–even pre-actinides, actinides and superheavy nuclei. The description of these correlated fission observables is based on the same PES for each fissioning nucleus. The evolution of the fission characteristics with increasing energy and isospin is shown to be related with the change of the minima in the PES (in the deformation space (β_L, β_H)) at scission. The minima on the PES (the maxima in the yields) are a direct result of the interplay between the LD energy, nucleus–nucleus interaction potential and shell correction energy.

For the electromagnetic-induced fission of even isotopes $^{222-230}$Th ($E^*_{\rm CN}(L=0) = 11\,$MeV), the transformation of the shape of the charge distribution from symmetric to asymmetric with increasing mass of a fissioning nucleus is reproduced. This transition leads to an increase in the neutron multiplicity at symmetry. The evolution from an one-peaked, to a three-peaked and then a two-peaked distribution is expected to occur across several mass units. This evolution is explained with the PES of scission configurations. The interplay between the LD surface energy and the nucleus–nucleus interaction potential is one of the main reasons for the appearance or disappearance of the asymmetric fission mode. This interplay of the macroscopic energies depends on the excitation energy and the isospin of the fissioning nucleus. The shell effects also affect indirectly, through the deformations of nuclei, the appearance of an asymmetric minimum in the driving potential. In our opinion, for a better understanding of the fission process, it is necessary to study in detail the relationship between the microscopic and macroscopic effects with increasing excitation energy.

It is demonstrated that the transition from the asymmetric to the symmetric fission regime, which occurs along the Th isotope chain, influences the average neutron multiplicity $\langle n\rangle$ and the average TKE $\langle$TKE$\rangle$. Because the observables $\langle n\rangle$ and $\langle$TKE$\rangle$ carry information about the symmetric–asymmetric fission transition, they can serve as markers to identify such transitions. The correlations between $\langle n \rangle$, $P_m(x)$, $\langle \mathrm{TKE} \rangle$ and the shapes of the charge/mass distributions can also be observed in the transition from asymmetric fission of ^{254,256}Fm (^{252}No) to the symmetric one of ^{258,260}Fm (^{258,262}No). It is interesting to study the correlations between these observables in the case of transition from unimodal to bimodal fission.

An important result is the expected conservation of the asymmetric shapes of the mass and charge distributions of the fission fragments at high enough excitation energies of the fissioning nuclei ^{226,228,230}Th. For the fissioning nucleus ^{226}Th at high excitation energies, the calculated charge distribution is well asymmetric but not three-equal-peaked one, which is desirable to test experimentally. As can be found, the three-equal-peaked charge distribution is not a specific property of the ^{226}Th nucleus. The shape of the charge (mass) distribution depends on the excitation energy. It is shown that macroscopic effects also play an important role in asymmetric fission at low and high excitation energies.

For the fissioning even–even nuclei $^{248-256}$Cf, the available experimental mass distributions in the spontaneous and thermal-neutron-induced fission are well described. As a global trend, the mass and charge yields are symmetrized with increasing excitation energy and decreasing isospin. In the fission of $^{248-256}$Cf at an excitation energy of about 55 MeV, the saturation of the symmetric component is found. As predicted, the charge distribution resulting from the fission of ^{248}Cf at excitation energies in the range between 25 MeV and 35 MeV shows both almost symmetric (Cd + Sn) and asymmetric (Ru + Xe) modes with relatively equal weights. In the fission of ^{248}Cf at $E^* = 45$ MeV and ^{252}Cf at $E^* = 55$–65 MeV, the difference is predicted between the shapes of mass (the pronounced asymmetric component) and charge (almost symmetric component Cd + Sn) yields. In the fission of $^{248-256}$Cf at low and high excitation energies, the mean proton and neutron numbers of heavy fragments are stable at around $\langle Z_H \rangle = 54$ and $\langle N_H \rangle = 80-86$. The same average values of the numbers of protons and neutrons in a heavy fission fragment appear in the fission of $^{222-230}$Th nuclei at low and high excitation energies.

For the fissioning even–even nuclei ^{254,256,258,260,264}Fm, ^{258,260,262,264}No and ^{262,264,266}Rf, the available experimental mass distributions in the spontaneous and thermal-neutron-induced fission are well described, except for ^{258}Fm(s.f.). In contrast to the experimental data, the calculated mass distribution for the spontaneous fission of ^{258}Fm is rather wide and has asymmetric bumps, even though the symmetric yields are extremely enhanced. However, the calculated spontaneous fission of ^{258}Fm shows a slightly asymmetric charge distribution. This difference between the calculated results and measured data is worth to be studied further experimentally and theoretically. In the case of the spontaneous fission of ^{260}Fm, the mass distribution is symmetric, but the charge distribution is asymmetric ($Z_L = 48$, $Z_H = 52$). For ^{260}No(s.f.), the mass distribution has asymmetric and symmetric peaks and the charge distribution has peaks at $Z_L = 50$ and $Z_H = 52$. For ^{266}Rf(s.f.), the mass yield is symmetric and the charge yield is asymmetric ($Z_L = 50$, $Z_H = 54$). For the thermal-neutron-induced fission of ^{257}Fm, the asymmetric charge distribution ($Z_L = 48$, $Z_H = 52$) is predicted. As is known, the reaction $^{257}\text{Fm}(n_{\text{th}}, f)$ has a symmetric mass yield.

Calculations for Fm isotopes suggest that a transition from an asymmetric to a symmetric mode occurs gradually but not as suddenly as displayed in Refs. 18 and 20. The fission of No isotopes may result in three-peaked distributions at the mass numbers $A = 250$ and $A = 256$. For the latter case, the experimental data seem to confirm this. The physical origins of these transitions are analyzed by looking at the components of the driving potential at scission. For heavy actinides, the shell effects play a dominant role by directly affecting the formations of minima in the driving potential or by indirectly facilitating the formation of a large number of fragmentations near $Z_i = 50$ and $N_i = 82$. In the case of the fission of Th isotopes, a monotonous variation of the macroscopic part of the driving potential and a rather weak role of the direct microscopic effects result in the change of the shape of the mass (charge) distribution over several mass units. For heavy actinides, this change is predicted to occur at a faster pace with increasing neutron number due to the large direct influence of the shell effects.

As demonstrated for the first time, mass yields become more symmetric with increasing excitation energy of the fissioning Fm and No. For ^{260,262,264}No (^{266}Rf), the asymmetric peaks at $Z_L = 50$ and $Z_H = 52$ ($Z_L = 50$ and $Z_H = 54$) are conserved with increasing E^*_{CN}. For ^{254,256}Fm, the shape of mass yields evolves rather slowly with increasing E^*_{CN}. The asymmetric component exists even at high excitation energy ($E^*_{\text{CN}} = 50\,\text{MeV}$) considered. In the fission of ^{258}Fm and ^{260}Fm, the symmetric component of the mass/charge distribution is enhanced much faster with excitation energy due to a rapid transition from strong shell-effect-governed to macroscopic-governed symmetric fission behavior. Once the shell effects are damped at high excitation energy, the fission is strictly driven by the competition between the LD binding energy and the interaction energy of two fragments. Any increase in the excitation energy goes toward accessing previously inaccessible asymmetric configurations. This, together with the smooth character of the macroscopic energies, mainly leads to a broadening of the mass distribution and saturation of the symmetric yields with increasing excitation energy. The saturation effect is worth to be studied experimentally. The presence of both single- and two-peaked mass yields in the fission of Fm and No isotopes is an ample test based on which the theoretical models can be better verified. One can hope that the results presented here will motivate further experimental activity.

The mass distributions of fission fragments of several fissioning nuclei ^{180,182,190,198}Hg* ($E^*_{\rm CN} \sim 33$–95 MeV) formed in the complete fusion reactions ^{36}Ar+^{144}Sm, ^{40}Ca+^{142}Nd, ^{36}Ar+^{154}Sm, α+^{194}Pt and ^{48}Ca+^{154}Sm are in good agreement with the experimental data. The low angular momentum L values contribute to the total yields leading to a rather smooth mass distribution, while the high L values contribute to the total yields preserving the microscopic shell effects. The combined microscopic and macroscopic effects preserve the structure of the distribution found usually at low excitation energies and angular momenta, while still weakly increasing the weight of symmetric configurations in the total yields. With increasing bombarding energy the symmetric yields increase up to a maximum at critical angular momentum, after which, as the excitation energy increases, the theoretical mass and charge distributions become wider. The charge distribution is predicted to be less affected by excitation energy and angular momentum than the mass distribution.

For the fissioning nuclei ^{180,182,190,198}Hg*, the asymmetric charge distributions are predicted up to high excitation energy. In contrast, for ^{202}Pb*, the symmetric charge distribution is obtained. The interesting cases found are the fissions of ^{182}Hg* at $E^*_{\rm CN} = 75$ MeV and ^{198}Hg* at $E^*_{\rm CN} = 49$ MeV. They exhibit a symmetric mass distribution (within the mass resolution of measurements), while the predicted charge distribution has a pronounced double-peaked structure. For the fissioning nuclei ^{180,182,190,198}Hg*, the average N_i/Z_i ratio of the fragments is not constant and depends on the excitation energy and angular momentum of a fissioning nucleus.

The study of the low-energy fission of actinides shows that the neutron excess ratio $\langle N_i\rangle/Z_i$ of fission fragments has a strong dependence on the fragment charge number Z_i, exhibiting a well-defined saw-tooth structure which is a direct result of the interplay between the neutron and proton shell closures. For example, if the neutron number in one of the fission fragments is close to the magic one, there is an interval of Z_i where N_i is almost unchangeable. In this case, the neutron excess ratio $\langle N_i\rangle/Z_i$ decreases with increasing Z_i. Furthermore, this structure is a property of the primary fragments rather than a consequence of the neutron emission from the fragments after post-separation. One cannot imply that the $\langle N_i \rangle / Z_i$ ratio of the initial fissioning CN is preserved in the primary fission fragments, but rather this assumption should be made only for high excitation energies, where the shell effects are considerably reduced.

To check two kinds of models in which the multi-chance fission has different contributions to the formation of the charge (mass) distribution, one can study the fission of ^{228}Th* produced in the transfer reaction ^{40}Ca+^{226}Ra or ^{32}S+^{226}Ra as a function of excitation energy. The comparison of different predictions of Refs. 25, 129 and 130 and of the improved scission-point model[69–78] with experimental data would shed light on the role of the multi-chance fission.

For the low-energy neutron-induced fission of ^{238}U nucleus, a possible explanation of the anomaly in charge yield of Mo/Sn fragments found by γ–γ coincidence spectroscopy[137] is presented. Unusually low measured yield of Mo/Sn pairs can be related with the small probability of generation of fragment spin in the scission-point DNS Mo + Sn with the spherical nucleus Sn.

Acknowledgments

The authors acknowledge the support of the Romania–JINR (Dubna) Cooperation Programme. The work of H. Paşca was supported by a Grant of the Ministry of Research, Innovation and Digitization, CNCS/CCCDI — UEFISCDI, Project No. PN-III-P1-1.1-PD-2019-0304 within PNCDI III. The work of G. G. Adamian, N. V. Antonenko and H. Paşca was supported by the Ministry of Science and Higher Education of the Russian Federation (Contract 075-10-2020-117).

References

1. H. J. Krappe and K. Pomorski, *Theory of Nuclear Fission* (Springer-Verlag, Berlin, 2012).
2. A. N. Andreyev, M. Huyse and P. Van Duppen, *Rev. Mod. Phys.* **85** (2013) 1541.
3. A. N. Andreyev, K. Nishio and K.-H. Schmidt, *Rep. Prog. Phys.* **81** (2018) 016301; K.-H. Schmidt and B. Jurado, *Rep. Prog. Phys.* **81** (2018) 106301.
4. K.-H. Schmidt *et al.*, *Nucl. Phys. A* **665** (2000) 221; K.-H. Schmidt, J. Benlliure and A. R. Junghans, *Nucl. Phys. A* **693** (2001) 169; C. Böckstiegel *et al.*, *Nucl. Phys. A* **802** (2008) 12.

5. U. Brosa, S. Grossmann and A. Müller, *Phys. Rep.* **197** (1990) 167.
6. A. N. Andreyev *et al.*, *Phys. Rev. Lett.* **105** (2010) 252502.
7. J. Elseviers *et al.*, *Phys. Rev. C* **88** (2013) 044321; L. Ghys *et al.*, *Phys. Rev. C* **90** (2014) 041301(R).
8. J. Randrup and P. Möller, *Phys. Rev. Lett.* **106** (2011) 132503; J. Randrup, P. Möller and A. J. Sierk, *Phys. Rev. C* **84** (2011) 034613; P. Möller, J. Randrup and A. J. Sierk, *Phys. Rev. C* **85** (2012) 024306.
9. M. Warda, A. Staszczak and W. Nazarewicz, *Phys. Rev. C* **86** (2012) 024601.
10. J. D. McDonnell, W. Nazarewicz, J. A. Sheikh, A. Staszczak and M. Warda, *Phys. Rev. C* **90** (2014) 021302(R).
11. G. Scamps and C. Simenel, *Phys. Rev. C* **100** (2019) 041602(R).
12. B. M. A. Swinton-Bland *et al.*, *Phys. Rev. C* **102** (2020) 054611.
13. E. Prasad *et al.*, *Phys. Lett. B* **811** (2020) 135941.
14. C. Schmitt *et al.*, *Phys. Rev. Lett.* **126** (2021) 132602.
15. A. V. Andreev, G. G. Adamian and N. V. Antonenko, *Phys. Rev. C* **86** (2012) 044315; A. V. Andreev, G. G. Adamian, N. V. Antonenko and A. N. Andreyev, *Phys. Rev. C* **88** (2013) 047604.
16. D. C. Hoffman *et al.*, *Phys. Rev. C* **41** (1990) 631.
17. D. C. Hoffman *et al.*, *J. Alloys Compd.* **213–214** (1994) 67.
18. E. K. Hulet *et al.*, *Phys. Rev. C* **21** (1980) 966.
19. S. Hofmann and G. Münzenberg, *Rev. Mod. Phys.* **72** (2000) 733.
20. E. K. Hulet *et al.*, *Phys. Rev. C* **40** (1989) 770.
21. I. V. Ryzhov *et al.*, *Phys. Rev. C* **83** (2011) 054603.
22. V. D. Simutkin, The title of thesis is Fragment Mass Distributions in Neutron-Induced Fission of ^{232}Th and ^{238}U from 10 to 60 MeV.
23. J. King *et al.*, *Eur. Phys. J. A* **53** (2017) 238.
24. R. Léguillon *et al.*, *Phys. Lett. B* **761** (2016) 125.
25. K. Hirose *et al.*, *Phys. Rev. Lett.* **119** (2017) 222501.
26. H. Naik, T. N. Nathaniel, A. Goswami, G. N. Kim, M. W. Lee, S. V. Suryanarayana, S. Ganesan, E. A. Kim, M.-H. Cho and K. L. Ramakumar, *Phys. Rev. C* **85** (2012) 024623 [*Erratum: Phys. Rev. C* **85** (2012) 039905]; H. Naik, V. K. Mulik, P. M. Prajapati, B. S. Shivasankar, S. V. Suryanarayana, K. C. Jagadeesan, S. V. Thakare, S. C. Sharma and A. Goswami, *Nucl. Phys. A* **913** (2013) 185; H. Naik, S. Mukerji, R. Crasta, S. V. Suryanarayana, S. C. Sharma and A. Goswami, *Nucl. Phys. A* **941** (2015) 16; H. Naik, S. Mukerji, S. V. Suryanarayana, K. C. Jagadeesan, S. V. Thakare and S. C. Sharma, *Nucl. Phys. A* **952** (2016) 100.
27. A. Deppman, E. Andrade-II, V. Guimarães, G. S. Karapetyan and N. A. Demekhina, *Phys. Rev. C* **87** (2013) 054604; A. Deppman, E. Andrade-II, V. Guimarães, G. S. Karapetyan, A. R. Balabekyan and N. A. Demekhina, *Phys. Rev. C* **88** (2013) 024608; A. Deppman *et al.*, *Phys. Rev. C* **88** (2013) 064609.
28. A. C. Berriman *et al.*, *Phys. Rev. C* **105** (2022) 064614.
29. D. Ramos *et al.*, *Phys. Rev. C* **97** (2018) 054612; D. Ramos *et al.*, *Phys. Rev. C* **99** (2019) 024615.
30. D. Ramos *et al.*, *Phys. Rev. Lett.* **123** (2019) 092503.

31. K. Nishio *et al.*, *Phys. Lett. B* **748** (2015) 89.
32. E. Prasad *et al.*, *Phys. Rev. C* **91** (2015) 064605.
33. I. Tsekhanovich *et al.*, *Phys. Lett. B* **790** (2019) 583.
34. A. A. Bogachev *et al.*, *Phys. Rev. C* **104** (2021) 024623.
35. E. M. Kozulin *et al.*, *Phys. Rev. C* **105** (2022) 014607.
36. D. Hilscher and H. Rossner, *Ann. Phys. (Fr.)* **17** (1992) 471.
37. H. Paşca, A. V. Andreev, G. G. Adamian and N. V. Antonenko, *Eur. Phys. J. A* **54** (2018) 104.
38. B. D. Wilkins, E. P. Steinberg and R. R. Chasman, *Phys. Rev. C* **14** (1976) 1832.
39. T. Matsuse, C. Beck, R. Nouicer and D. Mahboub, *Phys. Rev. C* **55** (1997) 1380.
40. S. J. Sanders, A. Szanto de Toledo and C. Beck, *Phys. Rep.* **311** (1999) 487.
41. S. Panebianco, J.-L. Sida, H. Goutte, J.-F. Lemaître, N. Dubray and S. Hilaire, *Phys. Rev. C* **86** (2012) 064601.
42. J.-F. Lemaître, S. Goriely, S. Hilaire and J.-L. Sida, *Phys. Rev. C* **99** (2019) 034612.
43. A. V. Andreev, G. G. Adamian, N. V. Antonenko, S. P. Ivanova and W. Scheid, *Eur. Phys. J. A* **22** (2004) 51; A. V. Andreev, G. G. Adamian, N. V. Antonenko and S. P. Ivanova, *Eur. Phys. J. A* **26** (2005) 327; A. V. Andreev, G. G. Adamian, N. V. Antonenko, S. P. Ivanova, S. N. Kuklin and W. Scheid, *Eur. Phys. J. A* **30** (2006) 579.
44. G. G. Adamian, N. V. Antonenko and W. Scheid, Clustering effects within the dinuclear model, in *Clusters in Nuclei*, Vol. **2**, ed. C. Beck, *Lecture Notes in Computer Science*, Vol. **848** (Springer-Verlag, Berlin, 2012), pp. 165–227.
45. N. Carjan, F. A. Ivanyuk, Yu. Ts. Oganessian and G. M. Ter-Akopian, *Nucl. Phys. A* **942** (2015) 97; N. Carjan, F. A. Ivanyuk and Yu. Ts. Oganessian, *Phys. Rev. C* **99** (2019) 064606.
46. M. Warda and A. Zdeb, *Phys. Scr.* **90** (2015) 114003.
47. J. Sadhukhan, S. A. Giuliani and W. Nazarewicz, *Phys. Rev. C* **105** (2022) 014619.
48. D. N. Poenaru, M. Ivascu, A. Sandulescu and W. Greiner, *Phys. Rev. C* **32** (1985) 572; D. N. Poenaru, R. A. Gherghescu and W. Greiner, *Phys. Rev. C* **73** (2006) 014608; D. N. Poenaru and R. A. Gherghescu, *Phys. Rev. C* **94** (2016) 014309.
49. C. Simenel and A. S. Umar, *Phys. Rev. C* **89** (2014) 031601(R).
50. G. Scamps, C. Simenel and D. Lacroix, *Phys. Rev. C* **92** (2015) 011602(R); Y. Tanimura, D. Lacroix and S. Ayik, *Phys. Rev. Lett.* **118** (2017) 152501.
51. N. Schunck and L. M. Robledo, *Rep. Prog. Phys.* **79** (2016) 116301.
52. A. Bulgac, P. Magierski, K. J. Roche and I. Stetcu, *Phys. Rev. Lett.* **116** (2016) 122504.
53. J. Zhao, B.-N. Lu, T. Niksić and D. Vretenar, *Phys. Rev. C* **92** (2015) 064315; J. Zhao, B.-N. Lu, T. Niksić, D. Vretenar and S.-G. Zhou, *Phys. Rev. C* **93** (2016) 044315.
54. P. Goddard, P. Stevenson and A. Rios, *Phys. Rev. C* **92** (2015) 054610; *Phys. Rev. C* **93** (2016) 014620.

55. P. Jachimovwicz, M. Kowal and J. Skalski, *Phys. Rev. C* **95** (2017) 014303.
56. K. Mazurek, P. N. Nadtochy, E. G. Ryabov and G. D. Adeev, *Eur. Phys. J. A* **53** (2017) 79.
57. P. Jachimovwicz, M. Kowal and J. Skalski, *At. Data Nucl. Data Tables* **138** (2021) 101393.
58. M. Caamaño *et al.*, *Phys. Rev. C* **92** (2015) 034606.
59. M. Caamaño and F. Farget, *Phys. Lett. B* **770** (2017) 72.
60. F. A. Ivanyuk, S. Chiba and Y. Aritomo, *Phys. Rev. C* **90** (2014) 054607.
61. R. N. Bernard, N. Pillet, L. M. Robledo and M. Anguiano, *Phys. Rev. C* **101** (2020) 044615.
62. Z. X. Ren, D. Vretenar, L. Niksić, P. W. Zhao, J. Zhao and J. Meng, *Phys. Rev. Lett.* **128** (2022) 172501.
63. E. Pellereau *et al.*, *Phys. Rev. C* **95** (2017) 054603.
64. A. Chatillon *et al.*, *Phys. Rev. C* **99** (2019) 054628.
65. A. Chatillon *et al.*, *Phys. Rev. Lett.* **124** (2020) 202502.
66. J.-F. Martin *et al.*, *Phys. Rev. C* **104** (2021) 044602.
67. A. I. Svirikhin *et al.*, *Phys. Part. Nucl. Lett.* **14** (2017) 571.
68. A. V. Isaev *et al.*, *Eur. Phys. J. A* **58** (2022) 108.
69. H. Paşca, A. V. Andreev, G. G. Adamian, N. V. Antonenko and Y. Kim, *Phys. Rev. C* **93** (2016) 054602.
70. H. Paşca, A. V. Andreev, G. G. Adamian and N. V. Antonenko, *Eur. Phys. J. A* **52** (2016) 369.
71. H. Paşca, A. V. Andreev, G. G. Adamian and N. V. Antonenko, *Phys. Lett. B* **760** (2016) 800.
72. H. Paşca, A. V. Andreev, G. G. Adamian and N. V. Antonenko, *Phys. Rev. C* **94** (2016) 064614.
73. H. Paşca, A. V. Andreev, G. G. Adamian and N. V. Antonenko, *Phys. Rev. C* **97** (2018) 034621.
74. H. Paşca, A. V. Andreev, G. G. Adamian and N. V. Antonenko, *Nucl. Phys. A* **969** (2018) 226.
75. H. Paşca, A. V. Andreev, G. G. Adamian and N. V. Antonenko, *Phys. Rev. C* **99** (2019) 064611
76. H. Paşca, A. V. Andreev, G. G. Adamian and N. V. Antonenko, *Phys. Rev. C* **101** (2020) 064604.
77. H. Paşca, A. V. Andreev, G. G. Adamian and N. V. Antonenko, *Phys. Rev. C* **104** (2021) 014604.
78. H. Pascą, A. V. Andreev, G. G. Adamian and N. V. Antonenko, *Phys. Rev. C* **107** (2023) 024603; Correlations between transition from symmetric to asymmetric fission and average neutron multiplicity and TKE, Unpublished manuscript (2021).
79. J. Maruhn and W. Greiner, *Z. Phys.* **251** (1972) 431.
80. G. Sauer, H. Chandra and U. Mosel, *Nucl. Phys. A* **264** (1976) 221.
81. M. Asghar, F. Caitucolli, B. Leroux, P. Perrin and G. Barreau, *Nucl. Phys. A* **368** (1981) 319.
82. G. G. Adamian *et al.*, *Int. J. Mod. Phys. E* **5** (1996) 191.

83. Sh. A. Kalandarov, G. G. Adamian, N. V. Antonenko and W. Scheid, *Phys. Rev. C* **82** (2010) 044603; **83** (2011) 054611; Sh. A. Kalandarov, G. G. Adamian, N. V. Antonenko, W. Scheid and J. P. Wieleczko, *Phys. Rev. C* **84** (2011) 064601.
84. R. Vandenbosch and J. R. Huizenga, *Nuclei Fission* (Academic Press, New York, 1973).
85. R. Vogt and J. Randrup, *Phys. Rev. C* **84** (2011) 044621.
86. W. Lang, H. G. Clerc, W. Wohlfarth, H. Schrader and K.-H. Schmidt, *Nucl. Phys. A* **345** (1980) 34; G. Siegert, H. Wollnik, J. Grief, R. Decker, G. Fiedler and B. Pfeiffer, *Phys. Rev. C* **14** (1976) 1864; G. Mariolopoulos, Ch. Hamelin, J. Blachot, J. P. Boucqet, B. Brissot, J. Crancon, H. Nifenecker and Ch. Ristori, *Nucl. Phys. A* **361** (1981) 213.
87. T. R. England and B. F. Rider, Evaluation and compilation of fission product yields 1993, LANL Report No. LA-UR-94-3106, Los Alamos National Laboratory (1994).
88. P. Schillebeeckx, C. Wagemans, A. J. Deruytter and R. Barthelemy, *Nucl. Phys. A* **545** (1992) 623.
89. R. Schmidt and H. Henschel, *Nucl. Phys. A* **395** (1983) 15.
90. C. Wagemans, E. Allaert, A. Deruytter, R. Barthélémy and P. Schillebeeckx, *Phys. Rev. C* **30** (1984) 218.
91. K. Nishio, Y. Nakagome, H. Yamamoto and I. Kimura, *Nucl. Phys. A* **632** (1998) 540.
92. K. Nishio, Y. Nakagome, I. Kanno and I. Kimura, *J. Nucl. Sci. Technol.* **32** (1995) 404.
93. C. Tsuchiya, Y. Nakagome, H. Yamana, H. Moriyama, K. Nishio, I. Kanno, K. Shin and I. Kimura, *J. Nucl. Sci. Technol.* **37** (2000) 941.
94. V. F. Apalin, Yu. N. Gritsyuk, I. E. Kutikov, V. I. Lebedev and L. A. Mikaelian, *Nucl. Phys. A* **71** (1965) 553.
95. B. C. Diven, H. C. Martin, R. F. Taschek and J. Terrell, *Phys. Rev.* **101** (1956) 1012.
96. N. E. Holden and M. S. Zucker, A reevaluation of the average prompt neutron emission multiplicity (Nubar) values from fission of uranium and transuranium nuclides, Report No. BNL-NCS-35513, Brookhaven National Laboratory (1985).
97. H. Zang, Z. Liu, S. Ding and S. Liu, *Nucl. Sci. Eng.* **86** (1984) 315.
98. R. Schmidt and H. Henschel, *Nucl. Phys. A* **395** (1983) 29.
99. A. I. Svirikhin, V. N. Dushin, M. L. Chelnokov, V. I. Chepigin, I. N. Izosimov, D. E. Katrasev, O. N. Malyshev, A. Minkova, A. G. Popeko, E. A. Sokol and A. V. Yeremin, *Eur. Phys. J. A* **44** (2010) 393; A. I. Svirikhin, A. V. Andreev, V. N. Dushin, M. L. Chelnokov, V. I. Chepigin, M. Gupta, A. V. Isaev, I. N. Izosimov, D. E. Katrasev, A. N. Kuznetsov, O. N. Malyshev, S. Mullins, A. G. Popeko, E. A. Sokol and A. V. Yeremin, *Eur. Phys. J. A* **48** (2012) 121; A. I. Svirikhin *et al.*, *Phys. Part. Nucl. Lett.* **16** (2019) 768.
100. Yu. A. Lazarev, *At. Energy Rev.* **15** (1977) 75.
101. S. N. Belenkiy, M. D. Skorokhvatov and A. V. Etenko, *At. Energiya* **55** (1983) 97 (in Russian).

102. D. A. Shaughnessy *et al.*, *Phys. Rev. C* **61** (2000) 044609.
103. D. A. Shaughnessy *et al.*, *Phys. Rev. C* **65** (2002) 024612.
104. J. P. Unik *et al.*, Fragment mass and kinetic energy distributions for fissioning systems ranging from mass 230 to 256, in *Physics and Chemistry of Fission*, Vol. II (IAEA, Vienna, 1974), pp. 19–45.
105. C. Romano *et al.*, *Phys. Rev. C* **81** (2010) 014607.
106. D. C. Hoffman *et al.*, *Phys. Rev. C* **21** (1980) 972.
107. J. P. Balagna *et al.*, *Phys. Rev. Lett.* **26** (1971) 145.
108. E. Cheifetz *et al.*, *Phys. Rev. C* **4** (1971) 1913.
109. F. Rejmund *et al.*, Isotopic fission-fragment yields measured at Lohengrin and deduced element yields, Report No. GANIL R 2010-01, http://hal.in2p3.fr/in2p3-00487186/fr/ (2010).
110. G. M. Ter-Akopian *et al.*, *Phys. Rev. C* **55** (1997) 1146.
111. C. E. Bemis *et al.*, *Phys. Rev. C* **15** (1977) 705.
112. K. F. Flynn *et al.*, *Phys. Rev. C* **5** (1972) 1725.
113. K. F. Flynn *et al.*, *Phys. Rev. C* **12** (1975) 1478.
114. R. M. Harbour *et al.*, *Phys. Rev. C* **8** (1973) 1488.
115. W. John *et al.*, *Phys. Rev. Lett.* **27** (1971) 45; R. C. Ragaini *et al.*, *Phys. Rev. C* **9** (1974) 399.
116. R. W. Lougheed, E. K. Hulet, K. J. Moody, J. F. Wild, R. J. Dougan, D. C. Hoffman, C. M. Gannett, R. A. Henderson and D. M. Lee, The spontaneous fission and decay properties of ^{261,262}Lr and the new isotope, ^{262}No, in *Proc. 3rd Chemistry Congr. North America: Nuclear Chemistry and Technology* (1988); R. W. Lougheed, E. K. Hulet, J. F. Wild, K. J. Moody, R. J. Dougan, C. M. Gannett, R. A. Henderson, D. C. Hoffmann and D. M. Lee, Discovery of 5-ms ^{262}No and the study of its spontaneous-fission decay properties, Nuclear Chemistry Division FY-1988 Annual Report No. UCAR-10062-88, Lawrence Livermore National Laboratory (1988).
117. M. R. Lane *et al.*, *Phys. Rev. C* **53** (1996) 2893.
118. J. Benlliure *et al.*, *Nucl. Phys. A* **628** (1998) 458.
119. M. G. Itkis *et al.*, *Sov. J. Nucl. Phys.* **52** (1990) 601; **53** (1991) 757.
120. G. N. Knyazheva *et al.*, *Phys. Rev. C* **75** (2007) 064602.
121. H. Q. Zhang, C. L. Zhang, C. J. Lin, Z. H. Liu, F. Yang, A. K. Nasirov, G. Mandaglio, M. Manganaro and G. Giardina, *Phys. Rev. C* **81** (2010) 034611.
122. K. V. Novikov *et al.*, *Phys. Rev. C* **102** (2020) 044605.
123. D. J. Hinde *et al.*, *Phys. Rev. C* **106** (2022) 064614.
124. A. K. Nasirov, K. Kim, G. Mandaglio, G. Giardina, A. I. Muminov and Y. Kim, *Eur. Phys. J. A* **49** (2013) 147.
125. G. Mandaglio, A. K. Nasirov, A. Anastasi, F. Curciarello, G. Fazio and G. Giardina, *Nucl. Phys. A* **979** (2018) 204.
126. A. K. Nasirov, B. M. Kayumov, G. Mandaglio, G. Giardina, K. Kim and Y. Kim, *Eur. Phys. J. A* **55** (2019) 29.
127. Ch. Schmitt and P. Moller, *Phys. Lett. B* **812** (2021) 136017.
128. M. Verriere, N. Schunck and D. Regnier, *Phys. Rev. C* **103** (2021) 054602.
129. K.-H. Schmidt, B. Jurado, C. Amouroux and C. Schmitt, *Nucl. Data Sheets* **131** (2016) 107.

130. P. Möller and C. Schmitt, *Eur. Phys. J. A* **53** (2017) 7.
131. D. G. Madland and J. R. Nix, *Nucl. Sci. Eng.* **81** (1982) 2113.
132. National Nuclear Data Center, Structure & decay, https://www.nndc.bnl.gov/ (2022).
133. P. Möller, A. J. Sierk, T. Ichikawa, A. Iwamoto, R. Bengtsson, H. Uhrenholt and S. Aberg, *Phys. Rev. C* **79** (2009) 064304.
134. A. J. Sierk, *Phys. Rev. C* **33** (1986) 2039.
135. A. S. Zubov, G. G. Adamian and N. V. Antonenko, *Phys. Part. Nucl.* **40** (2009) 847.
136. M. B. Chadwick *et al.*, *Nucl. Data Sheets* **112** (2011) 2887.
137. J. N. Wilson *et al.*, *Phys. Rev. Lett.* **118** (2017) 222501.
138. M. A. Kellett, O. Besillon and R. W. Mills, The JEFF-3.1/-3.1.1 radioactive decay data and fission yields sub-libraries, OECD Nuclear Energy Agency Report No. 6287 (2009).
139. H. Paşca, A. V. Andreev, G. G. Adamian, N. V. Antonenko and D. Lacroix, *Phys. Rev. C* **98** (2018) 014624.
140. J. R. Nix and W. J. Swiatecki, *Nucl. Phys. A* **71** (1965) 1.
141. J. O. Rasmussen, W. Nörenberg and H. J. Mang, *Nucl. Phys. A* **136** (1969) 465.
142. M. Zielinska-Pfabe and K. Dietrich, *Phys. Lett. B* **49** (1974) 123.
143. G. S. Popeko *et al.*, New data on the fragment angular momenta obtained for the ^{252}Cf(s.f.) with Gammasphere, in *Proc. Int. Conf. Fission and Properties of Neutron-Rich Nuclei*, eds. J. H. Hamilton and A. V. Ramayya (World Scientific, Singapore, 1998), p. 645.
144. S. Misicu, A. Sandulescu, G. M. Ter-Akopian and W. Greiner, *Phys. Rev. C* **60** (1999) 034613.
145. T. M. Shneidman, G. G. Adamian, N. V. Antonenko, S. P. Ivanova, R. V. Jolos and W. Scheid, *Phys. Rev. C* **65** (2002) 064302.
146. H. Pasca, Sh. A. Kalandarov, G. G. Adamian and N. V. Antonenko, *Phys. Rev. C* **96** (2017) 044611; *Nucl. Phys. A* **980** (2018) 143.
147. W. U. Schroeder and J. R. Huizenga, Damped nuclear reactions, in *Treatise on Heavy-Ion Science: Vol. 2: Fusion and Quasi-Fusion Phenomena*, ed. D. A. Bromley (Plenum Press, New York, 1984), pp. 115–726.
148. O. Litaize, L. Thulliez, O. Serot, A. Chebboubi and P. Tamagno, *EPJ Web Conf.* **169** (2018) 00012.

Generator coordinate method for nuclear octupole excitations: Status and perspectives*

E. F. Zhou and J. M. Yao †

School of Physics and Astronomy,
Sun Yat-sen University, Zhuhai 519082, P. R. China
†*yaojm8@sysu.edu.cn*

Strong octupole correlations have been observed in the low-lying states of atomic nuclei across various mass regions. In this paper, we provide an overview of Beyond Mean-Field (BMF) studies of nuclear octupole collective motions with Generator Coordinate Method (GCM) in combination with quantum-number projections that are implemented to restore the broken symmetries in nuclear mean-field states. We highlight recent developments within this framework and their applications to excitation spectra and electromagnetic transition rates in octupole-shaped nuclei and hypernuclei. We discuss the novel phenomena of nucleon clustering in light nuclei. Additionally, we explore the phase transition from octupole vibrations to rotational motions as spin increases in heavy nuclei. Lastly, we examine the status and future prospects of studies on octupole deformation effects in nuclear Schiff moments. These studies, along with the upper limits of atomic Electric Dipole Moment (EDM), impose stringent constraints on beyond-standard-model time-reversal-violating nucleon–nucleon interactions.

Keywords: Generator coordinate method; octupole deformation; Schiff moment.

1. Introduction

An atomic nucleus is a self-bound quantum many-body system composed of several to hundreds of nucleons. These nucleons distribute within a range of a few femtometers, approximately given by the formula $R \simeq 1.2A^{1/3}$ fm, where A is the mass number of the nucleus. With a specific combination of neutrons and protons (N, Z), an atomic nucleus can exhibit different equilibrium shapes characterized by nonzero values of dimensionless deformation

*To cite this article, please refer to its earlier version published in the *International Journal of Modern Physics E*, Volume 32(10), (49 pages), 2340011 (2023), DOI: 10.1142/S0218301323400116.

†Corresponding author.

parameters $\beta_{\lambda\mu}$. In mean-field approaches, these parameters are defined by the expectation values of multipole moment operators, represented as

$$\beta_{\lambda\mu} = \frac{4\pi}{3AR^{\lambda}} \langle \Phi(\mathbf{q}) | \hat{Q}_{\lambda\mu} | \Phi(\mathbf{q}) \rangle, \tag{1}$$

where $\hat{Q}_{\lambda\mu} = r^{\lambda} Y_{\lambda\mu}$ with $\mu = -\lambda, \ldots, \lambda$ and $Y_{\lambda\mu}$ stands for the spherical harmonic function of rank λ. The symbol $|\Phi(\mathbf{q})\rangle$ represents the wave function of a nuclear state in the intrinsic frame.

Most atomic nuclei exhibit either a spherical equilibrium shape, where $\beta_{\lambda\mu} = 0$, or intrinsic ellipsoidal shapes in their ground states. These ellipsoidal shapes can be prolate, indicated by $\beta_{20} > 0$, or oblate, indicated by $\beta_{20} < 0$. In rare instances, some atomic nuclei spontaneously break reflection symmetry by having a nonzero octupole deformation parameter, $\beta_{3\mu} \neq 0$. The presence of intrinsic octupole deformation in the nuclear ground state is inferred from the observation of low-lying negative-parity states or parity doublets connected with strong electric octupole ($E3$) transitions.[1,2]

The emergence of deformed equilibrium shapes in atomic nuclei is attributed to the strong residual nucleon–nucleon interaction between valence nucleons. This interaction can be parametrized as separable multipole-multipole interactions. In nuclei with neutron or proton numbers near 56, 88, 134 and 194, which are only slightly greater than the magic numbers 50, 82, 128 and 184, respectively, a strong octupole-octupole coupling occurs between pairs of orbitals with $\Delta\ell = \Delta j = 3$. These orbitals are located around the Fermi surface and include pairs such as $(d_{5/2}, h_{11/2})$, $(f_{7/2}, i_{13/2})$, $(g_{9/2}, j_{15/2})$ and $(h_{11/2}, k_{17/2})$. Notably, the isoscalar neutron-proton part[3] plays a significant role in driving atomic nuclei toward reflection-asymmetric octupole shapes in their ground states. Pairing correlation between nucleons also influences the occupation of nucleons on the orbitals around the Fermi surface, and thus, it has a significant impact on the equilibrium shapes. Whether an atomic nucleus exhibits octupole shapes in its ground state depends on the competition between the underlying shell structure and pairing correlation. Both aspects can be effectively addressed in self-consistent mean-field approaches.[4,5] Recent studies, based on Relativistic Hartree Bogoliubov (RHB) approaches,[6] Skyrme Hartree Fock (HF) + Bardeen-Cooper-Schrieffer (BCS)[7] and Skyrme Hartree Fock Bogoliubov (HFB) approaches,[8] have systematically explored atomic nuclei with nonzero octupole deformation β_{30} in their ground states. These studies reveal that the majority of octupole-deformed nuclei are indeed found near the intersection of neutron numbers $N \simeq$

88, 134 and 194 and proton numbers $Z \simeq 56$ and 88,[8] consistent with the finding with the macroscopic-microscopic nuclear model.[9]

Octupole collective motions in atomic nuclei manifest as low-lying octupole vibrational or rotational states. This phenomenon applies even to spherical nuclei with closed-shell configurations for neutrons and/or protons, such as ^{208}Pb,[10–13] where the first 3^- state is interpreted as an one-octupole-phonon state. This octupole-phonon state has been explored systematically in stable even–even nuclei with the methods of Generator Coordinate Method (GCM) + Parity Projection (PP)[12] and Quasiparticle Random Phase Approximation (QRPA).[14]

As the nucleon number departs from closed shells, atomic nuclei become deformed, leading to the observation of rotation bands built upon octupole vibrational states. The energy spectrum in this context can be effectively described by a phenomenological octupole collective Hamiltonian. For illustration purpose, we neglect the coupling of rotations and octupole vibrations. Under this approximation, the excitation energies of low-lying states in the collective model are given by[15]

$$E(J,K) = (n_0 + 1/2)\hbar\Omega_{30} + \sum_{\mu=1}^{3}(2n_\mu + 1 + |K_\mu|)\hbar\Omega_{3\mu} + \frac{\varepsilon}{2}[J(J+1) - K^2]. \tag{2}$$

Here, n_μ with $\mu = 0, 1, 2, 3$ correspond to the number of different types of octupole phonons, $K_\mu = 0, \pm1, \pm2, \ldots$ and K is the projection of the total angular momentum J along the intrinsic symmetric axis. The octupole vibrational states of type μ with the same value of the major quantum number N_μ defined as $N_\mu \equiv 2n_\mu + 1 + |K_\mu|$ are degenerate. The corresponding energy spectrum with multiple negative-parity bands is schematically depicted in Fig. 1. Notably, this energy spectrum has been observed in some isotopes, such as ^{152}Sm, where octupole bands with $K^\pi = 0^-$, 1^- and 2^- have been observed up to relatively high spin in the $(\alpha, 2n)$ data.

With an increase in the octupole-octupole coupling strength, the excitation energy of the octupole vibrational state decreases and eventually becomes degenerate with the ground state. Beyond a critical value, the nuclear system becomes unstable against octupole vibration, leading to the development of a static octupole-deformed shape in the ground state. As a consequence of this transition, the low-lying states become dominated by the rotational excitations of octupole-deformed shapes. This transition is illustrated in Fig. 2, which shows the shift from an octupole vibrator to an

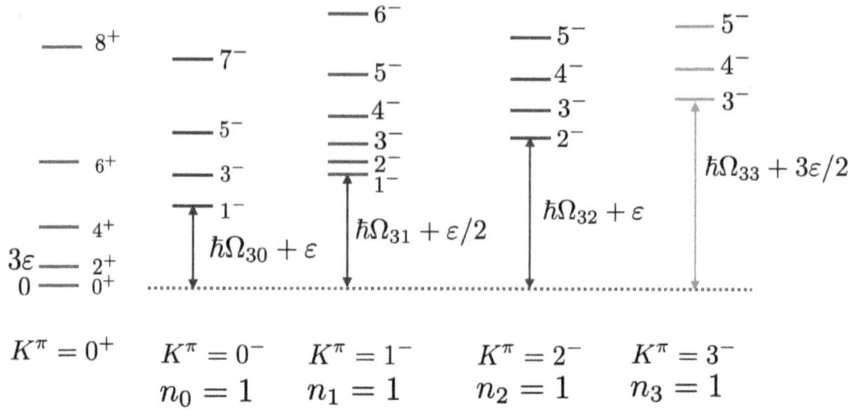

Fig. 1. Rotational bands in a collective model for an even–even deformed nucleus with different octupole fluctuations. The rotational bands are built on top of one-octupole-phonon states with various values of $K(=0,1,2,3)$. The labels for the bands include K^π (where K represents the angular momentum projection and π represents parity) and a nonzero oscillator number n_μ, which represents different types of octupole vibrations. This figure has been adapted from Ref. 15.

octupole rotor. In the rigid rotor with an octupole deformed shape, $\hbar\Omega_{30}=0$. Consequently, the positive and negative-parity states merge into one rotational band. To characterize this phase transition, the ratio of excitation energies, denoted as $R_{J/2}$, provides a signature. It is calculated as the ratio of the excitation energy for a specific angular momentum state J to that of the 2^+ state, i.e., $R_{J/2}=E_x(J)/E_x(2)$. As the transition progresses from an octupole vibrator to a rigid rotor, the staggering amplitude of the energy ratios R_{J2} decreases, as depicted in Fig. 2(c).

In many cases, the barrier height separating the two degenerate minimal states $|\Phi(\beta_{20},\beta_{30})\rangle$ and $|\Phi(\beta_{20},-\beta_{30})\rangle=\hat{P}|\Phi(\beta_{20},\beta_{30})\rangle$ (as illustrated in Fig. 2(c)) may not be high enough to eliminate the tunneling effect between them. In other words, the overlap $\langle\Phi(\beta_{20},\beta_{30})|\hat{P}|\Phi(\beta_{20},\beta_{30})\rangle$ of the two intrinsic wave functions with positive and negative values of β_{30}, connected by the space-reversal operator $\hat{P}$ is nonzero. This quantum tunneling effect leads to an energy shift between the positive and negative-parity states, resulting in staggering in the energy ratio. This phenomenon is observed in the low-spin states of all even–even octupole-deformed nuclei. It implies that the dynamical effect tends to obscure the transition pattern from an octupole vibrator to a rigid octupole rotor in nuclear ground states as a function of nucleon numbers. Strictly speaking, the so-called stable or permanent octupole-deformed shapes with $\langle\Phi|\hat{P}|\Phi\rangle=0$ are rarely realized in the ground

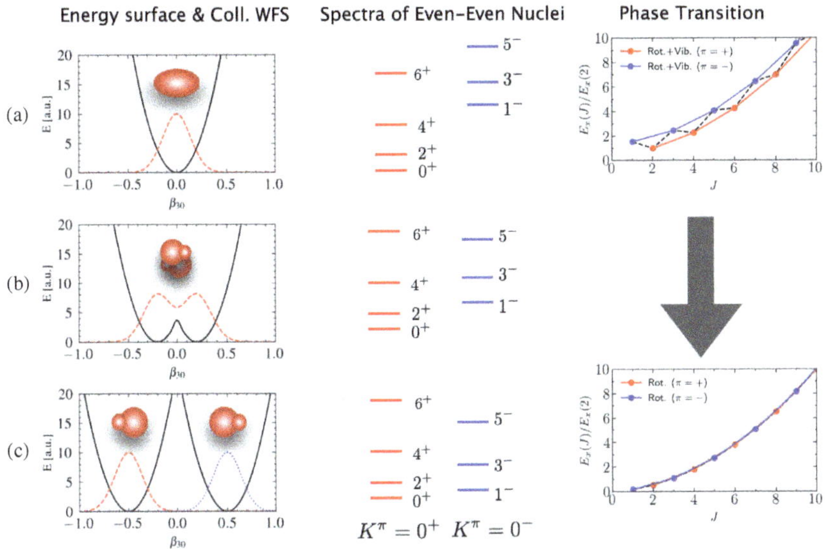

Fig. 2. The phase transition from an octupole vibrator to an octupole rotor in axially deformed even–even nuclei. (a) The octupole deformation β_{30} of the equilibrium shape in the ground state is zero. (b) The nonzero octupole deformation $|\beta_{30}|$ is developed in the ground state, but with a strong quantum tunneling effect. (c) The static octupole deformation $|\beta_{30}|$ is formed without tunneling between the intrinsic states with positive and negative values of β_{30}, resulting in a rigid octupole rotor. The barrier height separating the degenerate minima is set to infinity, resulting in an energy spectra characteristic of a rigid rotor. The corresponding energy spectra of $K^\pi = 0^+, 0^-$ and the excitation energy ratio $R_{J/2}$ as a function of the angular momentum J ($\hbar$) are also provided.

states of realistic nuclei. Even though strong $E3$ transitions are observed in their low-lying states, such as in ^{144,146}Ba[16–18] and ^{222,224,226}Ra,[19,20] which are claimed to have permanent octupole deformations. However, there is substantial evidence indicating that octupole shapes in heavy nuclei become stabilized with increasing rotation, leading to the merging of positive and negative-parity states into one rotation band.[21–23] This phenomenon highlights the intricate interplay between quantum tunneling effects and rotational motion in nuclear systems. It is worth mentioning that the onset of octupole deformation in atomic nuclei may also generate other exotic collective motions, such as multiple chiral doublet bands.[24,25]

Except for the nuclear models based on collective Hamiltonians[25–29] and IBM models,[30,31] the GCM, when combined with symmetry restoration techniques,[32,33] provides a powerful microscopic tool for studying nuclear collective motions associated with exotic shapes. In the context of a general mean-field wave function $|\Phi(\mathbf{q})\rangle$ with nonzero octupole deformation, both rotational symmetry and reflection symmetry are broken. Consequently, a nuclear state described by the mean-field wave function $|\Phi(\mathbf{q})\rangle$ does not possess a definite spin-parity J^π. Furthermore, if pairing correlation is considered by the introduction of quasiparticles in the mean-field state $|\Phi(\mathbf{q})\rangle$, nucleon numbers (N, Z) are not conserved. The broken symmetries of these mean-field wave functions with nonzero intrinsic deformations are restored using projections and finally mixed in the GCM. This Symmetry-Projected Generator Coordinate Method (SPGCM) has been extensively and successfully applied to describe nuclear low-lying states associated with octupole deformation. In the early studies, PP was implemented into GCM to restore the parities of nuclear states from the mean field calculations with the BKN force,[34] Skyrme[35–38] and Gogny forces.[39,40] In the recent decade, GCM + PP + Particle-Number Projection (PNP) + Angular-Momentum Projection (AMP) have been implemented into Covariant Density Functional Theory (CDFT)[22,41–43] and the HFB with the Gogny force.[44] Recently, PP + AMP has been implemented into RHB for nuclear states with exotic octupole shapes.[45] Within these SPGCM frameworks, the picture of rotation-induced phase transition from an octupole vibrator to an octupole-deformed rotor has been illustrated in ^{224}Ra[22] and ^{144}Ba.[23]

It is worth mentioning that recent theoretical studies on heavy-ion collisions with a hydrodynamic model have shown that the elliptic flow and triangular flow are sensitive to the nuclear quadrupole and octupole deformations of the colliding nuclei.[46–48] This indicates that flow measurements in heavy-ion collisions, especially when using isobaric systems, provide a complementary method for studying octupole collectivity in nuclear ground states.[49]

The emergence of strong octupole correlations in atomic nuclei may also have a significant impact on the studies of fundamental interactions and symmetries at nuclear level. The SPGCM study based on the CDFT has demonstrated that the effect of dynamical octupole correlation may quench the nuclear matrix element of neutrinoless double-beta decay in ^{150}Nd by about 7%.[50] See Ref. 51 for the review on the Beyond-Mean-Field (BMF) studies of neutrinoless double-beta decay. The nuclei with octupole deformations are also of great interest for the search of nonzero permanent atomic Electric Dipole Moment (EDM). The current upper limit provides a valuable constraint on the strengths of time-reversal (T) symmetry-violating interactions and serves as a test for beyond-standard models of Charge-Parity (CP) violation. The atomic EDM is proportional to the nuclear Schiff moment, which can be enhanced by several orders of magnitude due to octupole deformation.[52,53] Consequently, the sensitivity of atomic EDM experiments can be significantly increased when the corresponding atomic nucleus exhibits substantial octupole deformation in its ground state.

In this paper, we provide an overview of recent GCM studies on nuclear octupole collective motions. We address the role of octupole correlation in the formation of cluster structures in light nuclei, the staggering behavior observed in the low-spin parity-doublet states of medium-mass or heavy nuclei around ^{144}Ba and ^{224}Ra, and the enhancement of nuclear Schiff moments in odd-mass nuclei. Due to the page limit, lots of exciting developments[54–57] in the GCM will not be covered. For a general review of GCM studies on nuclear collective motions, readers can refer to Refs. 5, 58, 59 and 173.

The review is arranged as follows. In Sec. 2, we present the basic formulas for the SPGCM. For the sake of self-containment, we first introduce the general formulas of HFB approaches, which are usually employed to generate reference states for the SPGCM calculations. Afterward, we present the formulas for energy spectrum and electromagnetic multiple transition strengths. In Sec. 3, we highlight the SPGCM studies on nuclear collective motions. In Sec. 4, we introduce the status and perspectives of computing nuclear Schiff moments. A summary of the review is provided in Sec. 5.

2. The SPGCM

2.1. *The reference states with octupole deformations*

2.1.1. *The Hartree–Fock–Bogoliubov (HFB) state*

Both long-range collective correlations associated with deformations and short-range correlations associated with nucleon pairing can be simultaneously described with the HFB approach.[4] The nuclear wave function from the HFB approach is an approximation of the exact nuclear wave function in the intrinsic frame, known as the HFB wave function. This wave function is a general product of wave functions consisting of independently moving quasiparticles

$$|\Phi(\boldsymbol{q})\rangle = \prod_{k=1}^{M} \beta_k(\boldsymbol{q})|0\rangle, \tag{3}$$

where $|0\rangle$ denotes the particle vacuum. It is important to note that the state described by the HFB wave function is a quasiparticle vacuum, i.e., $\beta_k(\boldsymbol{q})|\Phi(\boldsymbol{q})\rangle = 0$. This state is referred to as the HFB state. The quasiparticle operators $\beta_k(\boldsymbol{q}), \beta_k^\dagger(\boldsymbol{q})$ are defined as linear combinations of the particle creation and annihilation operators $(c_p^\dagger, c_p)$,

$$\begin{pmatrix} \beta(\boldsymbol{q}) \\ \beta^\dagger(\boldsymbol{q}) \end{pmatrix} = \mathcal{W}^\dagger(\boldsymbol{q}) \begin{pmatrix} c \\ c^\dagger \end{pmatrix} = \begin{pmatrix} U^\dagger(\boldsymbol{q}) & V^\dagger(\boldsymbol{q}) \\ V^T(\boldsymbol{q}) & U^T(\boldsymbol{q}) \end{pmatrix} \begin{pmatrix} c \\ c^\dagger \end{pmatrix}, \tag{4}$$

where $\boldsymbol{q}$ labels different HFB states. The unitary transformation (4) is known as the Bogoliubov transformation with the $\mathcal{W}(\boldsymbol{q})$ being a $2M \times 2M$ matrix and it preserves the anticommutator relations. The unitarity requires that the matrices (U, V) satisfy the following relations:

$$U^\dagger U + V^\dagger V = 1, \quad UU^\dagger + V^* V^{\mathrm{T}} = 1, \tag{5}$$

$$U^{\mathrm{T}} V + V^{\mathrm{T}} U = 0, \quad UV^\dagger + V^* U^{\mathrm{T}} = 0. \tag{6}$$

The HFB wave function in (3) is defined by the Bogoliubov transformation coefficients (U, V), which are determined by the variational principle. According to Thouless' theorem,[60] any general product wave function $|\Phi\rangle$ which is not orthogonal to another product wave function $|\Phi^{(0)}\rangle$ can be expressed as

$$|\Phi(\mathcal{Z})\rangle = \mathcal{N} \exp\left(\sum_{k<k'} \mathcal{Z}_{kk'} \beta_k^\dagger \beta_{k'}^\dagger \right) |\Phi^{(0)}\rangle, \tag{7}$$

where the normalization constant $\mathcal{N} = \langle \Phi | \Phi^{(0)} \rangle$ and $\mathcal{Z}$ is a skew symmetric matrix. The HFB reference state $|\Phi^{(0)}\rangle$ is the vacuum of quasiparticle operator $\beta^{(0)}$, connecting to particle operators $(c_p^\dagger, c_p)$ with the Bogoliubov transformation coefficients $(U^{(0)}, V^{(0)})$.[a]

With the Thouless theorem (7), one can express the energy of HFB wave function $|\Phi\rangle$ in terms of the unknown matrix $\mathcal{Z}$,

$$E[\mathcal{Z}] = \frac{\langle \Phi(\mathcal{Z}) | \hat{H} | \Phi(\mathcal{Z}) \rangle}{\langle \Phi(\mathcal{Z}) | \Phi(\mathcal{Z}) \rangle}, \tag{8}$$

where $\hat{H}$ is the Hamiltonian operator for nuclear systems. The energy $E[\mathcal{Z}]$ is a function of $\mathcal{Z}$ to be determined in variational iteration methods.

Supposing the HFB wave function at nth step is $|\Phi^{(n)}\rangle$, the wave function $|\Phi^{(n+1)}\rangle$ at the $(n+1)$th step is determined by

$$|\Phi^{(n+1)}\rangle = \mathcal{N} \exp\left(\sum_{k<k'} \mathcal{Z}_{kk'}^{(n)} \beta_k^{(n)\dagger} \beta_{k'}^{(n)\dagger} \right) |\Phi^{(n)}\rangle. \tag{9}$$

[a] The proof of this theorem for HFB wave functions is provided in the textbook.[4]

The parameter matrix $G^{(n)}$ is defined as the gradient of the energy multiplied with a step size parameter η,

$$G^{(n)}_{kk'} \equiv -\eta \frac{\partial}{\partial \mathcal{Z}^*_{kk'}} E[\mathcal{Z}] = -\eta H^{20}_{kk'}, \quad G^{(n)}_{kk'} = -G^{(n)}_{k'k}. \tag{10}$$

To ensure the commutation relation for the new quasiparticle operator $\beta_k^{(n+1)}$ and guarantee the orthogonality relations for the new HFB wave function $(U^{(n+1)}, V^{(n+1)})$, one finds the following relations[4,61]:

$$\begin{aligned}\begin{pmatrix}\beta^{(n+1)\dagger}\\ \beta^{(n+1)}\end{pmatrix} &= \begin{pmatrix}[L^{-1}]^* & 0\\ 0 & L^{-1}\end{pmatrix}\begin{pmatrix}I & G^{(n)\dagger}\\ G^{(n)T} & I\end{pmatrix}\begin{pmatrix}\beta^{(n)\dagger}\\ \beta^{(n)}\end{pmatrix}\\ &= \begin{pmatrix}[L^{-1}]^* & [L^{-1}]^* G^{(n)\dagger}\\ L^{-1}G^{(n)T} & L^{-1}\end{pmatrix}\begin{pmatrix}\beta^{(n)\dagger}\\ \beta^{(n)}\end{pmatrix}\end{aligned} \tag{11}$$

and the coefficients are updated as follows:

$$U^{(n+1)} = (U^{(n)} + V^{(n)*}G^{(n)*})[L^{-1}]^\dagger, \tag{12a}$$

$$V^{(n+1)} = (V^{(n)} + U^{(n)*}G^{(n)*})[L^{-1}]^\dagger, \tag{12b}$$

where the L is determined by the Choleski decomposition

$$1 + G^{(n)T}G^{(n)*} = LL^\dagger \tag{13}$$

and $G^{(n)}$ by the energy gradient (10).

With the following two Bogoliubov transformations which relate the quasiparticle operators at the nth and $(n+1)$th iterations to the single-particle operators:

$$\begin{pmatrix}\beta^{(n)\dagger}\\ \beta^{(n)}\end{pmatrix} = \begin{pmatrix}U^{(n)T} & V^{(n)T}\\ V^{(n)\dagger} & U^{(n)\dagger}\end{pmatrix}\begin{pmatrix}c^\dagger\\ c\end{pmatrix}, \tag{14a}$$

$$\begin{pmatrix}\beta^{(n+1)\dagger}\\ \beta^{(n+1)}\end{pmatrix} = \begin{pmatrix}U^{(n+1)T} & V^{(n+1)T}\\ V^{(n+1)\dagger} & U^{(n+1)\dagger}\end{pmatrix}\begin{pmatrix}c^\dagger\\ c\end{pmatrix} \tag{14b}$$

and the unitary transformation between the two sets of quasiparticle operators,[4]

$$\begin{pmatrix}\beta^{(n+1)\dagger}\\ \beta^{(n+1)}\end{pmatrix} = \begin{pmatrix}\mathcal{U}^T & \mathcal{V}^T\\ \mathcal{V}^\dagger & \mathcal{U}^\dagger\end{pmatrix}\begin{pmatrix}\beta^{(n)\dagger}\\ \beta^{(n)}\end{pmatrix}, \tag{15}$$

one finds the expressions for the coefficients of the above unitary transformation,

$$\mathcal{U} = U^{(n)\dagger}U^{(n+1)} + V^{(n)\dagger}V^{(n+1)}, \tag{16a}$$

$$\mathcal{V} = V^{(n)T}U^{(n+1)} + U^{(n)T}V^{(n+1)}. \tag{16b}$$

Combining (11) and (15), one finds the expression for the inverse of the matrix L,

$$L^{-1} = \mathcal{U}^\dagger = U^{(n+1)\dagger}U^{(n)} + V^{(n+1)\dagger}V^{(n)}. \tag{17}$$

Therefore, the HFB wave function is updated iteratively based on Eqs. (10), (12) and (17). As the matrices G and H^{20} approach zero, the iteration converges with $|\Phi^{(n+1)}\rangle \to |\Phi^{(n)}\rangle$.

It is worth pointing out that the speed of convergence is sensitive to the choice of the step size, η. If η is too large, one can wildly oscillate around energy minima. If η is too small, then the minimization procedure becomes extremely slow. The proper size is determined by the inverse of the Hessian matrix, which is defined as the second derivative of the energy expectation value. Since the calculation of the Hessian matrix is usually time-consuming, the step size η is thus chosen as an empirical value around 10^{-3}. Meanwhile, one can improve the iteration by adding a momentum term, which can suppress oscillations in high-curvature directions. This method is also called the heavy-ball algorithm.[62,63] The momentum term becomes especially important in situations where the landscape is shallow and flat in some directions and narrow and steep in others. Moreover, one can also implement the conjugate gradient descent (GD) method. This method has been found to reduce the number of iterations by a factor of three to four.[61]

In order to generate a set of mean-field wave functions $|\Phi(\mathbf{q})\rangle$ with correct average nucleon numbers N_τ for neutrons and protons, and different intrinsic deformations $\beta_{\lambda\mu}$, such as quadrupole and octupole deformations, one needs to add the following constraints to the energy expectation value (8) in the minimization of energy:

$$\langle\Phi(\mathbf{q})|\hat{N}_\tau|\Phi(\mathbf{q})\rangle = N_\tau, \quad \langle\Phi(\mathbf{q})|\hat{Q}_{\lambda\mu}|\Phi(\mathbf{q})\rangle = q_{\lambda\mu}, \tag{18}$$

where $\lambda = 1, 2, 3$ and $\mu = 0$ (axial symmetry) with the constraint $q_{10} = 0$ to ensure the location of the center-of-mass coordinate at the origin. The total Routhian of the nucleus to be minimized becomes

$$\tilde{E}[\Phi] = \langle\Phi(\mathbf{q})|\hat{H} - \sum_{\tau=n,p} \lambda_\tau \hat{N}_\tau|\Phi(\mathbf{q})\rangle + \frac{1}{2}\sum_{\lambda=1,2,3} C_\lambda(\langle\Phi(\mathbf{q})|\hat{Q}_{\lambda 0}|\Phi(\mathbf{q})\rangle - q_{\lambda 0})^2, \tag{19}$$

where the Lagrange multipliers λ_τ is called chemical potential introduced in order to fix the average particle number, C_λ for deformations. The variational problem with the constraints (18) can be solved similarly with the gradient descent (GD) method, provided that the minimization is carried out within the hypersurface given by the constraints. For more details, see, for instance, Refs. 4, 63 and 64.

Alternatively, the HFB wave function can be solved with the iterative diagonalization method,[4] in which the following eigenvalue problem is solved:

$$\sum_q \begin{pmatrix} \tilde{h} & \Delta \\ -\Delta^* & -\tilde{h}^* \end{pmatrix}_{pq} \begin{pmatrix} U_{qk} \\ V_{qk} \end{pmatrix} = E_k \begin{pmatrix} U_{pk} \\ V_{pk} \end{pmatrix}, \tag{20}$$

where q runs over all single-particle basis and E_k is the so-called quasiparticle energy. The eigenvectors (U_{qk}, V_{qk}) are the coefficients of the Bogoliubov transformation. The matrix element of the Routhian $\tilde{h}$ for single-particle motions reads,

$$\tilde{h}^{(\tau)}_{pq}[\rho] = h_{pq}[\rho] - \lambda_\tau \delta_{pq} - \sum_{\lambda=1,2,3} C_\lambda (\mathrm{Tr}[\rho Q_{\lambda 0}] - q_{\lambda 0})(Q_{\lambda 0})_{pq}, \tag{21}$$

The matrix element of single-particle hamiltonian reads

$$h_{pq}[\rho] = t_{pq} + \sum_{rs} V^{pr}_{qs} \rho^{r}_{s}, \tag{22}$$

where $V^{pr}_{qs} \equiv \langle pr|V(1,2)|qs\rangle$ is the matrix element of two-body interaction. In Eq. (20), the matrix element of particle-particle (pairing) field is given by $\Delta^{pq} = \frac{1}{2}\sum_{rs} V^{pq}_{rs} \kappa_{rs}$. The one-body *normal* and *abnormal* density matrices ρ and κ (also known as pairing tensor) of the HFB state $|\Phi(\boldsymbol{q})\rangle$ are defined as

$$\rho^{p}_{q}(\boldsymbol{q};\boldsymbol{q}) \equiv \langle\Phi(\boldsymbol{q})|c^\dagger_p c_q|\Phi(\boldsymbol{q})\rangle = (V^* V^T)_{qp}, \tag{23a}$$

$$\kappa^{pq}(\boldsymbol{q};\boldsymbol{q}) \equiv \langle\Phi(\boldsymbol{q})|c^\dagger_p c^\dagger_q|\Phi(\boldsymbol{q})\rangle^* = (U^* V^T)^*_{pq}, \tag{23b}$$

$$\kappa_{rs}(\boldsymbol{q};\boldsymbol{q}) \equiv \langle\Phi(\boldsymbol{q})|c_s c_r|\Phi(\boldsymbol{q})\rangle = (V^* U^T)_{rs}. \tag{23c}$$

The solution to (20) defines the Bogoliubov transformation matrix $\mathcal{W}$ which diagonalizes the following matrix:

$$\begin{pmatrix} H^{11} & H^{20} \\ -H^{20*} & -H^{11*} \end{pmatrix} = \begin{pmatrix} U & V^* \\ V & U^* \end{pmatrix}^\dagger \begin{pmatrix} \tilde{h} & \Delta \\ -\Delta^* & -\tilde{h}^* \end{pmatrix} \begin{pmatrix} U & V^* \\ V & U^* \end{pmatrix} = \begin{pmatrix} E_k & 0 \\ 0 & -E_k \end{pmatrix}. \tag{24}$$

The convergence properties of the iterative diagonalization method depend sensitively on the types of constraints. In many cases, one cannot find convergence at all, in particular in regions of level crossings, where the iteration switches back and forth between different occupations. In contrast, the GD method does not show this kind of numerical instability.[64]

2.1.2. *The HFB state in canonical basis or BCS state*

According to the Bloch–Messiah theorem,[65] the Bogoliubov transformation matrix $\mathcal{W}(\boldsymbol{q})$ in (4) can always be decomposed into the product of three special matrices[4]

$$\mathcal{W} = \begin{pmatrix} D & 0 \\ 0 & D^* \end{pmatrix} \begin{pmatrix} \bar{U} & \bar{V} \\ \bar{V} & \bar{U} \end{pmatrix} \begin{pmatrix} C & 0 \\ 0 & C^* \end{pmatrix}. \tag{25}$$

In comparison with Eq. (4), one finds the relations $U = D\bar{U}C$ and $V = D^*\bar{V}C$, where D and C are $M \times M$ unitary matrices that transform particle operators and quasiparticle operators, respectively. The matrix D_{pl} transforms from the original basis p to the so-called canonical basis consisting of pairs $(l, \bar{l})$ of canonically conjugate states. The second transformation given by the real matrices $\bar{U}$ and $\bar{V}$,

$$\bar{U}_{ll'} = \begin{pmatrix} u_l & 0 \\ 0 & u_l \end{pmatrix} \delta_{ll'}, \quad \bar{V}_{ll'} = \begin{pmatrix} 0 & v_l \\ -v_l & 0 \end{pmatrix} \delta_{\bar{l}\bar{l}'} \tag{26}$$

is the Bogoliubov–Valatin transformation connecting the quasiparticle operators $(\alpha_l^\dagger, \alpha_{\bar{l}}^\dagger)$ in canonical basis with the single-particle operators $(a_l^\dagger, a_l)$ in HF basis

$$\alpha_l^\dagger = u_l a_l^\dagger - v_l a_{\bar{l}}, \tag{27a}$$

$$\alpha_{\bar{l}}^\dagger = u_l a_{\bar{l}}^\dagger + v_l a_l, \tag{27b}$$

where the index $l = 1, 2, \ldots, M/2$ and $u_l^2 + v_l^2 = 1$. The $v_l^2 \in [0, 1]$ is interpreted as the occupation probability of the lth single-particle state in canonical basis, in which the one-body density ρ is diagonal. The third matrix C in Eq. (25) describes a transformation in quasiparticle space from the operators α to the operators β in Eq. (4).

In the canonical basis or BCS wave function, the pairing field matrix element $\Delta_{k\bar{l}} = \Delta_k \delta_{kl}$ becomes diagonal, and the occupation probabilities v_k^2 of neutron $(\tau = n)$ or proton $(\tau = p)$ states are determined by the

$$v_k^2 = \frac{1}{2}\left(1 - \frac{\epsilon_k - \lambda_\tau}{\sqrt{(\epsilon_k - \lambda_\tau)^2 + \Delta_k^2}}\right), \tag{28}$$

where ϵ_k is the single-particle energy from the HF solution, the Fermi energy λ_τ determined by the condition $2\sum_{k>0}^{\tau} v_k^2 = N_\tau$, and the pairing gap Δ_l by the employed pairing force[4]

$$\Delta_k = -\frac{1}{2}\sum_l V_{l\bar{l}}^{k\bar{k}} \frac{\Delta_l}{\sqrt{(\epsilon_l - \lambda_\tau)^2 + \Delta_l^2}}. \tag{29}$$

2.1.3. *The HFB state with odd-number parity*

In the HFB approaches, the lowest-energy state of an odd-mass nucleus is described as a one-quasiparticle state,

$$|\Phi_{k_0}(\mathbf{q})\rangle = \beta_{k_0}^\dagger |\Phi(\mathbf{q})\rangle. \tag{30}$$

One can prove that this state is the vacuum of another type of quasiparticle operators $\gamma_{j=1,2,\ldots,M}$ defined as

$$\gamma_j = \begin{cases} \beta_j^\dagger, & j = k_0, \\ \beta_j, & \text{others} \end{cases} \tag{31}$$

and

$$\gamma_j |\Phi_{k_0}(\mathbf{q})\rangle = 0. \tag{32}$$

Since the quasiparticle operator $\beta_{k_0}^\dagger$ is a linear combination of single-particle creation and annihilation operators, it changes the number parity $(-1)^{N_\tau}$ of the state. Thus, the number parities of the wave functions $|\Phi(\mathbf{q})\rangle$ and $|\Phi_{k_0}(\mathbf{q})\rangle$ are different.

In the canonical basis, the wave function of an odd-mass nucleus can be constructed as

$$|\Phi_{k_0}(\mathbf{q})\rangle = \alpha_{k_0}^\dagger \prod_{k>0} (u_k + v_k a_k^\dagger a_{\bar{k}}^\dagger)|0\rangle, \tag{33}$$

where u_k and v_k are determined in the BCS calculation (28) and $\alpha_{k_0}^\dagger$ the quasiparticle creation operator (27b) defined in canonical basis. If the same-Fermi energy λ_τ is used for the odd-mass nucleus, the mean value of the particle-number operator, i.e., $\langle \Phi_{k_0}(\mathbf{q})|\hat{N}|\Phi_{k_0}(\mathbf{q})\rangle = N_\tau + u_{k_0}^2 - v_{k_0}^2 \neq N_\tau \pm 1$. Thus, one needs to readjust the Fermi energy. Here, there are different ways to construct mean-field reference states with odd-number parity corresponding to different levels of approximations. See Ref. 66 for details.

2.1.4. *Effective nuclear Hamiltonians*

The Hamiltonian operator of nuclear systems is usually composed of kinetic energy, two-body nuclear forces, and even three-body nuclear forces. In most applications, the effect of the three-body nuclear force is considered using a density-dependent two-body force. Particularly, phenomenological Hamiltonians or EDFs with parameters fitted to the properties of nuclear many-body systems are employed. These include nonrelativistic zero-range Skyrme forces, finite-range Gogny forces and relativistic effective Lagrangian densities.[5]

In the relativistic framework of CDFT, also known as the Relativistic Mean Field (RMF) approach, nucleons interact via the exchange of different effective meson fields or through different types of coupling vertices.[67–69] These approaches are commonly referred to as the meson-exchange model and point-coupling model, respectively. In both cases, an effective nuclear Hamiltonian is derived from the Lagrangian density. When applying the mean-field approximation, one obtains an Energy Density Functional (EDF) that depends on various four-component currents and their derivatives. Within this theoretical framework, a reflection-asymmetric CDFT was developed by expanding nucleons and mesons on the eigenfunctions of the two-center harmonic-oscillator potential.[70,71] This approach has been applied to investigate the shape evolution from spherical to octupole-deformed shapes in isotopes of Sm,[72] Th[73] and Ra.[74] Simultaneously, CDFT was developed using the standard (single-center) harmonic-oscillator basis for superheavy nuclei, incorporating axial octupole shape degrees of freedom,[75] as well as accommodating general shapes.[76]

It is worth mentioning that one may encounter singularity and self-interaction problems in the Energy Density Functional (EDF)-based BMF approaches.[77,78] Many efforts have been devoted to finding an energy functional free of these problems. These issues do not exist in Hamiltonian-based approaches. Remarkable progress has been achieved in the development of nuclear *ab initio* methods in the past decade, starting from Hamiltonians derived from chiral (Effective Field Theory (EFT)).[79] The obtained HFB state serves as a reference state for *ab initio* many-body approaches.[80] With the help of the (Similarity Renormalization Group (SRG))[81] and in-medium SRG,[82] one is able to derive effective nuclear interactions suitable for GCM calculations.[54,55] The implementation of symmetry-restoration methods into *ab initio* methods, starting from a realistic nuclear force, has become very attractive. Notably, the aforementioned singularity problems are not present in these *ab initio* frameworks.

2.2. *Symmetry restoration with projections*

Since symmetries are allowed to be violated during the variational process in the HFB approach, particle number, translational, rotational and parity symmetries are all broken in the octupole-deformed HFB states $|\Phi(\mathbf{q})\rangle$.

The wave function of a state with quantum-number projections is constructed as

$$|NZJK\pi;\mathbf{q}\rangle = \hat{P}^J_{MK}\hat{P}^N\hat{P}^Z\hat{P}^\pi|\Phi(\mathbf{q})\rangle. \tag{34}$$

Here, the generator coordinate $\mathbf{q}$ stands for the discretized deformation parameters $(\beta_{2\mu}, \beta_{3\mu})$ of the reference states $|\Phi(\mathbf{q})\rangle$ obtained from deformation-constrained self-consistent mean-field calculations using HF + BCS, HFB, relativistic RHB or CDFT. The $\hat{P}^J_{MK}$, $\hat{P}^{N,Z}$ and $\hat{P}^\pi$ are projection operators that extract components with the right angular momentum J, neutron number N, proton number Z and parity $\pi = \pm$.[4] The projection operators are defined as follows:

$$\hat{P}^J_{MK} = \frac{2J+1}{8\pi^2}\int d\Omega D^{J*}_{MK}(\Omega)\hat{R}(\Omega), \tag{35a}$$

$$\hat{P}^{N_\tau} = \frac{1}{2\pi}\int_0^{2\pi} d\varphi_\tau e^{i\phi_\tau(\hat{N}_\tau - N_\tau)}, \tag{35b}$$

$$\hat{P}^\pi = \frac{1}{2}(1+\pi\hat{P}), \tag{35c}$$

where the operator $\hat{P}^J_{MK}$ extracts the component of angular momentum along the intrinsic axis $\hat{z}$ given by K. The Wigner-D function is defined as $D^J_{MK}(\Omega) \equiv \langle JM|e^{i\varphi\hat{J}_z}e^{i\theta\hat{J}_y}e^{i\psi\hat{J}_z}|JK\rangle$ with $\Omega = (\varphi, \theta, \psi)$ representing the three Euler angles.

The energy of the projected state $|\Phi(\mathbf{q})\rangle$ is given by the ratio of the diagonal element of the Hamiltonian to that of norm kernels,

$$E^{NZJK\pi}(\mathbf{q}) = \frac{\mathscr{H}^{J\pi}_{\kappa,\kappa}}{\mathscr{N}^{J\pi}_{\kappa,\kappa}}, \tag{36}$$

where the hamiltonian kernel $\mathscr{H}^{J\pi}_{\kappa_a,\kappa_b}$ and norm kernel $\mathscr{N}^{J\pi}_{\kappa_a,\kappa_b}$ are given by

$$\mathscr{O}^{J\pi}_{\kappa_a,\kappa_b} = \langle NZJK_a\pi;\mathbf{q}_a|\hat{O}|NZJK_b\pi;\mathbf{q}_b\rangle = \langle\Phi(\mathbf{q}_a)|\hat{O}\hat{P}^J_{K_aK_b}\hat{P}^N\hat{P}^Z\hat{P}^\pi|\Phi(\mathbf{q}_b)\rangle \tag{37}$$

with the operator $\hat{O}$ representing $\hat{H}$ and 1, respectively. The matrix elements of a Hamiltonian operator $\hat{H}$ can be evaluated using the generalized Wick's theorem[83]

$$\begin{aligned}\mathscr{H}^{J\pi}_{\kappa_a,\kappa_b} = {} & \frac{2J+1}{8\pi^2}\int d\Omega D^J_{K_aK_b}(\Omega)\frac{1}{2\pi}\int_0^{2\pi} e^{i\varphi_n N_0}\frac{1}{2\pi}\int_0^{2\pi} e^{i\varphi_p Z_0}\\ & \times\frac{1}{2}[\langle\Phi(\mathbf{q}_a)|\hat{H}\hat{R}(\Omega)e^{-i\varphi_n\hat{N}}e^{-i\varphi_n\check{Z}}|\Phi(\mathbf{q}_b)\rangle\\ & +\pi\langle\Phi(\mathbf{q}_a)|\hat{H}\hat{R}(\Omega)e^{-i\varphi_n\hat{N}}e^{-i\varphi_n\check{Z}}\hat{P}|\Phi(\mathbf{q}_b)\rangle].\end{aligned} \tag{38}$$

In contrast to the Hamiltonian-based GCM, the hamiltonian kernel $\mathscr{H}^{J\pi}_{\kappa_a,\kappa_b}$ in the EDF-based GCM calculations is usually evaluated with the *mixed density* prescription, i.e., the hamiltonian overlap takes the same functional form of mean-field EDF, provided that all the densities ρ are replaced with the mixed densities $\tilde{\rho}$ and $\tilde{\rho}^P$,

$$\frac{\langle\Phi(\mathbf{q}_a)|\hat{H}\hat{R}(\Omega)e^{-i\varphi_n\hat{N}}e^{-i\varphi_p\check{Z}}|\Phi(\mathbf{q}_b)\rangle}{\langle\Phi(\mathbf{q}_a)|\hat{R}(\Omega)e^{-i\varphi_n\hat{N}}e^{-i\varphi_p\check{Z}}|\Phi(\mathbf{q}_b)\rangle} = E[\tilde{\rho},\nabla^2\tilde{\rho},\ldots], \tag{39a}$$

$$\frac{\langle\Phi(\mathbf{q}_a)|\hat{H}\hat{R}(\Omega)e^{-i\varphi_n\hat{N}}e^{-i\varphi_p\check{Z}}\hat{P}|\Phi(\mathbf{q}_b)\rangle}{\langle\Phi(\mathbf{q}_a)|\hat{R}(\Omega)e^{-i\varphi_n\hat{N}}e^{-i\varphi_p\check{Z}}\hat{P}|\Phi(\mathbf{q}_b)\rangle} = E[\tilde{\rho}^P,\nabla^2\tilde{\rho}^P,\ldots]. \tag{39b}$$

The matrix elements of mixed (transition) densities are defined as

$$\tilde{\rho}_{ij}(g) = \frac{\langle\Phi(\mathbf{q}_a)|c_j^\dagger c_i\hat{R}(\Omega)e^{-i\varphi_n\hat{N}}e^{-i\varphi_p\check{Z}}|\Phi(\mathbf{q}_b)\rangle}{\langle\Phi(\mathbf{q}_a)|\hat{R}(\Omega)e^{-i\varphi_n\hat{N}}e^{-i\varphi_p\check{Z}}|\Phi(\mathbf{q}_b)\rangle} = [V^*(\mathbf{q}_b,g)\mathbb{U}^{-1}V^T(\mathbf{q}_a)]_{ij}, \tag{40a}$$

$$\tilde{\rho}^P_{ij}(g) = \frac{\langle\Phi(\mathbf{q}_a)|c_j^\dagger c_i\hat{R}(\Omega)e^{-i\varphi_n\hat{N}}e^{-i\varphi_p\check{Z}}\hat{P}|\Phi(\mathbf{q}_b)\rangle}{\langle\Phi(\mathbf{q}_a)|\hat{R}(\Omega)e^{-i\varphi_n\hat{N}}e^{-i\varphi_p\check{Z}}\hat{P}|\Phi(\mathbf{q}_b)\rangle} = [V^{(p)*}(\mathbf{q}_b,g)\mathbb{U}^{(p)-1}V^T(\mathbf{q}_a)]_{ij} \tag{40b}$$

with the symbol g standing for $(\Omega,\varphi_n,\varphi_p)$ and $V^{(p)}_{qk}=(-1)^{\ell_q}V_{qk}$, where q is for the spherical Harmonic Oscillator (HO) basis $|q\rangle=|n\ell jm_j\rangle_q$. The matrix $\mathbb{U}$ is defined as[4,51]

$$\mathbb{U} = U^\dagger(\mathbf{q}_b,g)U(\mathbf{q}_a)+V^\dagger(\mathbf{q}_b,g)V(\mathbf{q}_a), \tag{41a}$$

$$\mathbb{U}^{(p)} = U^{(p)\dagger}(\mathbf{q}_b,g)U(\mathbf{q}_a)+V^{(p)\dagger}(\mathbf{q}_b,g)V(\mathbf{q}_a) \tag{41b}$$

with $U(\boldsymbol{q},g)$ and $V(\boldsymbol{q},g)$ defined by

$$U(\boldsymbol{q},g)=D(g)U(\boldsymbol{q}),\quad V(\boldsymbol{q},g)=D^*(g)V(\boldsymbol{q}), \tag{42}$$

where $D(g)$ is the product of the representation matrices of the rotation operators $\hat{R}(\Omega)\ e^{-i\varphi_n\hat{N}}e^{-i\varphi_p\check{Z}}$ in the HO basis.

2.3. *Configuration mixing with GCM*

In the SPGCM, the wave function $|\Psi^{J\pi}_\alpha\rangle$ of the αth state with spin-parity J^π is constructed as a linear combination of the projected states (34)[33]

$$|\Psi^{J\pi}_\alpha\rangle = \sum_{\kappa\in\{\mathbf{q},K\}} f^{J\pi\alpha}_\kappa|NZJK\pi;\mathbf{q}\rangle, \tag{43}$$

where the weight function $f_{\kappa}^{J\pi\alpha}$ is determined by the Hill–Wheeler–Griffin (HWG) equation[4,84]

$$\sum_{\kappa_b}[\mathscr{H}^{J\pi}_{\kappa_a,\kappa_b} - E^{J\pi}_{\alpha}\mathscr{N}^{J\pi}_{\kappa_a,\kappa_b}]f^{J\pi\alpha}_{\kappa_b} = 0, \tag{44}$$

where the hamiltonian kernel $\mathscr{H}^{J\pi}_{\kappa_a,\kappa_b}$ and norm kernel $\mathscr{N}^{J\pi}_{\kappa_a,\kappa_b}$ have been defined in (37).

The solution of the HWG equation (44) provides energy $E^{J\pi}_{\alpha}$ and weight function $f_{\kappa}^{J\pi\alpha}$ of the nuclear state. For even–even octupole-deformed nuclei built on the eigenstates of the simplex operator $\hat{S} = \hat{P}\hat{R}_y(\pi)$, the projected states with angular momentum J must necessarily have the natural parity $\pi = (-1)^J$. Therefore, for the states of the $K = 0$ band, only quantum numbers of angular momentum and parity such as $0^+, 1^-, 2^+, 3^-$, etc., are possible.

Since the basis functions $|NZJK\pi; \mathbf{q}\rangle$ are nonorthogonal to each other, one usually defines the collective wave function $g^{J\pi}_{\alpha}(K, \mathbf{q})$ as follows:

$$g^{J\pi}_{\alpha}(K, \mathbf{q}) = \sum_{\kappa'}(\mathscr{N}^{1/2})^{J\pi}_{\kappa,\kappa'}f^{J\pi\alpha}_{\kappa'}, \tag{45}$$

which fulfills the normalization condition

$$\sum_{K,\mathbf{q}}|g^{J\pi}_{\alpha}(K, \mathbf{q})|^2 = \sum_{\kappa}\sum_{\kappa'}f^{J\pi\alpha*}_{\kappa}\mathscr{N}^{J\pi}_{\kappa,\kappa'}f^{J\pi\alpha}_{\kappa'} = \langle\Psi^{J\pi}_{\alpha}|\Psi^{J\pi}_{\alpha}\rangle = 1. \tag{46}$$

The distribution of $g^{J\pi}_{\alpha}(K, \mathbf{q})$ over K and $\mathbf{q}$ reflects the contribution of each basis function (34) to the nuclear state $|\Psi^{J\pi}_{\alpha}\rangle$.

2.4. *Electromagnetic transition rates*

The probabilities of γ-ray transitions between nuclear states serve as an excellent probe to the details of nuclear structure. By combining experimentally measured half-lives, transition energies and partial gamma-ray and conversion line intensities (including branching ratios, mixing ratios and conversion coefficients), it becomes possible to determine the absolute γ-ray transition probabilities between nuclear states. These data offer a rigorous test of nuclear models.

With the wave functions of initial state $|\Psi^{J_i\pi_i}_{i}\rangle$ and final state $|\Psi^{J_f\pi_f}_{f}\rangle$, one can compute the γ-ray transition rate of the tensor operator $\hat{T}_{\lambda\mu}$ between these two states[4]

$$T_{i\to f}(T\lambda; J^{\pi_i}_i \to J^{\pi_f}_f) = 8\pi c\alpha\frac{(\lambda+1)}{\lambda[(2\lambda+1)!!]^2}\left(\frac{E_\gamma}{\hbar c}\right)^{2\lambda+1}B(T\lambda; J^{\pi_i}_i \to J^{\pi_f}_f), \tag{47}$$

where E_γ (in MeV) is the energy difference of the two states. Substituting the following constants:

$$c = 3.0 \times 10^{23}\,\text{fm/s}, \quad \hbar c = 197.327\,\text{fm}\cdot\text{MeV}, \quad \alpha = \frac{e^2}{4\pi\hbar c} = 1/137.035, \tag{48}$$

one finds the transition rate (in s^{-1})

$$T_{i\to f}(E\lambda; J_i^{\pi_i} \to J_f^{\pi_f}) = \alpha_\lambda E_\gamma^{2\lambda+1} B(E\lambda; J_i^{\pi_i} \to J_f^{\pi_f}), \tag{49}$$

where the constant α_λ for $\lambda = 1, 2, 3$ reads

$$\alpha_1 \simeq 1.59 \times 10^{15}, \quad \alpha_2 \simeq 1.23 \times 10^{9}, \quad \alpha_3 \simeq 5.71 \times 10^{2} \tag{50}$$

and the reduced transition strength $B(E\lambda)$ (in $e^2\text{fm}^{2\lambda}$) reads

$$B(E\lambda; J_i^{\pi_i} \to J_f^{\pi_f}) = \frac{1}{2J_i + 1} |M(E\lambda; J_i^{\pi_i} \to J_f^{\pi_f})|^2. \tag{51}$$

The reduced transition matrix element $M(E\lambda)$ is given by

$$\begin{aligned} &M(E\lambda; J_{\alpha_i}^{\pi_i} \to J_{\alpha_f}^{\pi_f}) \\ &\quad = \sum_{\mathbf{q}_i \mathbf{q}_f} f_{K_f}^{J_f \pi_f \alpha_f} f_{K_i}^{J_i \pi_i \alpha_i} \times \langle NZJ_f K_f \pi_f; \mathbf{q}_f || \hat{T}_\lambda || NZJ_i K_i \pi_i; \mathbf{q}_i \rangle, \end{aligned} \tag{52}$$

where the deformation-dependent reduced matrix element is determined by

$$\begin{aligned} &\langle NZJ_f K_f \pi_f; \mathbf{q}_f || \hat{T}_\lambda || NZJ_i K_i \pi_i; \mathbf{q}_i \rangle \\ &\quad = \delta_{1,\pi_f\pi_\lambda\pi_i} (2J_f + 1)(-1)^{J_f - K_f} \sum_{M'M''} \begin{pmatrix} J_f & \lambda & J_i \\ -K_f & M' & M'' \end{pmatrix} \\ &\qquad \times \langle \Phi(\mathbf{q}_f) | \hat{T}_{\lambda M'} \hat{P}_{M''K_i}^{J_i} \hat{P}^Z \hat{P}^N \hat{P}^\pi | \Phi(\mathbf{q}_i) \rangle. \end{aligned} \tag{53}$$

The nonzero value of the transition strength requires that the spin-parity quantum numbers of the initial and final states fulfill the following conditions:

$$\pi_f \pi_\lambda \pi_i = 1, \quad |J_f - J_i| \le \lambda \le J_f + J_i, \tag{54}$$

where the parity $\pi_\lambda = (-1)^\lambda$ for electric multipole transitions $\hat{T}_\lambda = r^\lambda Y_\lambda$ and $\pi_\lambda = (-1)^{\lambda+1}$ for magnetic multipole transitions $\hat{T}_\lambda = (g_s \mathbf{S} + \frac{2}{\lambda+1} g_\ell \mathbf{L}) \cdot (\nabla r^\lambda Y_\lambda)$ in units of nuclear magneton (μ_N). We note that free-space values of electric charges and $g_{s/\ell}$ factors are usually used in the calculations of full single-particle space. If a valence-space Hamiltonian is employed, effective values are used.

In the case that the intrinsic states $|\Phi(\mathbf{q})\rangle$ are restricted to have axial symmetry, the integration in (35a) over the two Euler angles (φ, ψ) can be carried out analytically. The reduced matrix element between the states with different deformations and $K_f = K_i = K$ is simplified as follows:

$$\begin{aligned}
&\langle NZJ_fK\pi_f; \mathbf{q}_f||\hat{T}_\lambda||NZJ_iK\pi_i; \mathbf{q}_i\rangle \\
&\quad = (-1)^{J_f-K}\delta_{1,\pi_f\pi_\lambda\pi_i}\frac{(2J_i+1)(2J_f+1)}{2}\sum_{\nu M}\begin{pmatrix} J_f & \lambda & J_i \\ -K & \nu & M \end{pmatrix} \\
&\qquad \times \int_0^\pi \sin\theta d\theta d^{J_i}_{MK}(\theta)\langle\Phi(\mathbf{q}_f)|\hat{T}_{\lambda\nu}e^{i\theta\hat{J}_y}\hat{P}^Z\hat{P}^N\hat{P}^{\pi_i}|\Phi(\mathbf{q}_i)\rangle. \qquad (55)
\end{aligned}$$

3. Nuclear Collective Motions with Octupole Shapes

3.1. *Reflection-asymmetric cluster structures in light nuclei*

Cluster structure appears in the excited states of many light nuclei, particularly in those of self-conjugate $4n$ nuclei close to the particle emission threshold, as illustrated by the Ikeda diagram.[85] Advances in nuclear theory have made it possible to study nucleon clustering phenomena in atomic nuclei, not only from the degrees of freedom of different clusters but also from the individual nucleons, with a full account of the Pauli exclusion principle. Cluster structures emerge automatically from the A-body wave function starting from realistic or effective nucleon–nucleon interactions.[86]

Self-consistent mean-field approaches, with all nucleons treated on the same footing, provide a good tool for understanding these phenomena. Previous studies in this context have shown that nucleons tend to form cluster structures in nuclear states at high spin,[87] isospin,[88] large deformation,[89–92] deep confining nuclear potential,[93] and with low density.[94] The GCM provides a BMF study of cluster structures in excited states of atomic nuclei by mixing mean-field reference states with both shell-model-like structure and cluster structure, with their weights determined by variational principles. As it will be illustrated below, the inclusion of reflection asymmetric shapes in the reference states is necessary for the reproduction of clustering structure in the light nuclei of concerned, including the Hoyle state of ^{12}C and molecular-like structure in ^{20}Ne. However, this breaks rotational symmetry and space-reversal symmetry of the nuclear Hamiltonian. These symmetries need to be restored to describe cluster structures in nuclear excited states characterized by the quantum numbers of spin and parity.

^{12}C is a well-known example of a nucleus with nucleon clustering structures. Its excited states are predominantly characterized by configurations involving three α particles, which can arrange themselves into various geometric shapes, such as triangular, linear, or others. Of particular importance is the second 0^+ state, commonly referred to as the Hoyle state, which plays a crucial role in nucleosynthesis. Researchers have extensively studied its cluster structure using various nuclear models, including ab initio methods.[95–98] In this context, our primary focus will be on investigations conducted within the SPGCM framework.

The nucleon clustering structure in ^{12}C has been extensively studied with the models of Antisymmetrized Molecular Dynamics (AMD)[99,100] and Fermionic Molecular Dynamics (FMD).[101] In both of these models, wave functions of configurations are constructed as Slater determinants of Gaussian wave packets. In the AMD model, parameters such as the center of the Gaussian wave packet, the width parameter, and the orientation of intrinsic spin for each nucleon are treated as variational parameters determined by the principles of variation. A study using this model with an effective nuclear interaction, including a zero-range three-body force and two-body interaction, has revealed that the ^{12}C ground state contains both sub-shell closure and 3α cluster components. Excited states in ^{12}C exhibit various 3α cluster structures. Additionally, the study highlights the importance of the α-cluster breaking component in the excited states.[100] In contrast, the FMD model treats the width parameter as a complex variational parameter, allowing it to vary for each single-particle state.[102] An effective interaction in the FMD model is derived from realistic Bonn or Argonne nucleon–nucleon interactions using the unitary correlation operator method. To simulate the effects of three-body correlations and genuine three-body forces, an additional two-body force is introduced, which includes momentum-dependence and spin-orbit forces with parameters adjusted to match binding energies and radii of finite nuclei. The configurations in the FMD model are generated through VAP calculations with constraints on radii and quadrupole deformation. Notably, the FMD model predicts that the Hoyle state in ^{12}C is a dilute self-bound gas of bosonic α particles, akin to a Bose–Einstein condensate.[103] It is also found that the spin-orbit force tends to break the α clusters, leading to a significant shell-model component in the FMD ground state. However, in the Hoyle state, α clustering remains dominant, albeit with a sizable component of shell-model nature.

In the past decade, researchers have also explored cluster structures in the low-lying states of ^{12}C using EDF-based GCM frameworks. For instance, Fukuoka *et al.* conducted configuration-mixing calculations of mean-field states with projections onto different parities and angular momenta.[104] In their approach, mean-field states were generated from Skyrme HF calculations, represented in three-dimensional coordinate-space allowing for various shapes. Initially, single-particle wave functions were prepared as Gaussian wave packets with random centers, evolving using the imaginary-time method. As the configurations evolved, various cluster structures emerged. To perform GCM calculations, configurations were selected based on a specific criterion: if the overlap between a given wave function and all selected configurations fell below a predefined cutoff value, that configuration was included in the GCM subspace. This approach allowed for a comprehensive exploration of cluster structures in ^{12}C. A similar concept has been applied in recent GCM studies, such as those related to neutrinoless double-beta decay.[56,105]

In recent years, extensive investigations into the low-lying states of ^{12}C have been carried out using the SPGCM. These studies involved mixing axial and reflection-asymmetric reference states generated from the RHB calculation.[43] The techniques of AMP, PNP and PP were implemented simultaneously to restore the broken symmetries in the reference states.

One common approach to revealing cluster structures in these states is by computing nucleon density profiles and electron-scattering form factors. For instance, the longitudinal Coulomb form factor $F_L(q)$, which describes transitions between initial $|\Psi^{J_i\pi_i}_{\alpha_i}\rangle$ and final $|\Psi^{J_f\pi_f}_{\alpha_f}\rangle$ states can be determined using the plane-wave Born approximation

$$F_L(q) = \frac{\sqrt{4\pi}}{Z}\int_0^\infty dr r^2 \rho^{J_f\alpha_f}_{J_i\alpha_i,L}(r) j_L(qr). \tag{56}$$

Here, q represents the momentum transfer for angular momentum L, $j_L(qr)$ is the spherical Bessel function of the first kind and $\rho^{J_f\alpha_f}_{J_i\alpha_i,L}(r)$ are the reduced transition densities of protons. These reduced transition densities can be expressed as follows[106]:

$$\rho^{J_f\alpha_f}_{J_i\alpha_i,L}(r) = (-1)^{J_f-J_i}\frac{2J_f+1}{2J_i+1}\sum_K \langle J_f 0 L K | J_i K\rangle \times \int d\hat{\mathbf{r}} \rho^{J_f J_i K 0}_{\alpha_f\alpha_i}(\mathbf{r}) Y^*_{LK}(\hat{\mathbf{r}}), \tag{57}$$

where $\rho_{\alpha_f\alpha_i}^{J_fJ_iK0}(\mathbf{r})$ stands for the pseudo-GCM density, which in the axial case is given by[106]

$$\rho_{\alpha_f\alpha_i}^{J_fJ_iK0}(\mathbf{r}) = \sum_{\mathbf{q},\mathbf{q}'} f_{\mathbf{q}'}^{J_f\pi_f\alpha_f} f_{\mathbf{q}}^{J_i\pi_i\alpha_i} \frac{2J_i+1}{2} \int_0^{\pi} d\theta \sin(\theta) d_{K0}^{J_i}(\theta)$$
$$\times \langle \Phi(\mathbf{q}')| \sum_{i=1}^{A} \delta(\mathbf{r}_i - \mathbf{r}) \hat{P}^N \hat{P}^Z e^{i\theta \widehat{J_y}} |\Phi(\mathbf{q})\rangle. \quad (58)$$

The quality of the theoretical description for nucleon density and transition density can be examined by looking at the elastic and inelastic form factors. To illustrate this, Fig. 3 displays the form factors of electron elastic and inelastic scattering, as well as nucleon densities of the predominant configurations in the first two 0^+ states of ^{12}C. The results of SPGCM + RHB calculation,[43] labeled as MR-EDF, are compared with those obtained from the calculations of SPGCM + AMD[100] and Tohsaki–Horiuchi–Schuck–Ropke

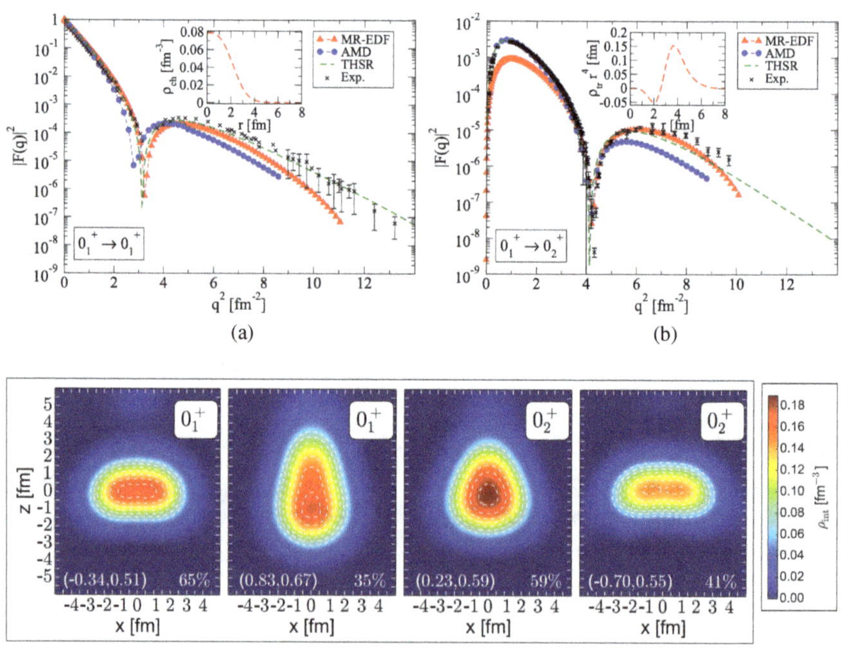

Fig. 3. The form factors of electron scatterings on ^{12}C for the (a) elastic $0_1^+ \to 0_1^+$ and (b) inelastic $0_1^+ \to 0_2^+$ processes from different calculations, in comparison with data. (c) The nucleon densities of the predominant configurations in the first two 0^+ states. Figure adapted from Ref. 43. Figure reprinted with permission from the American Physical Society.

(THSR) wave function.[107] It was found that the form factors are well reproduced by the MR-EDF calculation. In this calculation, the wave function of the 0_2^+ state is predominated by two configurations: a weakly prolate deformed configuration and a strongly oblate deformed one, as shown in Fig. 3(c). Both configurations have very large octupole deformations.

It is worth noting that the study conducted with the SPGCM + RHB framework[43] did not take into account triaxial deformation. To achieve the expected triangular distribution of the Hoyle state, axial symmetry needs to be broken. Actually, the constraint on multipole moments may not be an effective way for the study of cluster structure in atomic nuclei. Recent progress was made in the development of multi-dimensional RHB approach[76] in which atomic nuclei are allowed to have arbitrary geometric shapes. The techniques of PP combined with 3D Angular Momentum Projection (3DAMP) were employed to restore rotation symmetry in order to investigate the 3α cluster structure in ^{12}C.[108] In this study, a constraint on the triangular moment

$$\lambda_3 = \frac{4\pi}{3NR^3}\langle\Phi(\mathbf{q})|\hat{S}_3|\Phi(\mathbf{q})\rangle \tag{59}$$

was introduced to control the distance among the three α particles, arranged in an equilateral triangle, in the reference states. The corresponding operator $\hat{S}_3$ is defined as

$$\hat{S}_3 = (z^2 + \rho^2\cos^2\varphi)^{3/2}\cos(3\theta), \tag{60}$$

where ρ, φ and z represent the cylindrical coordinates, while θ is the polar angle on the x-z plane. Furthermore, to account for pairing correlations between nucleons, a separable pairing force of finite range was employed.[109] This approach allowed for a comprehensive exploration of the 3α cluster structure in ^{12}C with more degrees of freedom.

As seen from Fig. 4, spin-parity projection favors configurations with nonzero values of λ_3, suggesting the formation of 3α cluster structures. It is worth noting that the projected energy surface exhibits a significant softness. Thus, the extension of this study with the mixing of particle-number projected states with different values of the intrinsic triangular moment would provide a deeper insight into the 3α cluster structure in ^{12}C.

Another example nucleus with different cluster structures is ^{20}Ne. It has been investigated using AMD models, which were based either on spherical Gaussian basis functions (SphAMD)[112] or triaxially deformed (TriAMD) Gaussian basis functions.[113,114] These studies employed phenomenological

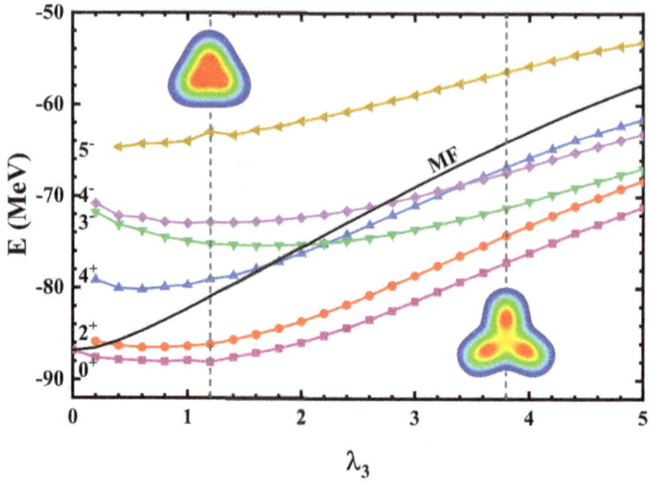

Fig. 4. (Color online) The energies of mean-field (black line) and spin-parity projected states with different J^π (scatter plot points) for ^{12}C as a function of the intrinsic triangular deformation parameter λ_3 defined in (59) from the RHB calculation. The insets display the density profiles of the intrinsic states at $\lambda_3 = 1.2$ and 3.8, indicated by two gray dashed lines. Taken from Ref. 108.

Hamiltonians or the Gogny D1S force, respectively. It was found in the SphAMD study that both positive and negative parity low-spin states are dominated by the diatomic molecular structures $\alpha+^{16}$O. With the consideration of triaxial deformation in the basis, the ground band with $K^\pi = 0^+$ turns out to have a mixed character of the deformed mean-field structure and the cluster structure. In contrast, the negative-parity state with $K^\pi = 0^-$ has an almost pure $\alpha+^{16}$O cluster structure. In particular, it was shown that the cluster structure moderates and the average distance between α and ^{16}O becomes smaller as the angular momentum increases.

In recent years, the cluster structures in ^{20}Ne were examined with the CDFT.[41,42] Figure 5 illustrates the mean-field energy surface of ^{20}Ne in the (β_{20}, β_{30}) deformation plane from the calculation using the relativistic PC-PK1.[115] Although a global energy minimum is found at the reflection-symmetric prolate shape $(\beta_2 = 0.5, \beta_3 = 0)$, this equilibrium shape is shown to be unstable against quantum fluctuations. A schematic representation of nuclear shapes corresponding to selected configurations along the optimal path on the energy surface highlights the progressive development of the reflection-asymmetric diatomic molecular structure $(\alpha+^{16}$O$)$ as the values of β_3 increase along the optimal path. This is consistent with the finding in the

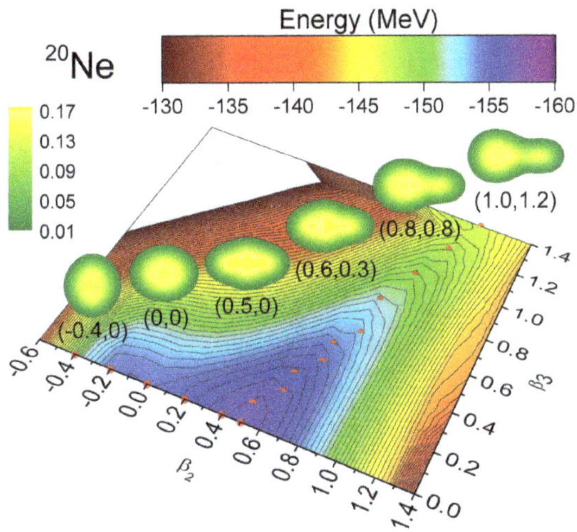

Fig. 5. (Color online) The mean-field energy surface of ^{20}Ne in two-dimensional (β_{20}, β_{30}) deformation plane, where the optimal configurations (along the valley) are indicated with red dots. The density profiles of several selected configurations are plotted (in fm^{-3}). Taken from Ref. 41.

TriAMD study based on the Gogny D1S force.[114] It is worth pointing out that only the configurations with nonzero value of β_3 contribute to the negative-parity states. Thus, the finding that the negative-parity states are predominated by the configuration with α+^{16}O structure does not necessary mean that the nucleus ^{20}Ne has a permanent octupole deformation. Actually, it has been shown in Refs. 41 and 42 that the collective wave functions of the negative-parity states are broadly distributed in the β_2-β_3 plane, confirming the picture of octupole vibration rather than the rotation with a permanent octupole deformation for the low-lying states of ^{20}Ne.

Figure 6 presents the low-energy spectra for ^{20}Ne obtained from different calculations. These calculations consider the dynamical octupole vibration effects and successfully reproduce both energy spectra and electric multipole transition strengths of low-lying parity-doublet bands. Compared to the results of calculations with the mixing of only prolate deformed configurations in Fig. 6 (e), the energy spectra become slightly stretched and $E2$ transition strengths are reduced in the calculations with octupole-deformed configurations. Moreover, it is shown in Fig. 6 that the energy spectra from the calculations using full configurations and optimal configurations are close to each other, and are more compressed than those of single-configuration calculation and the corresponding data. Among all the calculations, the $E2$ transition strengths in the ground state

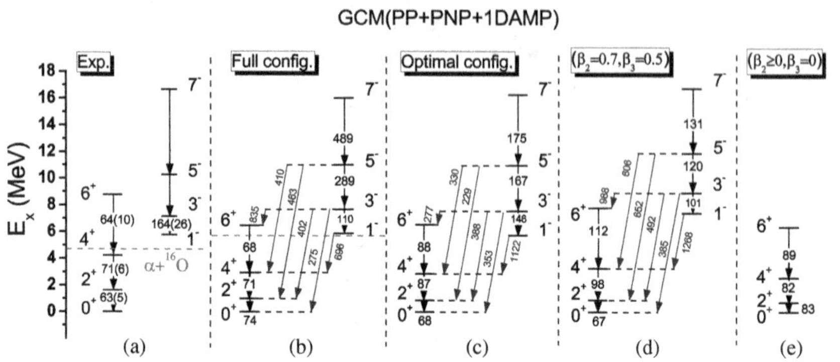

Fig. 6. (Color online) Energy spectra of ^{20}Ne from different calculations, in comparison with data. The optimal configurations in (c) are those indicated by red dots in Fig. 5. The single configuration with deformation parameters ($\beta_2 = 0.7, \beta_3 = 0.5$) in (d) corresponds to the energy minimum of the energy surface with AMP. Taken from Ref. 41.

band from the full configuration-mixing calculations are closest to the data. An in-depth analysis of the distributions of collective wave functions for the low-lying states of ^{20}Ne reveals that the positive-parity states are dominated by the α+^{12}C + α structure, and the negative-parity states by the α+^{16}O structure.

The low-lying states of hypernuclei offer a wealth of information about hyperon-nucleon interactions in nuclear medium and the impurity effect of hyperon in atomic nuclei. Extracting this valuable information from experimental data necessitates the use of accurate hypernuclear models. The SPGCM has been extended for hypernuclear spectroscopy based on different nuclear EDFs, including the Gogny force,[116,117] Skyrme force[118] or relativistic Lagrangian density.[119] Within this framework, the inclusion of dynamical octupole shapes may change significantly the interpretation of the spectroscopy of some hypernuclei, like $^{21}_{\Lambda}$Ne. Figure 7 displays the mean-field and projected energy curves as a function of the octupole deformation β_3 for the intrinsic states in ^{20}Ne and $^{21}_{\Lambda}$Ne. The quadrupole deformation parameter β_2 is fixed to the value around the energy minimum. Notably, the energies of the negative-parity states with the configuration $|^{20}\text{Ne}(K^\pi = 0^-)\rangle \otimes |\Lambda_s\rangle$ (blue dashed curves in Fig. 7(b)) are close to those with the configuration $|^{20}\text{Ne}(K^\pi = 0^+)\rangle \otimes |\Lambda_p\rangle$ (blue dashed curves in Fig. 7(c)),[110] where in these two types of configurations the Λ hyperon occupies either the first or second lowest-energy states. It turns out in the study[111] that these two configurations are strongly mixed due to the octupole correlations, producing the low-lying negative-parity states in Fig. 8. Moreover, the electric multipole transition strengths of the negative-parity states are significantly quenched with this configuration

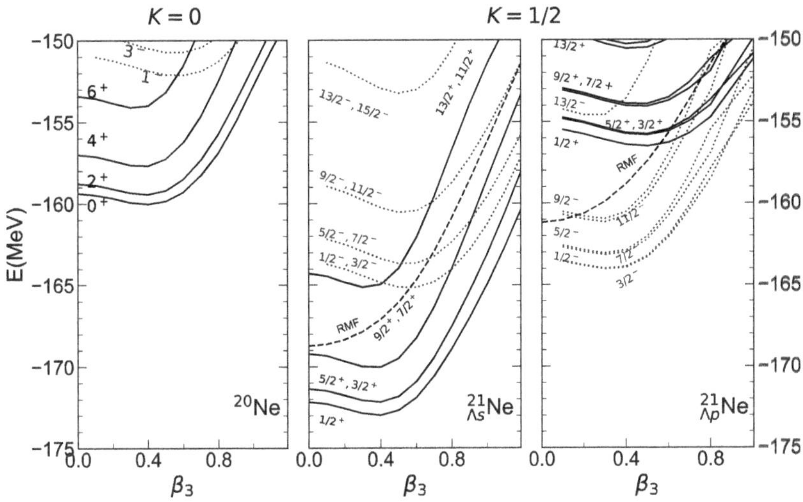

Fig. 7. The projected energy curves as a function of the octupole deformation β_3 of the intrinsic state ($\beta_2 = 0.60$ is fixed) in ^{20}Ne, $^{21}_{\Lambda_s}$Ne and $^{21}_{\Lambda_p}$Ne, respectively. The symbols Λ_s and Λ_p label the configurations where the Λ hyperon is put on the first and second lowest-energy orbits, respectively. The levels with positive (negative) parity are plotted with solid (dotted) curves. Taken from Ref. 110.

mixing effect. The density distribution of these two dominated configurations is also shown in Fig. 8. One can see that the negative-parity states in $^{21}_{\Lambda}$Ne are dominated by the mixing of the cluster structures $\alpha+^{16}\text{O} + \Lambda_s$ and $\alpha+^{12}\text{C} + \alpha + \Lambda_p$. In contrast, the positive-parity states are dominated by the $\alpha+^{12}\text{C} + \alpha + \Lambda_s$ structure, which are only slightly changed by the dynamical octupole effect.

From all the above studies, one may conclude that octupole shape is one of the relevant degrees of freedom for describing the clustering structure in ^{12}C and ^{20}Ne. However, the use of a constraint on the quadrupole and octupole moments might not be sufficient for the description of nucleon clustering structures. It is thus of importance to explore alternative ways to generate configurations with all possible cluster structures for the nuclei of interest within the SPGCM.

3.2. *Octupole vibrational states in near-spherical nuclei*

Nuclei with proton and/or neutron numbers near magic numbers often exhibit octupole correlations through vibrational motions. One of the most notable observations is the low-lying 3^- state, which is often interpreted as a

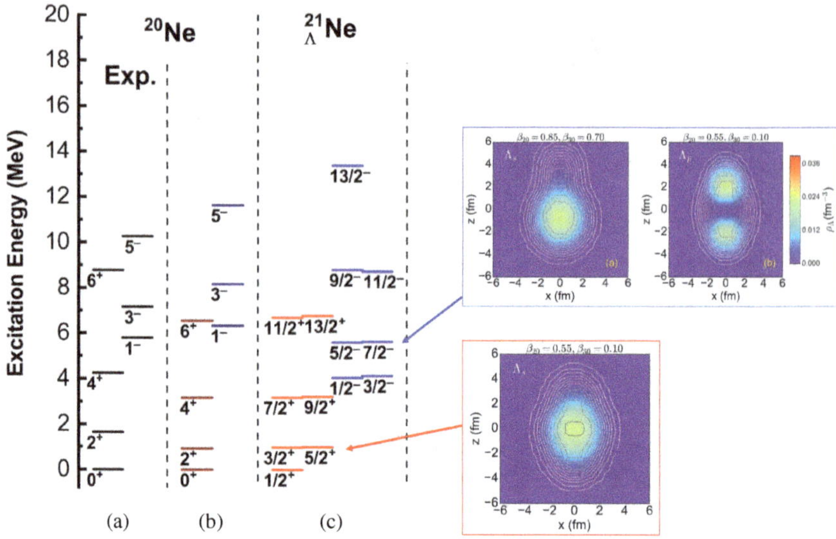

Fig. 8. The low-lying energy spectra of ^{20}Ne and $^{21}_{\Lambda}$Ne from the HyperGCM calculation with the mixing of quadrupole-octupole deformed configurations. The contour plots of the nucleon density and hyperon of the predominated configurations of positive and negative-parity states are also displayed. Results are taken from Ref. 111.

one-octupole-phonon state. The higher-lying states are dominated by the weak coupling between the octupole 3^- phonon and other single-particle excited states, or multi-phonon states with anharmonic characteristics. This picture has been illustrated through SPGCM studies that incorporate octupole shape fluctuation effects.[11,12,45]

The double-magic nucleus ^{208}Pb serves as a prime example of a near-spherical nucleus with intriguing low-lying octupole phonon states, which have been the focus of extensive research utilizing various nuclear models.[10] Numerous experiments have been conducted to search for multi-octupole-phonon states in ^{208}Pb.[121,122] A study in Ref. 11 extended the SPGCM to investigate the low-lying states of ^{208}Pb. This study employed a CDFT, with the generator coordinates chosen as the parameters of quadrupole deformation (β_2) and octupole deformation (β_3). The results, displayed in Fig. 9, showcase the low-lying energy spectra of ^{208}Pb. One can see from Fig. 9(c) that the multiples of two- and three-phonon states (labeled as 2ph and 3ph, respectively) appear at similar excitation energies to each other. The predicted spin-average of the excitation energies for the 2ph and 3ph states is 9.6 and 13.1 MeV, respectively, which is about twice and three times

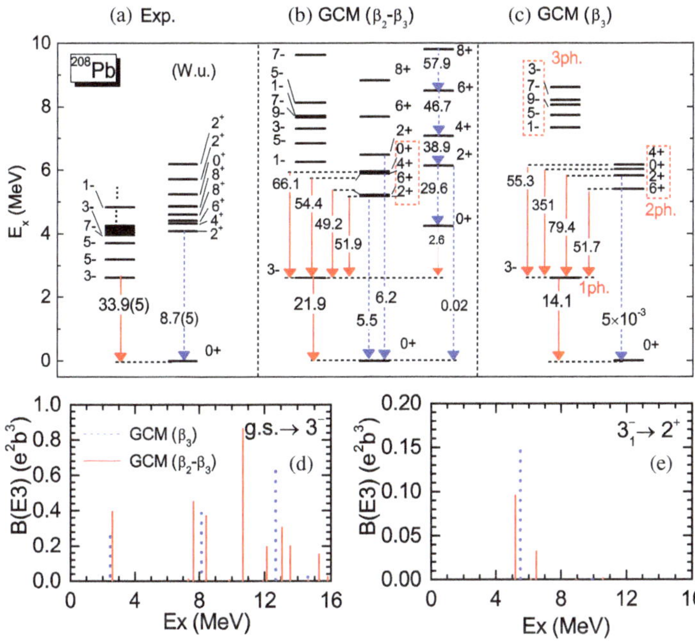

Fig. 9. (Color online) The low-lying energy spectra of ^{208}Pb obtained from the GCM calculation with the mixing of quadrupole-octupole (β_2-β_3) deformed configurations (b) or the mixing of only octupole (β_3) deformed configurations with $\beta_2 = 0$ (c), in comparison with the experimental data taken from Ref. 120. The panel (d) and (e) show the $E3$ transition strengths from the ground state to 3^- states and from the first 3^- state to 2^+ states as a function of the excitation energy of the final states. In (a)–(c), the red solid and the blue dashed lines indicate the $E3$ and $E2$ transition strengths (in W.u.), respectively. All the calculated excitation energies are scaled to the empirical excitation energy of the lowest 3^- state by dividing them by a factor of 1.78 and 1.64 in the panels (b) and (c), respectively. Figure adapted from Ref. 11. Figure reprinted with permission from the American Physical Society.

that of the one-octupole phonon state, 4.3 MeV. We note that the energy spectra in Fig. 9 are normalized to the data for the excitation energy of the 3_1^+ state. However, there are some energy displacements, indicating the existence of anharmonicity. The anharmonicity is shown to be stronger in the transition strengths. The $B(E3)$ value for the transition from the 2ph multiplets to the 3_1^+ state is much larger than twice the $B(E3)$ value from the 3_1^+ state to the ground state. Figure 9(b) shows that the inclusion of the quadrupole shape fluctuation slightly alters the excitation energies, and the transition strengths for the $3_1^+ \to 0_1^+$ and the $2_1^+ \to 0_1^+$ transitions are significantly improved. Figures 9(d) and 9(e) show that the E3 transitions become more fragmented after taking into account the interplay of the quadrupole and octupole shape fluctuations.

The low-lying 3^- states in ^{208}Pb turns out to have a significant impact on the cross-section of sub-barrier fusion reaction of ^{16}O+^{208}Pb, as shown in Fig. 10. The solid line is the fusion cross-sections (the upper panel) and the fusion barrier distribution (the lower panel) calculated with the CCFULL code[123] using the excitation energies and transition strengths from the SPGCM calculation as inputs. The fusion barrier distribution is defined as the second energy derivative of the product of the energy E and fusion cross-section σ_{fus}, that is, $d^2(E\sigma_{\text{fus}})/dE^2$. The results are compared to the results of calculation based on the two-phonon harmonic oscillator approximation and to the single-channel calculation. For the former, the 3_1^-, 2_1^+, $(3_1^-)^2$, $(2_1^+)^2$ and $3_1^- \otimes 2_1^+$ states are included within the harmonic oscillator coupling scheme.[124] It has been a long-standing problem that for this particular

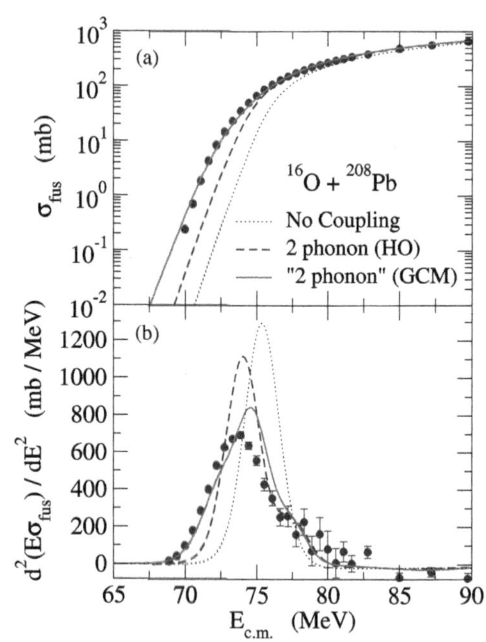

Fig. 10. The fusion cross-sections (upper panel) and the fusion barrier distributions (lower panel) for the ^{16}O+^{208}Pb system obtained with the semi-microscopic coupled-channels calculation with the coupling strengths from the SPGCM calculations (the solid line). The dashed and the dotted lines show the results of the two-phonon coupling in the harmonic oscillator limit and of the no-coupling limit, respectively. Figure adapted from Ref. 11. Figure reprinted with permission from the American Physical Society.

system the coupled-channels calculation with the harmonic oscillator couplings overestimates the height of the main peak in the barrier distribution[125] and underestimates the fusion cross-sections at energies below the Coulomb barrier. In contrast, the calculation using nuclear structure properties from the GCM calculation (see Fig. 9) yields a much lower peak in the fusion barrier distribution, leading to a better agreement with the experimental data both for the fusion cross-sections and for the barrier distribution. It turns out that the couplings between the 3_1^- and the 2_1^+ states, the two-phonon states and the excited negative-parity states, play an important role.

Nuclear octupole shapes may also play a significant role in heavy-ion collisions. The research by Zhang *et al.*[47] demonstrated that the triangular flow and anisotropy ϵ_3 in ^{96}Zr+^{96}Zr collisions are more pronounced than those in ^{96}Ru+^{96}Ru collisions. This discrepancy was attributed to the stronger octupole correlation in ^{96}Zr, a conclusion supported by recent RHB studies that include projections onto parity and angular momentum.[45]

3.3. *Transition from octupole vibrations to rotations*

The transition from octupole vibrational to rotational motions with the change of isospin or spin is a fascinating phenomenon in atomic nuclei. This transition can be deduced from the systematic evolution of energy spectra and electric multipole transition strengths, see Fig. 2.

Figure 11 provides a clear illustration of how the energy spectra and the energy ratio $R_{J/2}$ of low-spin states in ^{144}Ba change as the quadrupole β_2 and octupole β_3 deformation parameters increase. These results are obtained through symmetry-projection calculations based on different intrinsically deformed states. For cases where $\beta_3 = 0.1$, the amplitude of staggering in $R_{J/2}$ increases with the quadrupole deformation parameter β_2. Octupole-deformed nuclei typically have octupole deformation parameters ranging from 0.1 to 0.2, resulting in energy spectra that exhibit staggering in low-spin states, as discussed in the introduction.

It is worth noting that the 3^- state is predicted to be the first excited state for the weakly deformed state with $\beta_2 = 0$ and $\beta_3 = 0.1$ or 0.2. This observation aligns with findings in ^{208}Pb, as shown in Fig. 9. However, for states with larger octupole deformations, such as $\beta_3 = 0.2$ or 0.3, the odd–even parity states become interleaved, irrespective of the value of the quadrupole deformation β_2, and the characteristic staggering behavior disappears.

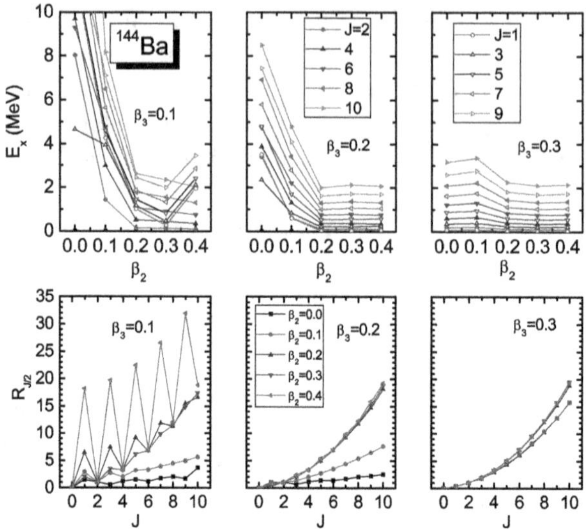

Fig. 11. The excitation energy of states projected onto particle number, angular momentum and parity based on an individual configuration for ^{144}Ba. (Lower panels) The ratio $R_{J/2}$ as a function of angular momentum $J(\hbar)$. Figure adapted from Ref. 23. Figure reprinted with permission from the American Physical Society.

Figure 12 displays the excitation energies of parity-doublets in ^{144}Ba from the SPGCM calculation using the PC-PK1, in comparison with data. It is shown that the staggering amplitude decreases gradually with the increase of angular momentum of the state. The distribution of the collective wave function shows that the predominant configuration of the positive-parity state is drifting to that of negative-parity state with the increase of the angular momentum. At the angular momentum $J \geq 8$, the distributions of the wave functions for the positive and negative states are close to each other, demonstrating clearly that the two bands of parity-doublet states merge into one band.

A similar rotation-induced shape transition picture is also observed in ^{224}Ra. As illustrated in Fig. 13(a), the energy surface and density distribution of the energy-minimal state of ^{224}Ra from the CDFT calculation[22] are shown. The global energy minimum is located at the state with deformation parameters $\beta_2 = 0.179$, $\beta_3 = 0.125, \beta_4 = 0.146$, which are slightly larger than the values $\beta_2 = 0.154, \beta_3 = 0.097, \beta_4 = 0.15$ reported in Ref. 19. The total nucleon distribution of this energy-minimum state is depicted in Fig. 13(b), revealing a well-developed pear shape. However, the energy difference between the energy minimum and the lowest

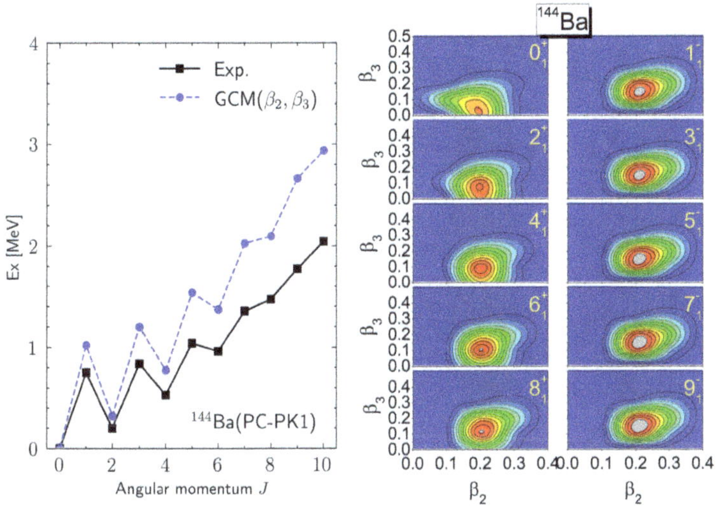

Fig. 12. The excitation energies (left panel) and the distributions of collective wave functions $|g_\alpha^{J\pi}(K=0,\mathbf{q})|^2$ (45) in the deformation parameters $\mathbf{q}(\beta_2, \beta_3)$ plane (right panel) for the parity-doublet states of ^{144}Ba.

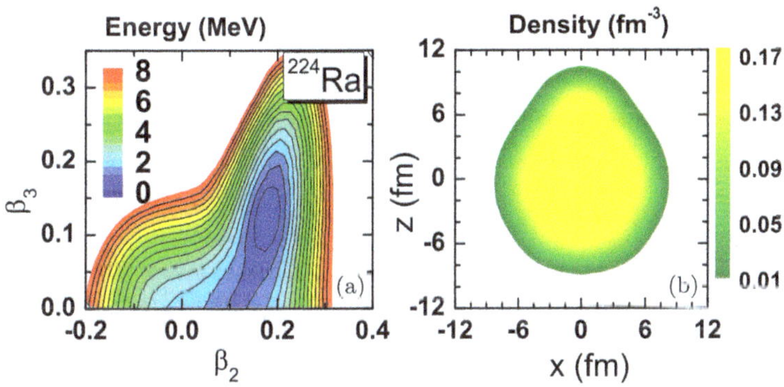

Fig. 13. (a) Mean-field energy surface of ^{224}Ra in the (β_{20}, β_{30}) plane and (b) distribution of nucleons for the energy-minimal state $|\Phi(\beta_2, \beta_3)\rangle$ in x-z plane from the RMF + BCS calculation using the PC-PK1 force.[115] Figure adapted from Ref. 22. Figure reprinted with permission from the American Physical Society.

reflection-symmetric configuration is not large enough to prevent shape fluctuations.

Figure 14 displays the low-lying parity doublets of ^{224}Ra. The spectra and the $E2$ and $E3$ transitions can be reproduced reasonably well using only the

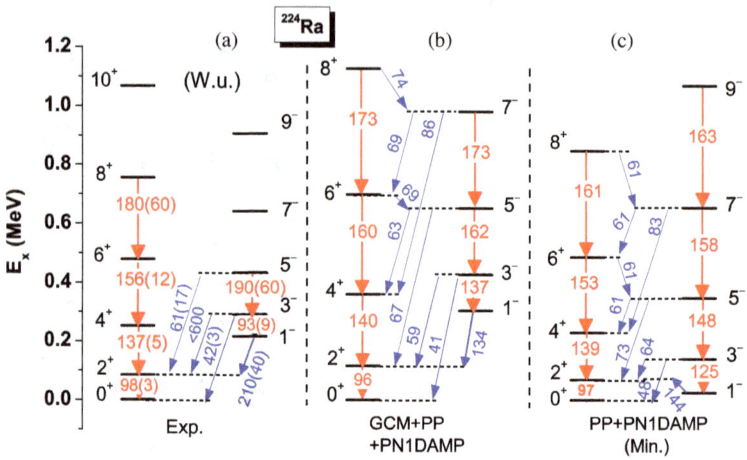

Fig. 14. (Color online) Low-lying energy spectra for ^{224}Ra from two different calculations, in comparison with data. The numbers on arrows are $E2$ (red color) and $E3$ (blue color) transition strengths (Weisskopf units). Figure adapted from Ref. 22. Figure reprinted with permission from the American Physical Society.

energy-minimum configuration. However, the predicted alternating parity rotation band as expected for a stable octupole shaped nucleus is not supported by the data. The energy displacement between the parity doublets can only be reproduced in the GCM calculation by mixing the configurations around the equilibrium shape, which does not affect appreciably the good description for the Eλ strengths. To understand the connection between spin-dependent parity splitting and shape evolution in a simple and intuitive way, the energy ratio $R_{J/2}$ and normalized staggering amplitude $S_{J/2}$ from the GCM calculations by mixing the octupole configurations at either $\beta_2 = 0.10$ or 0.15 are shown in Fig. 15(a), where the normalized staggering amplitude is defined as

$$S_{J/2} \equiv \left| E_x(J) - \frac{J+1}{2J+1} E_x(J-1) - \frac{J}{2J+1} E_x(J+1) \right| \Big/ E_x(2^+). \quad (61)$$

The gradually reduced energy staggering is shown in both cases. Figures 15(c) and 15(d) show that the projected energy surfaces with positive parity evolve evidently with increasing spin, presenting a transition from weakly deformed octupole shape to a large deformed one, which is consistent with the behavior of the spin-dependent collective potentials introduced phenomenologically in Ref. 126. This spin-controlled shape stabilization picture is demonstrated more clearly in the distributions of collective wave functions in Figs. 15(e)

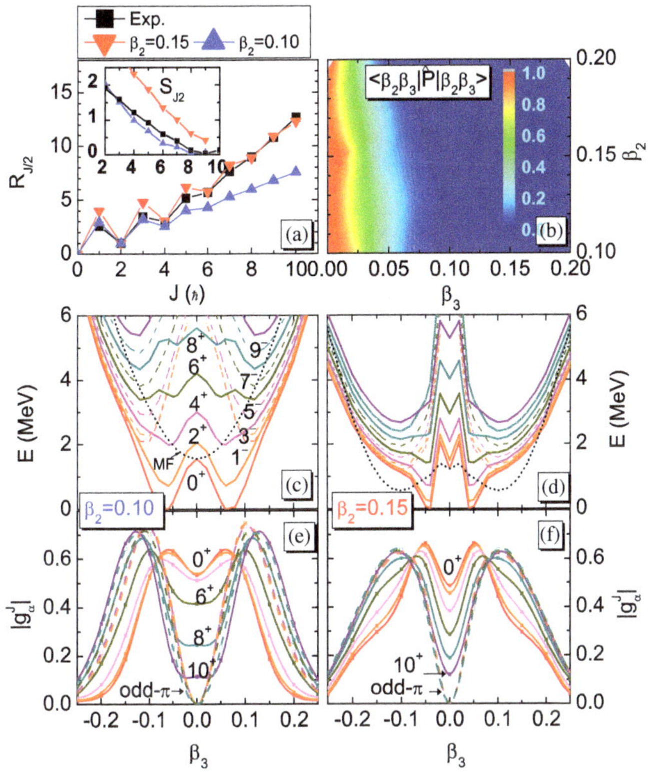

Fig. 15. (a) The energy ratio $R_{J/2} = E_x(J^\pi)/E_x(2^+)$ and (inset) staggering amplitude $S_{J/2}$ of (61) from the SPGCM calculation with the mixing of the configurations with different values of β_3, keeping the β_2 being fixed at 0.15 and 0.10, respectively. (b) Norm overlap $\langle\Phi(\beta_2, \beta_3|\hat{P}|\Phi(\beta_2, \beta_3)\rangle$ of the parity operator; (c) and (d) mean-field and projected energy curves (solid lines for even J, dashed lines for odd J) in ^{224}Ra. The mean-field energy surfaces in (c) and (d) are shifted down by 3.43 and 4.15MeV, respectively. (e) and (f) Absolute value of collective wave functions for parity-doublet states. The peak at $\beta_3 = 0.02$ in (d) is due to the collapse of pairing correlation between protons. Figure adapted from Ref. 22. Figure reprinted with permission from the American Physical Society.

and 15(f). With the increase of spin, the dominated configuration of positive-parity state drifts gradually from the weak octupole configurations ($\beta_3 \simeq 0.05$) to those with large octupole shapes ($\beta_3 \in [0.10, 0.15]$), for which the overlap $\langle\Phi|\hat{P}|\Phi\rangle \to 0$. The origin of spin-dependent parity splitting in low-spin states is related to the octupole shape stabilization of positive-parity states, the dominated shapes of which drift gradually to that of negative-parity ones.

4. Schiff Moments in Odd-Mass Nuclei and Octupole Deformation

The observation of nonzero permanent EDMs in elementary or composite particles, such as neutrons, nuclei, atoms or molecules, would signal the violation of T-symmetry. This violation also implies the violation of CP symmetry, as predicted by the CPT theorem.[127] CP violation is one of Sakharov's conditions for explaining the matter-antimatter asymmetry of the Universe.[128] However, CP violation within the Standard Model of particle physics cannot account for the observed asymmetry,[129,130] necessitating the existence of new CP-violating sources beyond the Standard Model. The EDM signals arising from PT-violating interactions allow for the differentiation of these much weaker CP-violating interactions from the dominant strong and electromagnetic interactions. Therefore, the study of EDMs in various systems has been at the forefront of particle physics, nuclear physics and atomic physics,[53] since the first attempt to measure the EDM of neutrons.[131]

The odd-mass atomic nuclei with static intrinsic octupole deformation or a soft octupole vibrational mode have an enhanced Schiff moment, which is important for searching for permanent EDM in atoms.[52] Over the past several decades, extensive research has been conducted on EDM at various levels. Figure 16 illustrates the connections from fundamental theory, including the Standard Model and beyond-standard-model physics, through a series of theoretical levels at different energy scales to experimentally accessible P-odd and T-odd observables in a variety of systems. More details can be found in the review papers.[52,53] Here, we provide a brief overview of the studies on nuclear Schiff moments, with a focus on the discrepancies among the predictions of different nuclear models and the impact of nuclear octupole deformation. In particular, we demonstrate the great potential of extending the SPGCM for calculating nuclear Schiff moments.

4.1. *Nuclear Schiff Moment (NSM)*

It was pointed out by Schiff[132] that for an atom composed of a point-like nucleus and nonrelativistic electrons, the effect of any nonzero nuclear EDM is completely screened by the atomic electrons, regardless of the magnitude of the external potential. As a result, the net atomic EDM is zero. However, this hindrance is lifted when the finite size of the nucleus is taken into account.

Let us consider the charge distribution $\rho(\mathbf{r})$ of an atomic nucleus inside an atom. The corresponding electrostatic potential at the position $\mathbf{R}$ (much

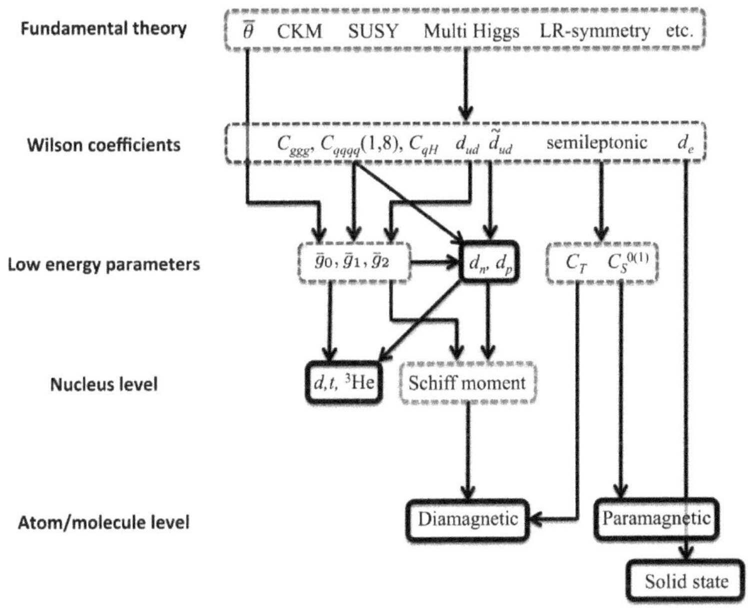

Fig. 16. The hierarchical diagram depicting the connections between a fundamental theory at a high-energy scale and an EDM in a measurable low-energy system. The dashed boxes represent levels that are dominated by theory, while the solid boxes identify systems that are the object of current and future experiments. Figure adapted from Ref. 53. Figure reprinted with permission from the American Physical Society.

larger than the nuclear size) is given by

$$\varphi(R) = \int \frac{e\rho(\mathbf{r})}{|\mathbf{R}-\mathbf{r}|} d^3r + \frac{1}{Z}(\mathbf{d}\cdot\nabla)\int \frac{\rho(\mathbf{r})}{|\mathbf{R}-\mathbf{r}|} d^3r, \tag{62}$$

where the second term is for the screen effect induced from the electrons of the atom. Only the dipole (PT-odd) term is kept, where $\mathbf{d}$ is the electric dipole moment of the atomic nucleus and Z is the proton number,

$$\mathbf{d} = \int d^3r \mathbf{r}\rho(\mathbf{r}), \quad Z = \int d^3r\rho(\mathbf{r}). \tag{63}$$

With the help of multipole expansion

$$\begin{aligned} \frac{1}{|\mathbf{R}-\mathbf{r}|} &= \sum_{\ell=0} \frac{r^\ell}{R^{\ell+1}} P_\ell(\cos\theta) \\ &= \frac{1}{R} - \mathbf{r}\cdot\nabla\frac{1}{R} + \frac{1}{2}(\mathbf{r}\cdot\nabla)^2\frac{1}{R} - \frac{1}{6}(\vec{r}\cdot\nabla)^3\frac{1}{R} + \cdots, \end{aligned} \tag{64}$$

the electrostatic potential $\varphi(R)$ is simplified as

$$\begin{aligned}\varphi(R) &= \frac{Ze}{R} - e\int d^3r\rho(\mathbf{r})\mathbf{r}\cdot\nabla\frac{1}{R} + \frac{e}{2}\int d^3r\rho(\mathbf{r})(\mathbf{r}\cdot\nabla)^2\frac{1}{R} - \frac{e}{6}\int d^3re\rho(\mathbf{r})\\ &\quad\times(\mathbf{r}\cdot\nabla)^3\frac{1}{R} + \frac{1}{Z}(\mathbf{d}\cdot\nabla)\int d^3r\frac{\rho(r)}{R} - \frac{1}{Z}(\mathbf{d}\cdot\nabla)\int d^3r\mathbf{r}\cdot\nabla\frac{1}{R}\\ &\quad+\frac{1}{Z}(\mathbf{d}\cdot\nabla)\frac{1}{2}\int d^3r(\mathbf{r}\cdot\nabla)^2\frac{1}{R} - \frac{1}{Z}(\mathbf{d}\cdot\nabla)\frac{1}{6}\int d^3r(\mathbf{r}\cdot\nabla)^3\frac{1}{R} + \cdots\\ &\equiv \varphi^{(\text{even})}(R) + \varphi^{(\text{odd})}(R),\end{aligned} \tag{65}$$

where the terms in the second and third lines are from the screen effect. One can see that the sub-leading term in the first line cancels the leading term in the second line, as shown in Refs. 132 and 133. The potential is divided into P-even and P-odd terms, where the P-odd terms are

$$\begin{aligned}\varphi^{(\text{odd})}(R) &= -\frac{e}{6}\int d^3r\rho(\mathbf{r})(\mathbf{r}\cdot\nabla)^3\frac{1}{R} + \frac{1}{Z}(\mathbf{d}\cdot\nabla)\frac{1}{2}\int d^3r(\mathbf{r}\cdot\nabla)^2\frac{1}{R} + \cdots\\ &= -\mathbf{S}\cdot\nabla\nabla^2\frac{1}{R} - \frac{1}{6}\sum_{i,j,k}Q_{ijk}\nabla_i\nabla_j\nabla_k\frac{1}{R} + \cdots.\end{aligned} \tag{66}$$

In the last equivalence, the following relation is used:

$$\begin{aligned}\nabla_i\nabla_j\nabla_k\frac{1}{R} &= \left[\nabla_i\nabla_j\nabla_k - \frac{1}{5}(\delta_{ij}\nabla_k + \delta_{ik}\nabla_j + \delta_{kj}\nabla_i)\nabla^2\right]\frac{1}{R}\\ &\quad+\frac{1}{5}(\delta_{ij}\nabla_k + \delta_{ik}\nabla_j + \delta_{kj}\nabla_i)\nabla^2\frac{1}{R}\end{aligned} \tag{67}$$

and the Schiff moment $\mathbf{S}$ is defined as

$$\mathbf{S} = \frac{1}{10}\left(\int e\rho(\mathbf{r})r^2\mathbf{r}d^3r - \frac{5}{3Z}\mathbf{d}\int\rho(\mathbf{r})r^2d^3r\right). \tag{68}$$

In the first quantization, the operator of Schiff moment is written as[133]

$$\hat{\mathbf{S}} = \hat{\mathbf{S}}^{(\text{c})} + \hat{\mathbf{S}}^{(\text{EDM})}, \tag{69}$$

where the first term is due to the charge distribution of the nucleus[b]

$$\hat{\mathbf{S}}^{(\text{c})} = \frac{1}{10}\sum_{p=1}^{Z}e\left(\hat{r}_p^2 - \frac{5}{3}R_c^2\right)\hat{\mathbf{r}}_p \tag{70}$$

[b] Here, $\hat{\mathbf{r}}_p$ stands for the coordinate operator for protons, instead of unit vector.

and the second term accounts for the contribution from the EDMs of neutron and proton inside the atomic nucleus,

$$\hat{\mathbf{S}}^{(\mathrm{EDM})} = \frac{1}{6}\sum_{N=1}^{A}\mathbf{d}_N(\hat{r}_N^2 - R_{ch}^2) + \frac{1}{5}\left[(\hat{\mathbf{r}}_N\cdot\mathbf{d}_N)\hat{\mathbf{r}}_N - \frac{1}{3}\mathbf{d}_N\hat{r}_N^2\right], \tag{71}$$

where R_c^2 is the mean squared radius of the nuclear charge distribution,

$$R_c^2 = \frac{1}{Z}\int d^3r\mathbf{r}^2\rho(\mathbf{r}) \tag{72}$$

and $\mathbf{d}_N$ is the EDM of a neutron ($N = n$) or proton ($N = p$),

$$\mathbf{d}_N = \int d^3r\mathbf{r}\rho_c^{(N)}(\mathbf{r}), \tag{73}$$

where $\rho_c^{(N)}(\mathbf{r})$ is the charge distribution inside the neutron or proton, $\mathbf{r}_N$ is the coordinate of nucleon. The sum $\sum_N$ runs over all nucleons, while $\sum_p$ is restricted to protons. It is shown in Eq. (71) that the $\mathbf{S}^{(\mathrm{EDM})}$ is zero if $\mathbf{d}_N = 0$. In the below discussion, we exclude this contribution to the Schiff moment.

The PT-violating electrostatic interaction $\hat{H}_{\mathrm{PT}}^{(eN)}$ between electrons and nuclear Schiff moment $\mathbf{S}$ in the atom is given by[134]

$$\hat{H}_{\mathrm{PT}}^{(eN)} = -4\pi e\sum_i \mathbf{S}\cdot\boldsymbol{\nabla}\delta^3(R_i), \tag{74}$$

which mixes atomic states of opposite parity and induces a T-violating EDM in the atomic ground state,[134]

$$d_A \equiv \langle\Psi_0^{(at)}|\hat{D}_z|\Psi_0^{(at)}\rangle \simeq 2\sum_{m\neq 0}\frac{\langle\Phi_0^{(at)}|\hat{H}_{\mathrm{PT}}^{(eN)}|\Phi_m^{(at)}\rangle\langle\Phi_m^{(at)}|\hat{D}_z|\Phi_0^{(at)}\rangle}{\mathcal{E}_0^A - \mathcal{E}_m^A}, \tag{75}$$

where $|\Psi_0^{(at)}\rangle$ is the perturbed atomic ground state with the mixture of parities due to $\hat{H}_{\mathrm{PT}}^{(eN)}$. The sum $\sum_m$ runs over a complete set of unperturbed atomic states $|\Phi_m^{(at)}\rangle$. The $|\Phi_0^{(at)}\rangle$ and $|\Phi_m^{(at)}\rangle$ denote the atomic ground state and excited states with $\mathcal{E}_0^A, \mathcal{E}_m^A$ being their energies and $\hat{D}_z = -e\hat{r}_z$ is the atomic electric dipole operator. The atomic states have been calculated with different atomic many-body approaches, including Dirac–Fock method,[134,135] the multiconfiguration Dirac–Hartree–Fock method,[136,137] RPA[134,135] and relativistic coupled-cluster (RCC) theory.[138–142] Here we quote the results by the Dirac–Fock method and RCC theory.

- In the Dirac–Fock calculation, atoms are treated as systems of closed-shell plus some valence electrons. The wave functions of electrons are determined by the single-particle relativistic HF equation. Correlations between the valence electrons were considered using the configuration-interacting method and the valence-core correlations were included using Many-Body Perturbation Theory (MBPT). The results for the three popular atoms ^{129}Xe, ^{199}Hg and ^{225}Ra are as follows[134,135]:

$$d_A(^{129}\text{Xe}) = 0.38 \times 10^{-17} S(e\,\text{fm}^3)^{-1}(e\,\text{cm}), \tag{76a}$$
$$d_A(^{199}\text{Hg}) = -2.8 \times 10^{-17} S(e\,\text{fm}^3)^{-1}(e\,\text{cm}), \tag{76b}$$
$$d_A(^{225}\text{Ra}) = -8.5 \times 10^{-17} S(e\,\text{fm}^3)^{-1}(e\,\text{cm}). \tag{76c}$$

- In the RCC calculation, the wave functions of atomic states are constructed as superposition of configurations truncated to singlets and doublets excitations on top of a reference state. The results of the RCC calculations for ^{129}Xe,[139] ^{199}Hg[138] and ^{225}Ra[140] are as follows:

$$d_A(^{129}\text{Xe}) = 0.336 \times 10^{-17} S(e\,\text{fm}^3)^{-1}(e\,\text{cm}), \tag{77a}$$
$$d_A(^{199}\text{Hg}) = -2.46 \times 10^{-17} S(e\,\text{fm}^3)^{-1}(e\,\text{cm}), \tag{77b}$$
$$d_A(^{225}\text{Ra}) = -6.79 \times 10^{-17} S(e\,\text{fm}^3)^{-1}(e\,\text{cm}). \tag{77c}$$

It is shown above that the atomic EDM d_A also depends on the Schiff moment S of atomic nucleus,

$$S \equiv \langle \Psi_0 | \hat{S}_z | \Psi_0 \rangle, \tag{78}$$

which is zero if parity is conserved in the nuclear state $|\Psi_0\rangle$. However, if there is a PT-violating nucleon–nucleon interaction $\hat{V}_{\text{PT}}$, parity will be violated in the nuclear state. Considering the fact that the $\hat{V}_{\text{PT}}$ is much weaker (if not zero) compared to the standard nuclear force, the wave function of parity-mixed nuclear ground state $|\Psi_0\rangle$ can be obtained with the first-order perturbation theory, i.e.,

$$|\Psi_0\rangle \simeq |\Phi_0\rangle + \sum_{i\neq 0} \frac{\langle \Phi_i | \hat{V}_{\text{PT}} | \Phi_0 \rangle}{E_0 - E_i} |\Phi_i\rangle. \tag{79}$$

Here, $|\Phi_0\rangle$ and $|\Phi_{i\neq 0}\rangle$ represent the wave functions of nuclear ground state and excited states of the Hamiltonian $\hat{H}_0$ with PT-even nuclear strong interaction. With this approximation, the Schiff moment S of the ground state $|\Psi_0\rangle$ becomes

$$S \simeq \sum_{i\neq 0} \frac{\langle \Phi_0 | \hat{S}_z | \Phi_i \rangle \langle \Phi_i | \hat{V}_{\text{PT}} | \Phi_0 \rangle}{E_0 - E_i} + \text{c.c.} \tag{80}$$

Since the PT-violating nuclear interaction $\hat{V}_{\rm PT}$ is a scalar and changes parity and $\hat{S}_z$ is a tensor operator of rank one, the intermediate excited states $|\Phi_i\rangle$ are required to have the same angular momentum as the ground state but with opposite parity, and the angular momentum should be nonzero. With these considerations, the atoms of odd-mass nuclei are chosen to search for permanent EDMs.

No matter which fundamental theory is considered for the CP-violating source at high-energy scale, as listed in Fig. 16, the PT-violating interaction $\hat{V}_{\rm PT}$ at the nuclear energy scale contains the contribution from the long-range one-pion exchange, as schematically depicted in Fig. 17. The standard PT-even πN coupling is described by the following Lagrangian density in the nonrelativistic framework:

$$\mathcal{L}_{\pi N}^{(\rm PT-even)} = -\frac{g_{\pi NN}}{2m_N}\bar{N}(\boldsymbol{\sigma}\cdot\boldsymbol{\nabla})\vec{\tau}\cdot\vec{\pi}N, \tag{81}$$

where N stands for nucleon field, $g_{\pi NN} = m_N g_A/f_\pi$ for the coupling constant with the pion decay constant $f_\pi \simeq 92.4\,\text{MeV}$, and axial-vector coupling $g_A \simeq 1.27$. The nonrelativistic Lagrangian for the PT-odd πN coupling is[52]

$$\mathcal{L}_{\pi N}^{(\rm PT-odd)} = \bar{N}[\bar{g}_0\vec{\tau}\cdot\vec{\pi} + \bar{g}_1\pi^0 + \bar{g}_2(3\tau_3\pi^0 - \vec{\tau}\cdot\vec{\pi})]N, \tag{82}$$

where $\bar{g}_{i=0,1,2}$ are the unknown Low-Energy Constants (LECs) for the isoscalar, isovector and isotensor types of PT-violating πN interactions. With the above two Lagrangian densities, one can derive the PT-violating two-body interaction $\hat{V}_{\rm PT}(\mathbf{r}_{12})$ in the coordinate-space representation[143–146]

$$\hat{V}_{\rm PT}(\mathbf{r}_{12}) = g_{\pi NN}(\bar{g}_0\mathcal{V}_0^{(\rm PT-odd)} + \bar{g}_1\mathcal{V}_1^{(\rm PT-odd)} + \bar{g}_2\mathcal{V}_2^{(\rm PT-odd)}), \tag{83}$$

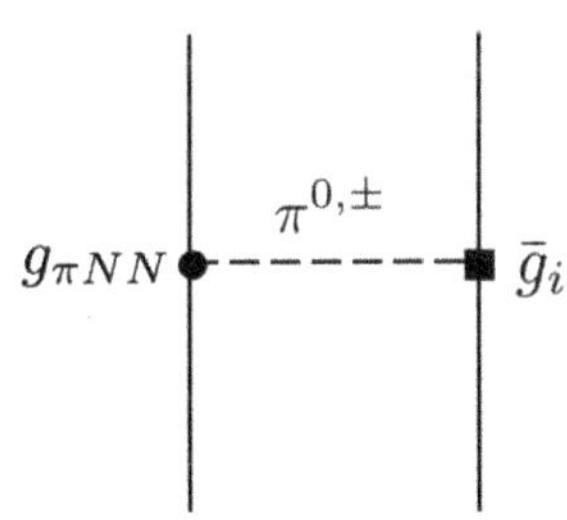

Fig. 17. The long-range 1π-exchange diagram contributing to the leading-order PT-odd nucleon–nucleon interactions. The solid and dashed lines represent nucleon and pion, respectively. The filled circle represents the standard PT-even πN interaction with the Low-Energy Constant $g_{\pi NN}$, while the square stands for the PT-violating πN interactions of isoscalar, isovector and isotensor types with the LECs $\bar{g}_{0,1,2}$, respectively.

where the PT-odd potentials $\mathcal{V}_i^{(\mathrm{PT-odd})}(\mathbf{r}_{12})$ are given by

$$\mathcal{V}_0^{(\mathrm{PT-odd})}(\mathbf{r}_{12}) = -\frac{m_\pi^2}{8\pi m_N}(\vec{\tau}_1 \cdot \vec{\tau}_2)(\boldsymbol{\sigma}_{12}^- \cdot \hat{\mathbf{r}}_{12})\frac{e^{-x}}{x}\left(1+\frac{1}{x}\right), \tag{84a}$$

$$\mathcal{V}_1^{(\mathrm{PT-odd})}(\mathbf{r}_{12}) = +\frac{m_\pi^2}{16\pi m_N}[(\tau_{1z}+\tau_{2z})(\boldsymbol{\sigma}_{12}^- \cdot \mathbf{r}_{12})$$
$$+(\tau_{1z}-\tau_{2z})(\boldsymbol{\sigma}_{12}^+ \cdot \hat{\mathbf{r}}_{12})]\frac{e^{-x}}{x}\left(1+\frac{1}{x}\right), \tag{84b}$$

$$\mathcal{V}_2^{(\mathrm{PT-odd})}(\mathbf{r}_{12}) = -\frac{m_\pi^2}{8\pi m_N}(3\tau_{1z}\tau_{2z} - \vec{\tau}_1 \cdot \vec{\tau}_2)(\boldsymbol{\sigma}_{12}^- \cdot \hat{\mathbf{r}}_{12})\frac{e^{-x}}{x}\left(1+\frac{1}{x}\right). \tag{84c}$$

Here, $\mathbf{r}_{12} \equiv \mathbf{r}_1 - \mathbf{r}_2$ is for the relative coordinate of two interacting nucleons, and $\boldsymbol{\sigma}_{12}^{\pm} \equiv \boldsymbol{\sigma}_1 \pm \boldsymbol{\sigma}_2$. Two dimensionless quantities $x \equiv m_\pi r_{12}$ and $\hat{\mathbf{r}}_{12} \equiv \mathbf{r}_{12}/r_{12}$ are introduced. The third-component of isospin τ_z is $+1$ for neutrons and -1 for protons, m_π and m_N are the masses of pion and nucleon, respectively.

Substituting the PT-violating nuclear interaction (83) into (80), one can express the Schiff moment in terms of the LECs $\bar{g}_i$,

$$S = g_{\pi NN}(a_0\bar{g}_0 + a_1\bar{g}_1 + a_2\bar{g}_2), \tag{85}$$

where the coefficients $a_{i=0,1,2}$ contain all the nuclear-structure information from theoretical calculations and thus vary with nuclear models.

4.2. *Calculations of NSM*

According to Eq. (80), the nuclear Schiff moment S may be enhanced in nuclei with a low-lying excited state having a similar energy and the same angular momentum as the ground state but with opposite parity. This type of low-energy structure is expected to exist in nuclei with intrinsic octupole deformation or a soft octupole vibration mode, which could lead to strongly enhanced nuclear Schiff moments by a factor of up to two to three orders of magnitude.[145,147,148] Therefore, the Schiff moments of odd-mass nuclei with strong octupole correlations have garnered significant attention in the nuclear theory community. However, determining Schiff moments presents a significant challenge because they are sensitive to the excitation energies and wave functions of parity-doublet states. The nucleus ^{225}Ra serves as a notable example of octupole-deformed nuclei that are of great interest to experimentalists.[53] The energy splitting between the parity doublets (ground state with spin-parity $1/2^+$ and excited state $1/2^-$) is approximately 55 keV,[149] as shown in Fig. 18.

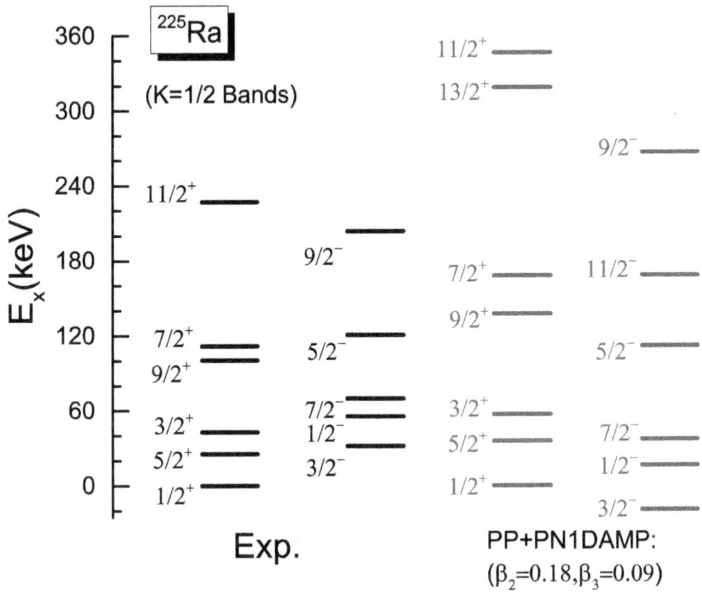

Fig. 18. The energy spectra of low-lying parity-doublet states with $K = 1/2$ in ^{225}Ra from the PP+PN1DAMP calculation based on the quadrupole-octupole deformed state with deformation parameters ($\beta_2 = 0.18, \beta_3 = 0.09$) by the CDFT, in comparison with corresponding data.[120] The results are taken from Ref. 66.

A variety of nuclear models based on different levels of approximation have been employed to investigate the low-lying states and Schiff moments (80) of nuclei that are relevant for the measurements of atomic EDMs. These models include the simple individual-particle model (IPM) with a phenomenological Woods–Saxon (WS) potential,[150] and the IPM + RPA with core-polarization corrections determined from RPA method,[151] the self-consistent mean-field approach of Skyrme HFB/HF + BCS[152–154] and the SHFB + QRPA with core-polarization corrections by QRPA,[155] the collective models of Particle-Rotor Model (PRM)[156] and Quadrupole-Octupole Vibrator (QOV),[157] and the valence-space Interacting Shell Model (ISM) with different effective interactions and model truncation schemes.[158–161] In the following, we sketch the basic ideas of the three types of popularly used nuclear models for the Schiff moments.

- The collective particle-core coupling model or PRM, in which the wave function of ground state with a small amount of mixing of the opposite-parity component due to the PT-odd nuclear force $V_{\rm PT}$ is approximated as[156]

$$|\Psi\rangle = \frac{1}{\sqrt{2}}[(1+\alpha)|JMK\rangle + (1-\alpha)|JM-K\rangle], \tag{86}$$

where the mixing amplitude α is determined by

$$\alpha = \frac{\langle JMK(-)|V^{PT}|JMK(+)\rangle}{E^+ - E^-} \tag{87}$$

and $E^+ - E^-$ is the energy splitting between the parity-doublet states. The wave function of $|JMK\rangle$ is a product of collective rotation part and intrinsic part,

$$|JMK\rangle = \left(\frac{2J+1}{8\pi^2}\right)^{1/2} D^J_{MK}(\varphi,\vartheta,\psi)\varphi_K^{(A)}\chi_c^{(A)}, \tag{88}$$

where D^J_{MK} is the Wigner-D function, $\chi_c^{(A)}$ is the wave function of the quadrupole and octupole deformed (reflection asymmetric) nuclear core in the intrinsic frame. The $\varphi_K^{(A)}$ stands for the wave function of the unpaired nucleon in the intrinsic frame, with an angular momentum projection of K along the symmetric axis, usually determined by the Schrodinger equation with a reflection-asymmetric single-particle plus pairing Hamiltonian in the particle-core coupling models.[156,162]
For simplicity, one can use the following expression to estimate the Schiff moment[148,163]:

$$S \approx 1 \times 10^{-4} \frac{J}{J+1}\beta_2\beta_3^2 ZA^{2/3}\frac{\text{keV}}{E^- - E^+} e\eta\,\text{fm}^3, \tag{89}$$

where β_2 and β_3 are the quadrupole and octupole deformation parameters of the nucleus in intrinsic frame, η is the dimensionless strength constant (a function of the LECs $\bar{g}_i$[151]) of the one-body mean-field potential corresponding to the PT-violating nuclear force,[145,164]

$$\langle a|\hat{U}_{\text{PT}}|b\rangle = \sum_{\epsilon_c<\epsilon_F}\langle ac|\hat{V}_{\text{PT}}|bc\rangle \simeq \frac{G}{\sqrt{2}}\frac{\eta}{2m_N}\langle a|(\boldsymbol{\sigma}\cdot\nabla)\rho_t|b\rangle. \tag{90}$$

Here G is the Fermi constant, ρ_t is the total density of neutrons and protons.

It is seen from Eq. (89) that the Schiff moment S is proportional not only to $ZA^{2/3}$, but also to quadrupole deformation β_2 and the square of octupole deformation parameter β_3.[148] This implies that the nuclear Schiff moment can be greatly enhanced by octupole deformation.[147,157,165] Thus, the atoms with an odd–even heavy octupole-correlated nuclei are of particular interest for the experimental search for permanent atomic EDMs. Indeed, it is shown in Fig. 19 that the a_i values of ^{225}Ra are overall larger than those of ^{129}Xe and ^{199}Hg by two orders of magnitude.

- The self-consistent mean-field approaches, in which all nucleons are treated at the same foot in the mean-field approximation,[145,152,153] various Skyrme EDFs are usually used. The Schiff moment in the mean-field approximation is given by the product of the intrinsic expectation value

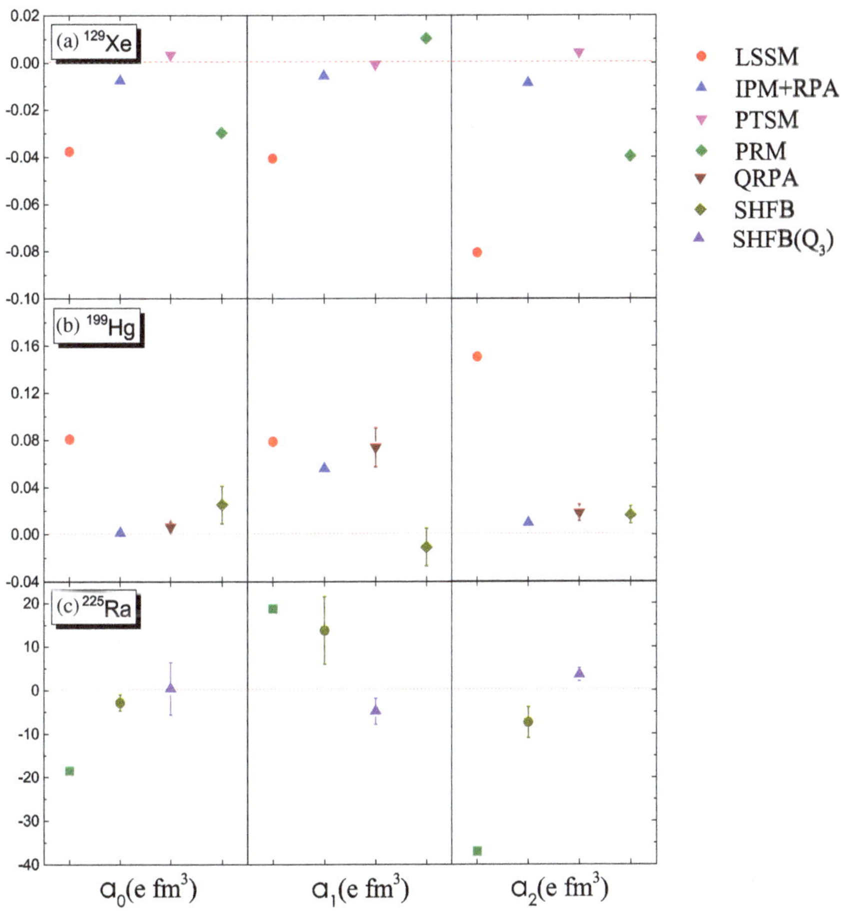

Fig. 19. The coefficients (a_0, a_1, a_2) of nuclear Schiff moments for ^{129}Xe, ^{199}Hg and ^{225}Ra from different model calculations, including the IPM + RPA,[151] SHFB,[152,153] SHFB(Q_3),[154] PRM,[156] LSSM,[160] SHFB + QRPA[155] and PTSM.[159]

of $\hat{S}_z$ and $\hat{V}_{\rm PT}$,

$$S \approx -2\frac{J}{J+1} \quad \langle\Phi(\mathbf{q})|\hat{S}_z|\Phi(\mathbf{q})\rangle \quad \frac{\langle\Phi(\mathbf{q})|V_{\rm PT}|\Phi(\mathbf{q})\rangle}{E^- - E^+}, \tag{91}$$

where the $J/(J+1)$ comes from the transformation from intrinsic frame to laboratory frame. $|\Phi(\mathbf{q})\rangle$ is the parity-violating wave function of mean-field energy-minimal state. Since parity is usually violated in the mean-field states, there is no way to calculate the energy of $E_\pm$. Thus, the energy difference $E^- - E^+$ of two opposite parity states but with same angular momentum is taken from the data. The mean-field potential $\hat{U}_{\rm PT}$ under the zero-range approximation ($m_\pi \to \infty$) is simplified as[145]

$$\begin{aligned}\hat{U}_{PT}(\boldsymbol{r}) \simeq -\frac{g_{\pi NN}}{2m_\pi^2 m_N}\Bigg\{&\sum_{i=1}^{A}\boldsymbol{\sigma}_i\tau_{z,i}\cdot[(\bar{g}_0+2\bar{g}_2)\boldsymbol{\nabla}\rho_1(\boldsymbol{r})-\bar{g}_1\boldsymbol{\nabla}\rho_0(\boldsymbol{r})]\\ &+\frac{1}{2}\sum_{i=1}^{A}\boldsymbol{\sigma}_i\cdot[(-3\bar{g}_0+\bar{g}_1\tau_{z,i})\boldsymbol{J}_0(\boldsymbol{r})+(\bar{g}_1+\bar{g}_0\tau_{z,i}-4\bar{g}_2\tau_{z,i})\boldsymbol{J}_1(\boldsymbol{r})]\Bigg\},\end{aligned} \tag{92}$$

where the isoscalar and isovector densities $\rho_0 = \rho_n + \rho_p$ and $\rho_1 = \rho_n - \rho_p$, and the "spin-orbit" current $\boldsymbol{J}_{i=0,1}(\boldsymbol{r})$ in the Skyrme EDF. The intrinsic expectation value of $\hat{V}_{\rm PT}$ is finally given by the expectation value of $\hat{U}_{PT}$. Since the summation $\sum_i$ runs over all the nucleons, core polarization effect is automatically taken into account.

- The BMF approaches in which the ground state and excited states are calculated in the laboratory frame with the configuration mixing effect. Recently, *ab initio* no-core shell model (NCSM) framework has been applied to calculate EDMs of several light nuclei using chiral two- and three-body interactions and a PT-violating Hamiltonian based on a one-meson-exchange model.[166] Application of this kind of studies to the Schiff moments of heavy nuclei are very challenging. So far, only the valence-space ISM[160,161] based on different approximations have been applied to nuclear Schiff moments, including the Pair-Truncated Shell Model (PTSM),[158,159] where many-body configurations are composed of nucleon pairs with specific angular momenta. In these studies, the two-body matrix elements of PT-violating nucleon–nucleon interactions are evaluated explicitly.

Given the remarkable success of the SPGCM in describing low-lying states of octupole-deformed even–even nuclei, there is considerable interest in extending this framework to investigate nuclear Schiff moments. To achieve this, it's essential to adapt the GCM for odd-mass nuclei, which has already been developed based on HFB states utilizing Skyrme[167] and Gogny[168] forces. However, these studies have not yet incorporated octupole deformation and parity projection. More recently, progress has been made in the SPGCM based on the CDFT for quadrupole-octupole deformed odd-mass nuclei.[66] As shown in Fig. 18, this approach successfully reproduces the main features of the energy spectrum for parity doublets with $K = 1/2$ in ^{225}Ra, with the exception that the negative-parity band is lower in energy than the positive-parity one which is different from the data. This discrepancy is expected to be resolved by incorporating the mixing of deformed states with different values of (β_2, β_3).

Table 1 summarizes the coefficients $a_{0,1,2}$ of nuclear Schiff moments obtained from various models for nuclei of experimental interest. Figure 19 provides a comparison of these coefficients for three prominent isotopes

Table 1. The coefficients a_i ($e$fm^3) of nuclear Schiff moments from the calculations of different nuclear models.

Isotopes	a_0	a_1	a_2	Nuclear models
^{153}Eu	−9.62	47.3	−25.53	PRM[148]
^{129}Xe	−0.038	−0.041	−0.081	LSSM[160]
^{129}Xe	−0.008	−0.006	−0.009	IPM + RPA[151]
^{129}Xe	0.003	−0.001	0.004	PTSM[159]
^{129}Xe	−0.03	0.01	−0.04	PRM[163]
^{199}Hg	0.080	0.078	0.15	LSSM[160]
^{199}Hg	0.0004	0.055	0.009	IPM + RPA[151,169]
^{199}Hg	[0.002, 0.010]	[0.057, 0.090]	[0.011, 0.025]	SHFB + QRPA[155]
^{199}Hg	[0.009, 0.041]	[−0.027, +0.005]	[0.009, 0.024]	SHFB[153]
^{221}Rn	−0.04(10)	−1.7(3)	0.67(10)	SHFB(Q_3)[154]
^{223}Rn	−62	62	−100	PRM[156]
^{223}Rn	−0.08(8)	−2.4(4)	0.86(10)	SHFB(Q_3)[154]
^{223}Ra	−25	25	−50	PRM[156]
^{223}Fr	−31	31	−62	PRM[156]
^{223}Fr	−0.02	−0.02	−0.04	QOV[157]
^{223}Fr	0.07(20)	−0.8(7)	0.05(40)	SHFB(Q_3)[154]
^{225}Ra	−18.6	18.6	−37.2	PRM[156]
^{225}Ra	[−1.0, −4.7]	[6.0, 21.5]	[−3.9, −11.0]	SHFB[152]
^{225}Ra	0.2(6)	−5(3)	3.3(1.5)	SHFB(Q_3)[154]
^{227}Ac	−26	129	−69	PRM[148]
^{229}Pa	−1.2(3)	−0.9(9)	−0.3(5)	SHFB(Q_3)[154]
^{235}U	−7.8	38.7	−20.7	PRM[148]
^{237}Np	−15.6	77.4	−41.4	PRM[148]

^{129}Xe, ^{199}Hg and ^{225}Ra. It is evident that the values of a_i exhibit significant discrepancies, varying by up to two to three orders of magnitude and sometimes even having different signs. This substantial variation in a_i introduces considerable uncertainty into the Schiff moments and consequently, into the allowed values for the LECs characterizing Parity-and-Time (PT)-violating nucleon–nucleon interactions.

In a Skyrme HFB study,[152] it was observed that finite-range effects reduce the direct matrix elements of PT-odd interactions (and the corresponding a_i values) from the zero-range limit by a factor of two or three. Furthermore, the different choices of Skyrme EDFs introduce an uncertainty of a similar magnitude. Minimizing the discrepancy among these models remains a primary objective in the nuclear theory community. Achieving this goal involves understanding the assumptions and approximations made by each model and quantifying the potential errors associated with them.[52]

Recent research has unveiled strong correlations between the intrinsic Schiff moments $\langle\Phi(\mathbf{q})|\hat{S}_z|\Phi(\mathbf{q})\rangle$ of odd-mass nuclei and the octupole moments $Q_{30} = \langle\Phi(\mathbf{q})|\hat{Q}_{30}|\Phi(\mathbf{q})\rangle$ of neighboring even–even nuclei.[154] With the correlation relation and available data on octupole moments Q_{30}, it has become possible to determine the intrinsic Schiff moments. This correlation significantly reduces systematic uncertainties originating from nuclear EDFs. Furthermore, linear correlation relations have been identified between the coefficients a_i and Q_{30}. Nevertheless, despite these correlations, substantial uncertainties persist in the resulting a_i values due to the significant uncertainty in the data of Q_{30}.

4.3. *Constraints on the LECs of PT-violating nuclear force*

The best experimental upper limits (95% C.L.) on the ground state atomic EDMs of the three atoms ^{129}Xe,[170] ^{199}Hg[171] and ^{225}Ra,[172]

$$|d_A(^{129}\text{Xe})| \le 1.4 \times 10^{-27}(e\,\text{cm}), \tag{93a}$$
$$|d_A(^{199}\text{Hg})| \le 7.4 \times 10^{-30}(e\,\text{cm}), \tag{93b}$$
$$|d_A(^{225}\text{Ra})| \le 1.4 \times 10^{-23}(e\,\text{cm}) \tag{93c}$$

and the coefficients a_i for the nuclei ^{129}Xe, ^{199}Hg and ^{225}Ra from Ref. 52,

$$S(^{129}\text{Xe}) = g_{\pi NN}(-0.008\bar{g}_0 - 0.006\bar{g}_1 - 0.009\bar{g}_2)(e\,\text{fm}^3), \tag{94a}$$
$$S(^{199}\text{Hg}) = g_{\pi NN}(+0.01\bar{g}_0 \pm 0.02\bar{g}_1 + 0.02\bar{g}_2)(e\,\text{fm}^3), \tag{94b}$$
$$S(^{225}\text{Ra}) = g_{\pi NN}(-1.5\bar{g}_0 + 6.0\bar{g}_1 - 4.0\bar{g}_2)(e\,\text{fm}^3), \tag{94c}$$

provide constraints on the allowed values for the LECs $\bar{g}_i$ of PT-violating nuclear force.

To determine the ranges of these values, we assume that the atomic EDMs d_A are only contributed to by the long-range PT-violating nuclear force $V_{\rm PT}$, cf. Fig. 17, and sample the values of $\bar{g}_{0,1,2}$ approximately 10^9 times. The samples that can reproduce atomic EDMs below their corresponding upper limits are displayed in Figs. 20 and 21. The results for the two choices of $a_1(^{199}\text{Hg}) = \pm 0.02$ are compared. It is observed that there are correlations between the three LECs, especially in the case of $a_1(^{199}\text{Hg}) = -0.02$. Quantitatively, the upper limits of the three atomic EDMs provide constraints on the $\bar{g}_i$, which, for the case

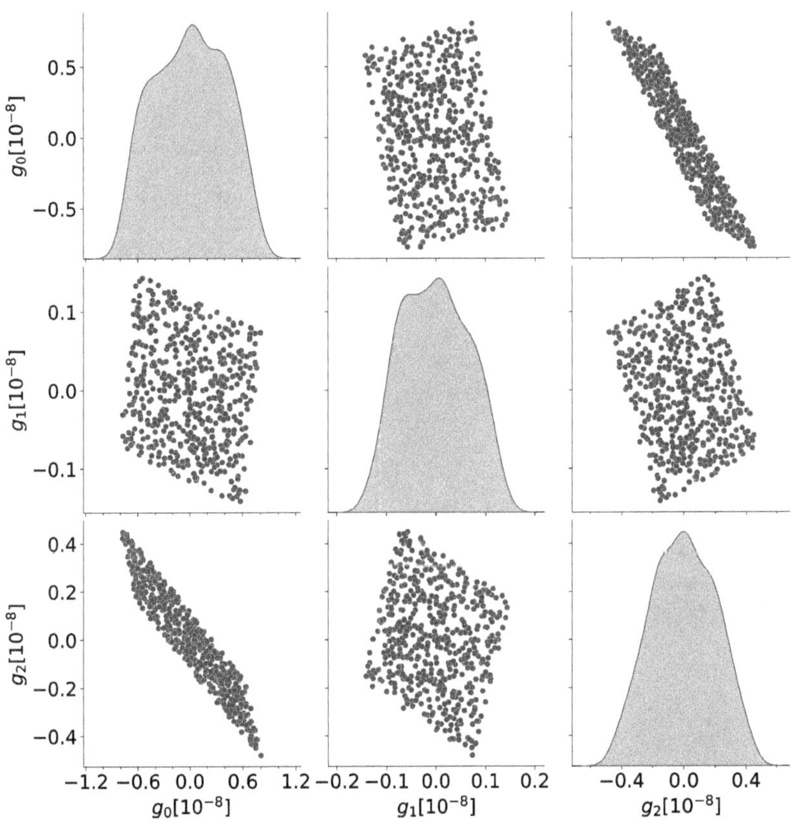

Fig. 20. The distribution of the allowed values for the LECs $\bar{g}_{0,1,2}[10^{-8}]$ of the CP-violating nuclear force constrained by the latest measurements on the EDMs of ^{129}Xe, ^{199}Hg and ^{225}Ra and the theoretical calculations for the atomic factors[134,135] and coefficients in the nuclear Schiff moments,[52] where $a_1(^{199}\text{Hg}) = 0.02$ for ^{199}Hg is used. See text for details.

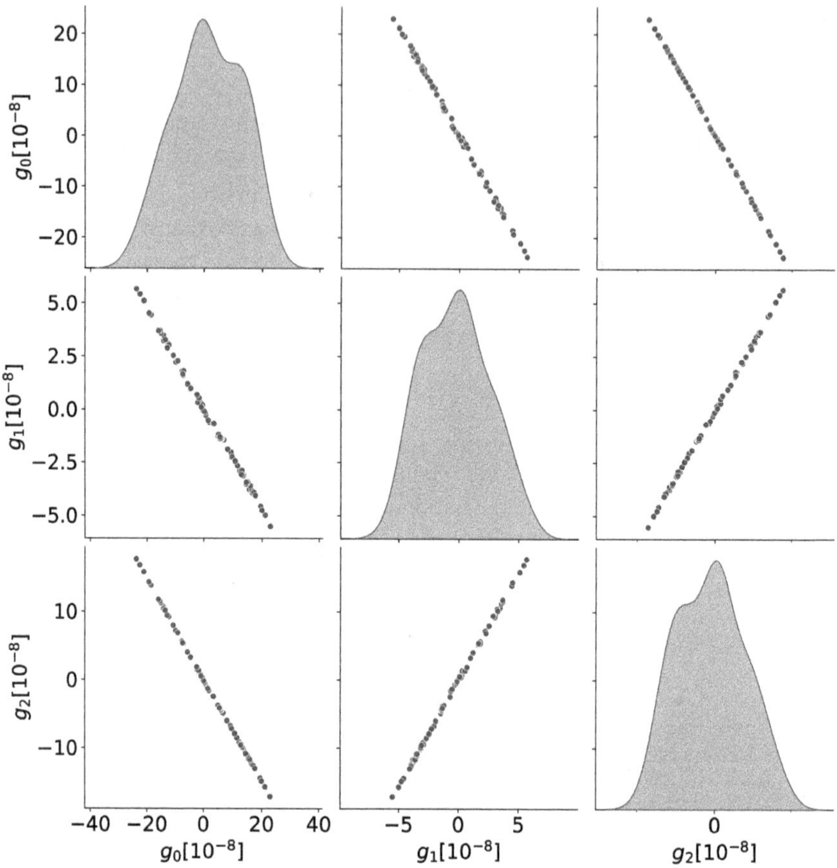

Fig. 21. Same as Fig. 20, but with $a_1 = -0.02$ for ^{199}Hg.

$a_1(^{199}\text{Hg}) = 0.02$, are

$$\bar{g}_0 \in [-0.78, 0.80] \times 10^{-8}, \tag{95a}$$

$$\bar{g}_1 \in [-0.14, 0.14] \times 10^{-8}, \tag{95b}$$

$$\bar{g}_2 \in [-0.48, 0.45] \times 10^{-8} \tag{95c}$$

and for the case $a_1(^{199}\text{Hg}) = -0.02$,

$$\bar{g}_0 \in [-24, 23] \times 10^{-8}, \tag{96a}$$

$$\bar{g}_1 \in [-5.5, 5.6] \times 10^{-8}, \tag{96b}$$

$$\bar{g}_2 \in [-17, 17] \times 10^{-8}. \tag{96c}$$

It is evident that changing the value of the coefficient a_1 for the nuclear Schiff moment of ^{199}Hg from -0.02 to $+0.02$ reduces the ranges of $\bar{g}_i$ by approximately one order of magnitude. This underscores the importance of accurate calculations of nuclear Schiff moments for all the nuclei of interest. Furthermore, it can be observed that the LEC $\bar{g}_1$ associated with the isovector type of PT-violating πN coupling is more tightly constrained than those of isoscalar and isotensor types.

5. Summary and Perspectives

Octupole deformation enriches the structure of atomic nuclei by introducing novel collective excitation modes, the description of which has garnered significant attention in the pursuit of a profound microscopic theory. In this paper, we introduce the symmetry-projected generator coordinate method (SPGCM) for studying nuclear octupole collective motions. This method is based on reference states with different reflection-asymmetric shapes obtained from self-consistent mean-field calculations. The success of SPGCM has been demonstrated through applications to phenomena such as nucleon clustering in light nuclei, and the phase transition from octupole vibrations to rotational motions in heavy nuclei. The SPGCM has also provided important inputs to understand structure of hypernuclei and sub-barrier fusion reactions.

The atoms with octupole-shaped odd-mass nuclei have been of great experimental interest to search for permanent atomic electric dipole moments, as the signal is proportional to nuclear Schiff moment which could be enhanced by several orders of magnitude if the nuclei possess intrinsic octupole deformation. We have reviewed the current status of studies on nuclear Schiff moments. Significant discrepancies in nuclear Schiff moments among different phenomenological nuclear models and the impact of these uncertainties on the determination of the low energy constants of PT-violating nucleon–nucleon interactions have been discussed in detail. Given the success of the SPGCM discussed above, we have shown the great potential for extending this framework to study the low-lying states of odd-mass nuclei and nuclear Schiff moments with octupole deformation, even though numerous challenges need to be overcome. The combination of SPGCM with modern nuclear EDFs, especially the CDFT where the parameters of the time-odd currents (important for odd-mass and rotating nuclei) share the same coupling constants with those of time-even densities, or with the ab initio method of in-medium similarity renormalization group starting from nuclear interactions derived from chiral effective field theory, will offer a compelling approach for determining nuclear Schiff moments in the future.

Acknowledgments

We are grateful to Y. Fu, H. Hergert, C.F. Jiao, G. Li, Z.P. Li, B.N. Lv, J. Meng, H. Mei, P. Ring, Y.T. Rong, L.J. Wang and X.Y. Wu for collaboration and fruitful discussions on the works reviewed in this paper. This work is supported in part by the National Natural Science Foundation of China (Grant Nos. 12375119 and 12141501) and the Guangdong Basic and Applied Basic Research Foundation (2023A1515010936).

ORCID

J. M. Yao https://orcid.org/0000-0001-9505-1852

References

1. I. Ahmad and P. A. Butler, *Ann. Rev. Nucl. Part. Sci.* **43** (1993) 71.
2. P. A. Butler and W. Nazarewicz, *Rev. Mod. Phys.* **68** (1996) 349.
3. M. Chen, T. Li, J. Dobaczewski and W. Nazarewicz, *Phys. Rev. C* **103** (2021) 034303.
4. P. Ring and P. Schuck, *The Nuclear Many-Body Problem* (Springer-Verlag, New York, 1980).
5. M. Bender, P.-H. Heenen and P.-G. Reinhard, *Rev. Mod. Phys.* **75** (2003) 121.
6. S. E. Agbemava, A. V. Afanasjev and P. Ring, *Phys. Rev. C* **93** (2016) 044304.
7. S. Ebata and T. Nakatsukasa, *Phys. Scr.* **92** (2017) 064005.
8. Y. Cao, S. E. Agbemava, A. V. Afanasjev, W. Nazarewicz and E. Olsen, *Phys. Rev. C* **102** (2020) 024311.
9. P. Möller, R. Bengtsson, B. G. Carlsson, P. Olivius, T. Ichikawa, H. Sagawa and A. Iwamoto, *Atom. Data Nucl. Data Tabl.* **94** (2008) 758.
10. B. A. Brown, *Phys. Rev. Lett.* **85** (2000) 5300.
11. J. M. Yao and K. Hagino, *Phys. Rev. C* **94** (2016) 011303.
12. L. M. Robledo, *Eur. Phys. J. A* **52** (2016) 300.
13. P. Van Isacker and M. Rejmund, *Phys. Rev. Res.* **4** (2022) L022031.
14. B. M. Loc, N. L. Anh, P. Papakonstantinou and N. Auerbach, *Phys. Rev. C* **108** (2023) 024303.
15. J. M. Eisenberg and W. Greiner, *Nuclear Theory* (North-Holland Publishing Company, Amsterdam, 1970).
16. D. R. Phillips *et al.*, *J. Phys. G* **48** (2021) 072001.
17. B. Bucher *et al.*, *Phys. Rev. Lett.* **116** (2016) 112503.
18. B. Bucher *et al.*, *Phys. Rev. Lett.* **118** (2017) 152504.
19. L. P. Gaffney *et al.*, *Nature* **497** (2013) 199.
20. P. A. Butler *et al.*, *Phys. Rev. Lett.* **124** (2020) 042503.

21. E. Garrote, J. L. Egido and L. M. Robledo, *Phys. Lett. B* **410** (1997) 86.
22. J. M. Yao, E. F. Zhou and Z. P. Li, *Phys. Rev. C* **92** (2015) 041304.
23. Y. Fu, H. Wang, L. J. Wang and J. M. Yao, *Phys. Rev. C* **97** (2018) 024338.
24. C. Liu, S. Y. Wang, R. A. Bark, S. Q. Zhang, J. Meng, B. Qi, P. Jones, S. M. Wyngaardt, J. Zhao, C. Xu, S.-G. Zhou, S. Wang, D. P. Sun, L. Liu, Z. Q. Li, N. B. Zhang, H. Jia, X. Q. Li, H. Hua, Q. B. Chen, Z. G. Xiao, H. J. Li, L. H. Zhu, T. D. Bucher, T. Dinoko, J. Easton, K. Juhász, A. Kamblawe, E. Khaleel, N. Khumalo, E. A. Lawrie, J. J. Lawrie, S. N. T. Majola, S. M. Mullins, S. Murray, J. Ndayishimye, D. Negi, S. P. Noncolela, S. S. Ntshangase, B. M. Nyakó, J. N. Orce, P. Papka, J. F. Sharpey-Schafer, O. Shirinda, P. Sithole, M. A. Stankiewicz and M. Wiedeking, *Phys. Rev. Lett.* **116** (2016) 112501.
25. Y. Y. Wang, S. Q. Zhang, P. W. Zhao and J. Meng, *Phys. Lett. B* **792** (2019) 454.
26. J. L. Egido and L. M. Robledo, *Nucl. Phys. A* **494** (1989) 85.
27. Z. P. Li, B. Y. Song, J. M. Yao, D. Vretenar and J. Meng, *Phys. Lett. B* **726** (2013) 866.
28. S. Y. Xia, H. Tao, Y. Lu, Z. P. Li, T. Niksic and D. Vretenar, *Phys. Rev. C* **96** (2017) 054303.
29. P. Zhao and Z. Li, *Int. J. Mod. Phys. E* **27** (2018) 1830007.
30. K. Nomura, R. Rodríguez-Guzmán and L. M. Robledo, *Phys. Rev. C* **92** (2015) 014312.
31. K. Nomura, arXiv:2211.00279.
32. J. A. Sheikh, J. Dobaczewski, P. Ring, L. M. Robledo and C. Yannouleas, *J. Phys. G* **48** (2021) 123001.
33. J. M. Yao, *Symmetry Restoration Methods* (Springer-Nature, 2022).
34. S. Marcos, H. Flocard and P. H. Heenen, *Nucl. Phys. A* **410** (1983) 125.
35. P. Bonche, S. J. Krieger, M. S. Weiss, J. Dobaczewski, H. Flocard and P.-H. Heenen, *Phys. Rev. Lett.* **66** (1991) 876.
36. J. Skalski, P.-H. Heenen, P. Bonche, H. Flocard and J. Meyer, *Nucl. Phys. A* **551** (1993) 109.
37. P.-H. Heenen, J. Skalski, P. Bonche and H. Flocard, *Phys. Rev. C* **50** (1994) 802.
38. P. H. Heenen, A. Valor, M. Bender, P. Bonche and H. Flocard, *Eur. Phys. J. A* **11** (2001) 393.
39. J. L. Egido and L. M. Robledo, *Nucl. Phys. A* **524** (1991) 65.
40. R. Rodríguez-Guzmán, L. M. Robledo and P. Sarriguren, *Phys. Rev. C* **86** (2012) 034336.
41. E. F. Zhou, J. M. Yao, Z. P. Li, J. Meng and P. Ring, *Phys. Lett. B* **753** (2016) 227.
42. P. Marević, J. P. Ebran, E. Khan, T. Nikšić and D. Vretenar, *Phys. Rev. C* **97** (2018) 024334.
43. P. Marević, J.-P. Ebran, E. Khan, T. Nikšić, and D. Vretenar, *Phys. Rev. C* **99** (2019) 034317.
44. R. N. Bernard, L. M. Robledo and T. R. Rodríguez, *Phys. Rev. C* **93** (2016) 061302.

45. Y. T. Rong, X. Y. Wu, B. N. Lu and J. M. Yao, *Phys. Lett. B* **840** (2023) 137896.
46. G. Giacalone, J. Jia and C. Zhang, *Phys. Rev. Lett.* **127** (2021) 242301.
47. C. Zhang and J. Jia, *Phys. Rev. Lett.* **128** (2022) 022301.
48. B. Bally, M. Bender, G. Giacalone and V. Somà, *Phys. Rev. Lett.* **128** (2022) 082301.
49. B. Bally *et al.*, arXiv:2209.11042 [nucl-ex].
50. J. M. Yao and J. Engel, *Phys. Rev. C* **94** (2016) 014306.
51. J. M. Yao, J. Meng, Y. F. Niu and P. Ring, *Prog. Part. Nucl. Phys.* **126** (2022) 103965.
52. J. Engel, M. J. Ramsey-Musolf and U. van Kolck, *Prog. Part. Nucl. Phys.* **71** (2013) 21.
53. T. Chupp, P. Fierlinger, M. Ramsey-Musolf and J. Singh, *Rev. Mod. Phys.* **91** (2019) 015001.
54. J. M. Yao, B. Bally, J. Engel, R. Wirth, T. R. Rodríguez and H. Hergert, *Phys. Rev. Lett.* **124** (2020) 232501.
55. M. Frosini, T. Duguet, Jean-P. Ebran, B. Bally, H. Hergert, T. R. Rodríguez, R. Roth, J. M. Yao and V. Somà, *Eur. Phys. J. A* **58** (2022) 64.
56. X. Zhang, W. Lin, J. M. Yao, C. F. Jiao, A. M. Romero, T. R. Rodríguez and H. Hergert, *Phys. Rev. C* **107** (2023) 024304.
57. M. Matsumoto, Y. Tanimura and K. Hagino, arXiv:2308.13233 [nucl-th].
58. L. M. Robledo, T. R. Rodríguez and R. R. Rodríguez-Guzmán, *J. Phys. G* **46** (2018) 013001.
59. T. Niksic, D. Vretenar and P. Ring, *Prog. Part. Nucl. Phys.* **66** (2011) 519.
60. D. J. Thouless, *Nucl. Phys.* **21** (1960) 225.
61. J. L. Egido, J. Lessing, V. Martin and L. M. Robledo, *Nucl. Phys. A* **594** (1995) 70.
62. W. Ryssens, M. Bender and P. H. Heenen, *Eur. Phys. J. A* **55** (2019) 93.
63. B. Bally, A. Sánchez-Fernández and T. R. Rodríguez, *Eur. Phys. J. A* **57** (2021) 69 [Erratum-ibid. **57** (2021) 124].
64. J. L. Egido, H. J. Mang and P. Ring, *Nucl. Phys. A* **334** (1980) 1.
65. C. Bloch and A. Messiah, *Nucl. Phys.* **39** (1962) 95.
66. E. F. Zhou *et al.*, in preparation (2023).
67. D. Vretenar, A. V. Afanasjev, G. A. Lalazissis and P. Ring, *Phys. Rep.* **409** (2005) 101.
68. J. Meng, H. Toki, S. G. Zhou, S. Q. Zhang, W. H. Long and L. S. Geng, *Prog. Part. Nucl. Phys.* **57** (2006) 470.
69. J. Meng, L. S. Song and J. M. Yao, *Int. J. Mod. Phys. E* **26** (2017) 1740020.
70. L. S. Geng, J. Meng and H. Toki, *Chin. Phys. Lett.* **24** (2007) 1865.
71. Z. Li, S. Chen, M. Zhou, Y. Chen and Z. Li, arXiv:2309.03461 [nucl-th].
72. W. Zhang, Z. P. Li, S. Q. Zhang and J. Meng, *Phys. Rev. C* **81** (2010) 034302.
73. J.-Y. Guo, P. Jiao and X.-Z. Fang, *Phys. Rev. C* **82** (2010) 047301.
74. L. F. Yu, P. W. Zhao, S. Q. Zhang and J. Meng, arXiv:1211.0601 [nucl-th].
75. H. Abusara, A. V. Afanasjev and P. Ring, *Phys. Rev. C* **85** (2012) 024314.
76. B.-N. Lu, E.-G. Zhao and S.-G. Zhou, *Phys. Rev. C* **85** (2012) 011301.
77. M. Bender, T. Duguet and D. Lacroix, *Phys. Rev. C* **79** (2009) 044319.

78. T. Duguet, M. Bender, K. Bennaceur, D. Lacroix and T. Lesinski, *Phys. Rev. C* **79** (2009) 044320.
79. S. Weinberg, *Nucl. Phys. B* **363** (1991) 3.
80. H. Hergert, *Front. Phys.* **8** (2020) 379.
81. S. K. Bogner, R. J. Furnstahl and A. Schwenk, *Prog. Part. Nucl. Phys.* **65** (2010) 94.
82. H. Hergert, S. K. Bogner, T. D. Morris, A. Schwenk and K. Tsukiyama, *Phys. Rep.* **621** (2016) 165.
83. R. Balian and E. Brezin, *Nuovo Cimento B* **64** (1969) 37.
84. D. L. Hill and J. A. Wheeler, *Phys. Rev.* **89** (1953) 1102.
85. K. Ikeda, N. Takigawa and H. Horiuchi, *Prog. Theor. Phys. Suppl.* **E68** (1968) 464.
86. M. Freer, H. Horiuchi, Y. Kanada-En'yo, D. Lee and U.-G. Meißner, *Rev. Mod. Phys.* **90** (2018) 035004.
87. T. Ichikawa, J. A. Maruhn, N. Itagaki and S. Ohkubo, *Phys. Rev. Lett.* **107** (2011) 112501.
88. P. W. Zhao, N. Itagaki and J. Meng, *Phys. Rev. Lett.* **115** (2015) 022501.
89. P. Arumugam, B. K. Sharma, S. K. Patra and Raj K. Gupta, *Phys. Rev. C* **71** (2005) 064308.
90. J. P. Ebran, E. Khan, T. Niksic and D. Vretenar, *Phys. Rev. C* **87** (2013) 044307.
91. J. M. Yao, N. Itagaki and J. Meng, *Phys. Rev. C* **90** (2014) 054307.
92. M. Matsumoto and Y. Tanimura, *Phys. Rev. C* **106** (2022) 014307.
93. J. P. Ebran, E. Khan, T. Niksic and D. Vretenar, *Nature* **487** (2012) 341.
94. M. Girod and P. Schuck, *Phys. Rev. Lett.* **111** (2013) 132503.
95. E. Epelbaum, H. Krebs, D. Lee and U.-G. Meissner, *Phys. Rev. Lett.* **106** (2011) 192501.
96. E. Epelbaum, H. Krebs, T. A. Lahde, D. Lee and U.-G. Meissner, *Phys. Rev. Lett.* **109** (2012) 252501.
97. T. Otsuka, T. Abe, T. Yoshida, Y. Tsunoda, N. Shimizu, N. Itagaki, Y. Utsuno, J. Vary, P. Maris and H. Ueno, *Nat. Commun.* **13** (2022) 2234.
98. S. Shen, S. Elhatisari, T. A. Lähde, D. Lee, B.-N. Lu and U.-G. Meißner, *Nat. Commun.* **14** (2023) 2777.
99. Y. Kanada-En'yo, *Phys. Rev. Lett.* **81** (1998) 5291.
100. Y. Kanada-En'yo, *Prog. Theor. Phys.* **117** (2007) 655.
101. H. Feldmeier and J. Schnack, *Rev. Mod. Phys.* **72** (2000) 655.
102. T. Neff and H. Feldmeier, *Nucl. Phys. A* **738** (2004) 357.
103. M. Chernykh, H. Feldmeier, T. Neff, P. von Neumann-Cosel and A. Richter, *Phys. Rev. Lett.* **98** (2007) 032501.
104. Y. Fukuoka, S. Shinohara, Y. Funaki, T. Nakatsukasa and K. Yabana, *Phys. Rev. C* **88** (2013) 014321.
105. A. M. Romero, J. M. Yao, B. Bally, T. R. Rodríguez and J. Engel, *Phys. Rev. C* **104** (2021) 054317.
106. J. M. Yao, M. Bender and P.-H. Heenen, *Phys. Rev. C* **91** (2015) 024301.
107. Y. Funaki, *Phys. Rev. C* **92** (2015) 021302.
108. K. Wang and B.-N. Lu, *Commun. Theor. Phys.* **74** (2022) 015303.

109. Y. Tian, Z. Y. Ma and P. Ring, *Phys. Lett. B* **676** (2009) 44.
110. H. Xia, X. Wu, H. Mei and J. Yao, *Sci. China Phys. Mech. Astron.* **62** (2019) 42011.
111. H. Xia, X. Wu, H. Mei and J. Yao, *Sci. China Phys. Mech. Astron.* **66** (2023) 252011.
112. Y. Kanada-Enyo and H. Horiuchi, *Prog. Theor. Phys.* **93** (1995) 115.
113. Y. Taniguchi, M. Kimura and H. Horiuchi, *Prog. Theor. Phys.* **112** (2004) 475.
114. M. Kimura, *Phys. Rev. C* **69** (2004) 044319.
115. P. W. Zhao, Z. P. Li, J. M. Yao and J. Meng, *Phys. Rev. C* **82** (2010) 054319.
116. M. Isaka, M. Kimura, A. Doté and A. Ohnishi, *Phys. Rev. C* **83** (2011) 054304.
117. M. Isaka, H. Homma, M. Kimura, A. Doté and A. Ohnishi, *Phys. Rev. C* **85** (2012) 034303.
118. J.-W. Cui, X.-R. Zhou, L.-X. Guo and H.-J. Schulze, *Phys. Rev. C* **95** (2017) 024323.
119. H. Mei, K. Hagino and J. M. Yao, *Phys. Rev. C* **93** (2016) 011301.
120. National Nuclear Data Center, NuDat 2 Database (2020), https://www.nndc.bnl.gov/nudat2.
121. M. Yeh, P. E. Garrett, C. A. McGrath, S. W. Yates and T. Belgya, *Phys. Rev. Lett.* **76** (1996) 1208.
122. K. Vetter *et al.*, *Phys. Rev. C* **58** (1998) R2631.
123. K. Hagino, N. Rowley and A. T. Kruppa, *Comput. Phys. Commun.* **123** (1999) 143.
124. K. Hagino and N. Takigawa, *Prog. Theor. Phys.* **128** (2012) 1061.
125. C. R. Morton, A. C. Berriman, M. Dasgupta, D. J. Hinde, J. O. Newton, K. Hagino and I. J. Thompson, *Phys. Rev. C* **60** (1999) 044608.
126. R. V. Jolos and P. von Brentano, *Phys. Rev. C* **49** (1994) R2301.
127. G. Luders, *Kong. Dan. Vid. Sel. Mat. Fys. Med.* **28N5** (1954) 1.
128. A. D. Sakharov, *Pisma Zh. Eksp. Teor. Fiz.* **5** (1967) 32.
129. D. E. Morrissey and M. J. Ramsey-Musolf, *New J. Phys.* **12** (2012) 125003.
130. M. Dine and A. Kusenko, *New J. Phys.* **76** (2004) 125003.
131. J. H. Smith, E. M. Purcell and N. F. Ramsey, *Phys. Rev.* **108** (1957) 120.
132. L. I. Schiff, *Phys. Rev.* **132** (1963) 2194.
133. V. V. Flambaum, I. B. Khriplovich and O. P. Sushkov, *Sov. Phys. JETP* **60** (1984) 873.
134. V. A. Dzuba, V. V. Flambaum, J. S. M. Ginges and M. G. Kozlov, *Phys. Rev. A* **66** (2002) 012111.
135. V. A. Dzuba, V. V. Flambaum and S. G. Porsev, *Phys. Rev. A* **80** (2009) 032120.
136. L. Radziute, G. Gaigalas, P. Jonsson and J. Biero, *Phys. Rev. A* **90** (2014) 012528.
137. L. Radžiūtė, G. Gaigalas, P. Jönsson and J. Bieroń, *Phys. Rev. A* **93** (2016) 062508.
138. K. V. P. Latha, D. Angom, B. P. Das and D. Mukherjee, *Phys. Rev. Lett.* **103** (2009) 083001 [Erratum-ibid. **115** (2015) 059902].
139. Y. Singh, B. K. Sahoo and B. P. Das, *Phys. Rev. A* **89** (2014) 030502.
140. Y. Singh and B. K. Sahoo, *Phys. Rev. A* **92** (2015) 022502.

141. B. K. Sahoo, *Phys. Rev. D* **95** (2017) 013002.
142. B. K. Sahoo and B. P. Das, *Phys. Rev. Lett.* **120** (2018) 203001.
143. W. C. Haxton and E. M. Henley, *Phys. Rev. Lett.* **51** (1983) 1937.
144. P. Herczeg, *Prog. Part. Nucl. Phys.* **46** (2001) 413.
145. J. Engel, M. Bender, J. Dobaczewski, J. H. De Jesus and P. Olbratowski, *Phys. Rev. C* **68** (2003) 025501.
146. C. M. Maekawa, E. Mereghetti, J. de Vries and U. van Kolck, *Nucl. Phys. A* **872** (2011) 117.
147. J. Engel, J. L. Friar and A. C. Hayes, *Phys. Rev. C* **61** (2000) 035502.
148. V. V. Flambaum and H. Feldmeier, *Phys. Rev. C* **101** (2020) 015502.
149. R. G. Helmer, M. A. Lee, C. W. Reich and I. Ahmad, *Nucl. Phys. A* **474** (1987) 77.
150. V. V. Flambaum, I. B. Khriplovich and O. P. Sushkov, *Nucl. Phys. A* **449** (1986) 750.
151. V. F. Dmitriev, R. A. Sen'kov and N. Auerbach, *Phys. Rev. C* **71** (2005) 035501.
152. J. Dobaczewski and J. Engel, *Phys. Rev. Lett.* **94** (2005) 232502.
153. S. Ban, J. Dobaczewski, J. Engel and A. Shukla, *Phys. Rev. C* **82** (2010) 015501.
154. J. Dobaczewski, J. Engel, M. Kortelainen and P. Becker, *Phys. Rev. Lett.* **121** (2018) 232501.
155. J. H. de Jesus and J. Engel, *Phys. Rev. C* **72** (2005) 045503.
156. V. Spevak, N. Auerbach and V. V. Flambaum, *Phys. Rev. C* **56** (1997) 1357.
157. N. Auerbach, V. F. Dmitriev, V. V. Flambaum, A. Lisetskiy, R. A. Sen'kov and V. G. Zelevinsky, *Phys. Rev. C* **74** (2006) 025502.
158. N. Yoshinaga, K. Higashiyama, R. Arai and E. Teruya, *Phys. Rev. C* **87** (2013) 044332.
159. E. Teruya, N. Yoshinaga, K. Higashiyama and K. Asahi, *Phys. Rev. C* **96** (2017) 015501.
160. K. Yanase and N. Shimizu, *Phys. Rev. C* **102** (2020) 065502.
161. K. Yanase, N. Shimizu, K. Higashiyama and N. Yoshinaga, arXiv:2210.08498.
162. N. Auerbach, V. V. Flambaum and V. Spevak, *Phys. Rev. Lett.* **76** (1996) 4316.
163. V. V. Flambaum and V. A. Dzuba, *Phys. Rev. A* **101** (2020) 042504.
164. O. P. Sushkov, V. V. Flambaum and I. B. Khriplovich, *Sov. Phys. JETP* **60** (1984) 11.
165. V. V. Flambaum and V. G. Zelevinsky, *Phys. Rev. C* **68** (2003) 035502.
166. P. Froese and P. Navrátil, *Phys. Rev. C* **104** (2021) 025502.
167. B. Bally, B. Avez, M. Bender and P. H. Heenen, *Phys. Rev. Lett.* **113** (2014) 162501.
168. M. Borrajo and J. L. Egido, *Phys. Lett. B* **764** (2017) 328.
169. V. F. Dmitriev and R. A. Sen'kov, *Phys. Atom. Nucl.* **66** (2003) 1940.
170. N. Sachdeva *et al.*, *Phys. Rev. Lett.* **123** (2019) 143003.
171. B. Graner, Y. Chen, E. G. Lindahl and B. R. Heckel, *Phys. Rev. Lett.* **116** (2016) 161601 [Erratum-ibid. **119** (2017) 119901].
172. M. Bishof *et al.*, *Phys. Rev. C* **94** (2016) 025501.
173. C. W. Wong, *Phys. Report.* **15** (1975) 283.

Microscopic formulation of the interacting boson model for reflection asymmetric nuclei*

Kosuke Nomura

Department of Physics, University of Zagreb, Zagreb, HR 10000, Croatia

knomura@phy.hr

Reflection asymmetric, octupole shapes in nuclei are a prominent aspect of nuclear structure, and have been recurrently studied over the decades. Recent experiments using radioactive-ion beams have provided evidence for stable octupole shapes. A variety of nuclear models have been employed for the related theoretical analyses. We review recent studies on the nuclear octupole shapes and collective excitations within the interacting boson model. A special focus is placed on the microscopic formulation of this model by using the mean-field method that is based on the framework of nuclear density functional theory. As an illustrative example, a stable octupole deformation, and a shape phase transition as a function of nucleon number that involves both quadrupole and octupole degrees of freedom are shown to occur in light actinides. Systematic spectroscopic studies indicate enhancement of the octupole collectivity in a wide mass region. Couplings between the octupole and additional degrees of freedom are incorporated in a microscopic manner in the boson system, and shown to play a crucial role in the description of the related intriguing nuclear structure phenomena such as the shape coexistence.

Keywords: Octupole deformation; interacting boson model; mean-field model; density functional theory; shape phase transition; shape coexistence.

1. Introduction

Reflection asymmetric, octupole deformation of the atomic nuclei presents a theme of great interest in nuclear structure physics.[1,2] The octupole deformation is expected to emerge in those mass regions in the chart of nuclides in which a coupling occurs between the single-particle orbitals that differ by $\Delta\ell = \Delta j = 3\hbar$ (ℓ and j are quantum numbers of single-particle state) and are opposite in parity. Empirical signatures of the octupole deformation

*To cite this article, please refer to its earlier version published in the *International Journal of Modern Physics E*, Volume 32(10), (20 pages), 2340001 (2023), DOI: 10.1142/S0218301323400013.

are low-energy negative-parity states forming an alternating-parity band with the positive-parity yrast states, and enhanced electric dipole and octupole transitions within the band.

The existence of low-lying negative-parity states that are associated with the reflection asymmetric deformation was recognized in the 1950s. Since then the topic has been recurrently pursued. Recent experiments using radioactive-ion beams have identified evidence for stable octupole shapes, e.g., in light actinides[3–5] and lanthanides.[6,7] Besides that, theoretical studies that are to support and predict the occurrence of nuclear octupole states have been extensively conducted using a variety of nuclear structure models. An extensive list of the relevant experimental and theoretical studies is given in the review articles, e.g., Refs. 1 and 2. In addition, the presence of static octupole correlations in the nucleus enhances an atomic electric dipole moment (EDM). The observation of a nonzero EDM would imply the violation of the CP symmetry, hence is important for exploring new physics beyond the standard model of elementary particles.[8]

In this paper, we review recent theoretical investigations on the reflection asymmetric nuclear shapes and related collective states within the interacting boson model (IBM).[9] A special focus is placed on the microscopic formulation of this model in terms of the nuclear mean-field method. Within this theoretical scheme, a stable octupole shape emerges in a characteristic set of nuclei in which octupole deformation is empirically suggested to occur. A transition between the quadrupole and octupole states as functions of the nucleon number is discussed in the context of the quantum phase transition in nuclear shapes.[10] Coupling between single-particle and collective octupole degrees of freedom is shown to play an important role in determining the low-lying band structure of the odd-mass nuclei. Furthermore, the octupole correlations are shown to be relevant for understanding the nature of the low-energy excited 0^+ states, which are often considered as a signature of shape coexistence.[11,12]

2. Interacting Boson Model for Reflection Asymmetric Nuclei

The IBM, proposed by Arima and Iachello,[9,13] has been remarkably successful in reproducing low-energy quadrupole collective states in a large number of medium-heavy and heavy nuclei. Its basic assumption is that the multi-fermion dynamics of the nuclear surface deformation is simulated in terms of the monopole, s, and quadrupole, d, bosons, which represent, from a microscopic point of view,[9,14,15] collective S and D pairs of valence nucleons corresponding to spin and parity 0^+ and 2^+, respectively. This corresponds to a drastic truncation of the full shell-model configuration space which becomes prohibitively large for heavy and open-shell nuclei. The model is also intimately related to the group theory. A prominent feature is the emergence of the dynamical symmetry, i.e., if the Hamiltonian is written as a specific combination of Casimir operators then the Hamiltonian is associated with a certain intrinsic structure and is exactly solvable. The sd-IBM, for instance, constitutes the bosonic algebra U(6), and from it appear three subalgebras U(5), SU(3) and O(6), which correspond to vibrational, rotational, and γ-unstable limits of the quadrupole mode, respectively.

To study negative-parity states, the IBM was extended[16–19] so that the boson model space should contain, in addition to the positive-parity s and d bosons, the negative-parity bosons such as the dipole, p, and octupole, f, bosons representing the spin and parity 1^- and 3^- nucleon pairs, respectively. The $spdf$- and sdf-boson models have been employed in phenomenological studies on the quadrupole–octupole coupled collective excitations in realistic cases in actinides,[20] rare-earth[21] and lanthanides.[22] From a group theoretical point of view, the sdf- and $spdf$-boson models constitute, respectively, the boson algebras U(13) and U(16). The group structures including the dynamical symmetries of the quadrupole and octupole states have been analyzed in this respect.[17,18,23,24]

Besides numerous successes in reproducing observed collective spectra, the IBM should have its microscopic basis on nucleonic degrees of freedom, and attempts have been made to establish a link between the IBM and more microscopic nuclear structure models. A standard method, referred to as the Otsuka–Arima–Iachello (OAI) mapping,[14,15] is such that a seniority-based shell-model state given in terms of the S, D, ... nucleon pairs is mapped onto the equivalent s, d, ... boson state. The number of bosons is set equal to that of the valence nucleons. The OAI method has been shown to be valid in limited realistic cases of moderately deformed nuclei, i.e., nearly spherical and γ-unstable ones.[25]

More recently, making use of the fact that the potential energy surface of a given nuclear mean-field model can be simulated by that of the boson system, the IBM Hamiltonian has been shown[26] to be derived in the general situations of quadrupole collective mode, covering the nearly spherical vibrational [U(5)],[26,27] strongly deformed rotational [SU(3)],[28] γ-unstable [O(6)][26,27] and triaxially deformed states.[29] The method was extended further to address various problems in nuclear structure, including the shape coexistence,[30] octupole deformation and collectivity,[31–33] description of odd-mass nuclei[34] and nuclear β decays.[35,36] In what follows, we outline the above-mentioned method and discuss its applications to those nuclear properties in which octupole correlations are expected to be relevant.

3. Mean-Field Derivation of the Interacting Boson Model

3.1. *Mean-field calculations*

Among contemporary theoretical approaches, the nuclear density functional theory (DFT)[37–42] allows for a global and accurate prediction of bulk nuclear matter and intrinsic properties of finite nuclei, including the mass, radii, deformations, etc., and collective excitations over the entire region of the nuclear chart. Its basic implementation is in the self-consistent mean-field (SCMF) calculations, in which an energy density functional (EDF) is constructed as a functional of one-body density matrices that correspond to a single product state. Both nonrelativistic[37,41] and relativistic[38–40] EDFs have been successfully applied to numerous studies on nuclear structure phenomena.

The starting point is a set of constrained SCMF calculations[43] for each nucleus to obtain the potential energy surface, that is, the total mean-field energy as a function of the relevant shape degrees of freedom. Here, the constraints are on the axially-symmetric intrinsic quadrupole Q_{20} and octupole Q_{30} moments, which are related to the deformation variables β_2 and β_3, respectively.

As an illustrative example, Fig. 1 shows the mean-field potential energy surfaces for the axially-symmetric even–even nuclei $^{222-232}$Th. The SCMF calculations are performed within the relativistic Hartree–Bogoliubov (RHB) method[38,39] based on the density-dependent point-coupling (DD-PC1)[44] interaction and the separable pairing force of finite range.[45] Variation of the topology of energy surface as a function of nucleon number gives a qualitative interpretation of nuclear shape evolution. The global (equilibrium) minimum occurs at nearly spherical mean-field configuration $(\beta_2, \beta_3) \approx (0.05, 0.0)$ for ^{222}Th. In the neighboring isotope ^{224}Th a nonzero octupole minimum with $\beta_3 \approx 0.1$ is seen. The octupole deformation becomes most pronounced for ^{226}Th. Note that the occurrence of the stable octupole minimum is inferred by the behaviors of the single-particle levels near the Fermi energies. At the deformations that correspond to the octupole global minimum, level densities of the single proton and neutron orbits were indeed shown to be lower near the Fermi energies.[46] For heavier nuclei ^{228,230}Th, the energy surface becomes softer along the β_3 direction. For ^{232}Th, the octupole minimum disappears, but the potential is substantially soft in the β_3 deformation.

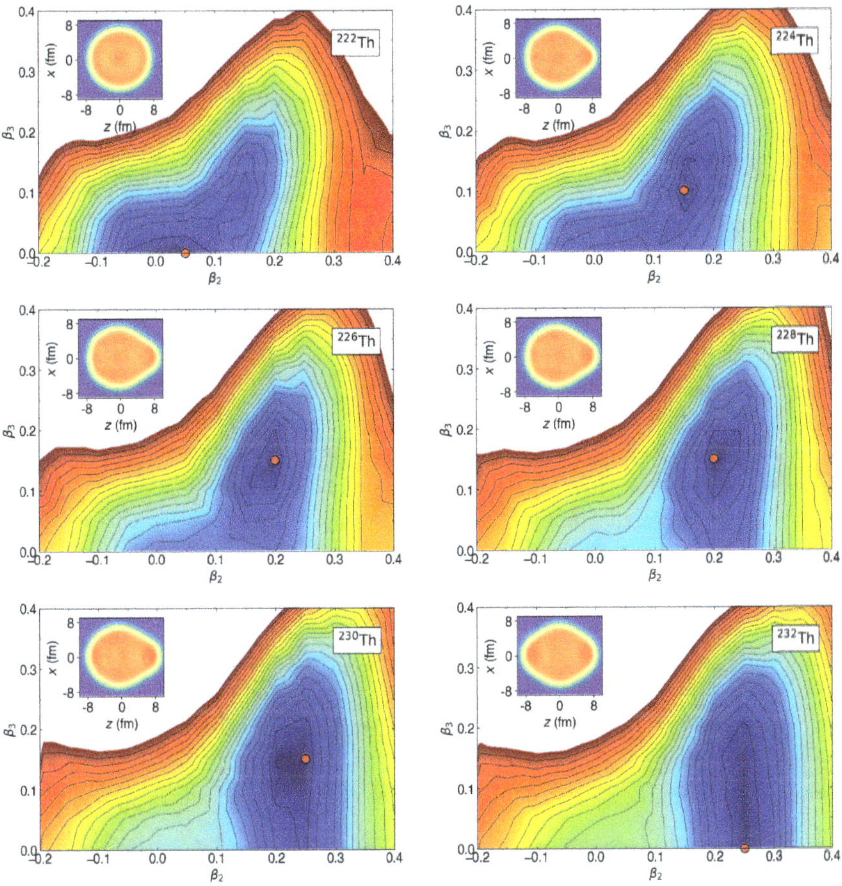

Fig. 1. Potential energy surfaces for $^{222-232}$Th in terms of the axially-symmetric quadrupole β_2 and octupole β_3 deformations, calculated within the relativistic Hartree–Bogoliubov method using the functional DD-PC1 and separable pairing force. The energy surface is plotted up to 10 MeV from the global minimum, indicated by the solid red circle. The energy difference between the neighboring contour lines is 0.5 MeV. The calculated intrinsic nucleon density in the x–z plane is also shown for each nucleus.

Figure 1 also shows total intrinsic nucleon density in the x–z plane for each nucleus, computed within the RHB method with the constraints on those β_2 and β_3 deformations corresponding to the global minimum. One finds the intrinsic nucleon densities resembling a completely spherically symmetric shape for ^{222}Th, a reflection asymmetric shape for $^{224-230}$Th, and a reflection symmetric shape for ^{232}Th.

3.2. *Mapping onto the boson system*

Let us turn to the boson system. We employ the simplest version of the sdf-boson model, where no distinction is made between proton and neutron bosons. A form of the sdf-IBM Hamiltonian appropriate for realistic calculations[31,32] reads

$$\hat{H}_{\mathrm{B}} = \epsilon_d \hat{n}_d + \epsilon_f \hat{n}_f + \kappa_2 \hat{Q} \cdot \hat{Q} + \kappa_2' \hat{L}_d \cdot \hat{L}_d + \kappa_3 : \hat{V}_3^{\dagger} \cdot \hat{V}_3 : . \quad (1)$$

$\hat{n}_d = d^{\dagger} \cdot \tilde{d}$ and $\hat{n}_f = f^{\dagger} \cdot \tilde{f}$ are number operators for d and f bosons, with ϵ_d and ϵ_f being single-d and -f boson energies relative to the s boson one, respectively. Note that $\tilde{d}_{\mu} = (-1)^{-\mu} d_{-\mu}$, $\tilde{f}_{\mu} = (-1)^{3-\mu} f_{-\mu}$, and the dot $(\cdot)$ means scalar product. The third term in Eq. (1) is the quadrupole–quadrupole interaction with the strength parameter κ_2, and $\hat{Q} = s^{\dagger}\tilde{d} + d^{\dagger}s + \chi_d [d^{\dagger} \times \tilde{d}]^{(2)} + \chi_f [f^{\dagger} \times \tilde{f}]^{(2)}$ is the quadrupole operator. χ_d and χ_f are dimensionless parameters. The fourth term, with the strength κ_2', is specifically required for strongly axially deformed nuclei,[28] and $\hat{L}_d$ is the angular momentum operator for d bosons $\hat{L}_d = \sqrt{10}[d^{\dagger} \times \tilde{d}]^{(1)}$. The last term, expressed in the normal-ordered form, represents the octupole–octupole interaction with the strength parameter κ_3, and the operator $\hat{V}_3^{\dagger} = s^{\dagger}\tilde{f} + \chi_3 [d^{\dagger} \times \tilde{f}]^{(3)}$, with χ_3 being another dimensionless parameter. For details, see Ref. 32.

A given boson model Hamiltonian can be related to a certain geometrical structure by using the boson coherent state.[47–49] For the sdf-IBM, if axial symmetry is assumed the coherent state $|\phi\rangle$ is expressed as

$$|\phi\rangle = (n!)^{-1/2} (\lambda^{\dagger})^n |0\rangle, \lambda^{\dagger} = (1 + \tilde{\beta}_2^2 + \tilde{\beta}_3^2)^{-1/2} (s^{\dagger} + \tilde{\beta}_2 d_0^{\dagger} + \tilde{\beta}_3 f_0^{\dagger}). \quad (2)$$

Here, n denotes the number of bosons, and $|0\rangle$ is the boson vacuum, i.e., the inert core. The amplitudes $\tilde{\beta}_2$ and $\tilde{\beta}_3$ are considered as the boson analogs of β_2 and β_3 deformations in the geometrical model. The energy surface of the boson system, denoted by $E_{\mathrm{IBM}}(\tilde{\beta}_2, \tilde{\beta}_3)$, is obtained by taking an expectation value $\langle\phi|\hat{H}_{\mathrm{B}}|\phi\rangle$, which also depends on the parameters of the Hamiltonian $\hat{H}_{\mathrm{B}}$ (1). To a good approximation, the bosonic deformation variables $(\tilde{\beta}_2, \tilde{\beta}_3)$ are connected to the fermionic counterparts (β_2, β_3) through the relations[26,32,47]

$$\tilde{\beta}_2 = c_2 \beta_2, \quad \tilde{\beta}_3 = c_3 \beta_3 \quad (3)$$

with c_2 and c_3 being constants of proportionality. Typical values for these constants are ~ 5. The fact that the bosonic deformations are larger than the fermionic ones reflects that the IBM is built on the limited configuration (valence) space while the SCMF model considers all nucleons in the nucleus.

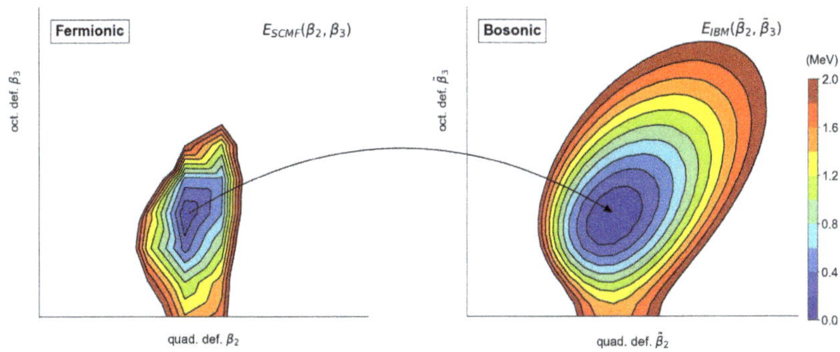

Fig. 2. A schematic illustration of the mapping of the fermionic potential energy surface $E_{\rm SCMF}(\beta_2, \beta_3)$ onto the bosonic counterpart $E_{\rm IBM}(\tilde{\beta}_2, \tilde{\beta}_3)$.

Together with the relations in (3), the SCMF and IBM potential energy surfaces are equated to each other:

$$E_{\rm SCMF}(\beta_2, \beta_3) \sim E_{\rm IBM}(\tilde{\beta}_2, \tilde{\beta}_3). \tag{4}$$

This procedure represents a mapping of the fermionic energy surface onto the bosonic one, as schematically illustrated in Fig. 2. The parameters of the Hamiltonian $\hat{H}_{\rm B}$, and the proportionality constants c_2 and c_3 for the deformation variables are determined by the mapping procedure. In other words, the IBM parameters are derived so that the bosonic energy surface becomes similar to the fermionic one as much as possible. Note that the relation (4) is an approximate equality that should be satisfied within a limited range of the (β_2, β_3) space, that is, in the vicinity of the global minimum. This is based on the fact that the low-energy quadrupole and octupole collective states are predominantly accounted for by the SCMF solutions near the global minimum.[26,27,46,50]

In addition, since the parameter κ_2' for the rotational term $\hat{L}_d \cdot \hat{L}_d$ [see Eq. (1)] cannot be uniquely determined from the comparison of the potential energy surfaces, it is derived separately,[28] from the comparison of the intrinsic moment of inertia in the boson system to the fermionic one computed in the cranking approximation.

Having determined the parameters for $\hat{H}_{\rm B}$ by the mapping procedure, the resultant IBM Hamiltonian is numerically diagonalized, resulting in the energies and wave functions of the excited states of both parities. It should be noted that no phenomenological adjustment is made in this procedure, that is, the *sdf*-IBM Hamiltonian is determined only by using inputs provided by the microscopic calculations within the EDF framework, which is also not specifically adjusted to the observed octupole states.

At this point, it is worth mentioning that the octupole shapes and collective excitations have also been extensively studied by a number of beyond-mean-field approaches within the EDF framework, including the symmetry projected generator coordinate method (see, e.g., Refs. 41, 51, and an extensive list of references are therein) and the quadrupole–octupole collective model.[50,52,53]

3.3. *Octupole shape phase transitions in Th isotopes*

Figure 3 shows the calculated energy spectra for the even-spin positive-parity and odd-spin negative-parity yrast states for $^{220-232}$Th. Agreement between the calculated and experimental[54] energy spectra is good. As one sees in Fig. 3(a), the overall decrease in energy of the positive-parity levels indicates a transition from vibrational to rotational nuclear structures. The negative-parity levels, on the other hand, exhibit a characteristic parabolic behavior with the minimum energy at ^{226}Th. [see Fig. 3(b)]. Around ^{226}Th, bands with both parities are close to each other, and appear to form an approximate alternating-parity band as expected for a stable octupole state.

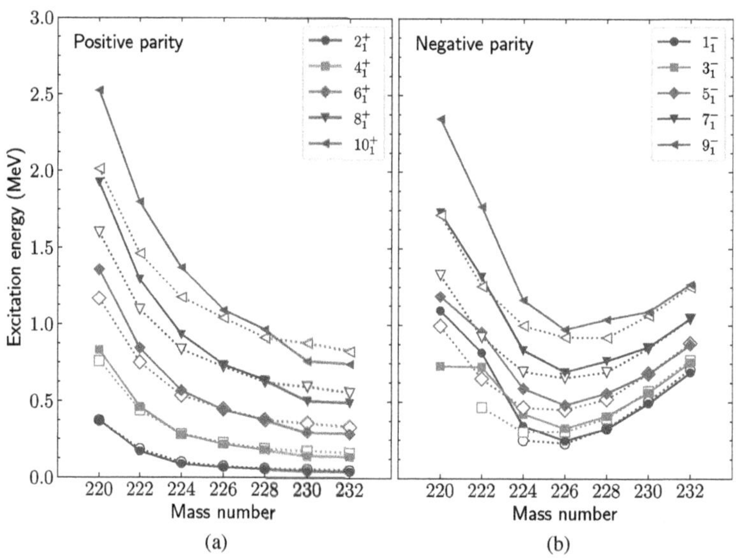

Fig. 3. Evolution of the excitation energies for the (a) positive- and (b) negative-parity yrast states for $^{220-232}$Th. Solid (open) symbols represent the calculated (experimental) values, and are connected by solid (broken) lines.

For $A > 226$, the negative-parity levels keep rising with mass number, while the positive-parity ones gradually decrease. Toward the nucleus ^{232}Th, the positive- and negative-parity bands are almost decoupled from each other, and appear to form separate $\Delta I = 2$ rotational bands.

To identify the occurrence of the alternating-parity band structure in the Th isotopes, it is convenient to study the signature splitting:

$$S(I) = \frac{[E(I+1) - E(I)] - [E(I) - E(I-1)]}{E(2_1^+)}. \tag{5}$$

$E(I)$ stands for the excitation energies of the $I = 1_1^-, 2_1^+, 3_1^-, 4_1^+, \ldots$ states. The quantity $S(I)$ is sensitive to the splitting between the positive- and negative-parity rotational bands. For an ideal alternating-parity band, the positive- and negative-parity bands should appear with an equal energy splitting, hence $S(I) \approx 0$. Nonzero $S(I)$ value would indicate a deviation from the pure alternating-parity band. This situation is illustrated in Fig. 4. For $^{220-226}$Th, the $S(I)$ value is nearly equal to zero. For $^{228-232}$Th, in turn, the odd–even spin staggering occurs, and the deviation from the limit $S(I) = 0$ becomes even larger with mass number.

The systematic behavior of the calculated energy spectra for the yrast bands with both parities, point to the occurrence of a type of quantum phase transition[10] along the chain of Th isotopes, that is, the structural evolution from nearly spherical to stable octupole shapes toward ^{226}Th, and to an octupole vibrational state around ^{230,232}Th associated with the octupole-soft potential. The behavior of the calculated spectroscopic properties reflects, to a large extent, the variation of the topology of the SCMF energy surface in

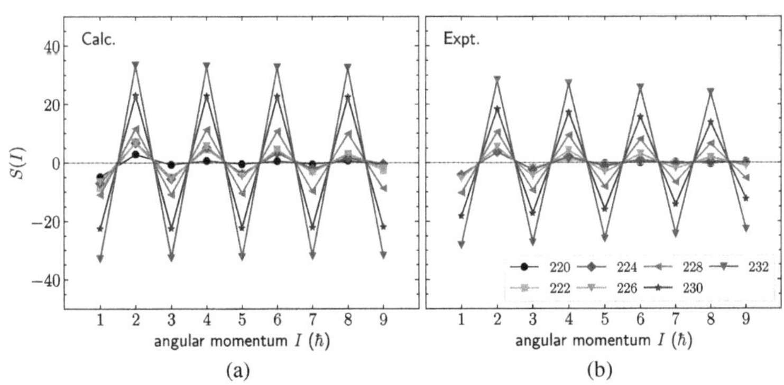

Fig. 4. Calculated (a) and experimental[54] (b) signature splittings $S(I)$ between the even-I positive-parity and odd-I negative-parity bands for $^{220-230}$Th as functions of the angular momentum.

the $\beta_2 - \beta_3$ plane, as was shown in Fig. 1. In addition, the Th isotopes appear to be an ideal case that exhibits a stable octupole deformation, as well as a shape phase transition that involves quadrupole and octupole modes.

4. Global Study

The octupolarity is suggested to be enhanced in several different mass regions, corresponding to the neutron and proton numbers (N, Z) $\approx$ (134, 88), (88, 56) (56, 56), (56, 34), (34, 34), ..., at which the $\Delta\ell = \Delta j = 3\hbar$ coupling between the single-particle orbitals can occur.

Figure 5 shows the calculated 3_1^- excitation energies for the nuclei in those characteristic mass regions in which octupole deformation is expected to occur. The plot has been made as a compilation of the results[33,51,55–57] obtained from the sdf-IBM that is based on the Hartree–Fock–Bogoliubov (HFB) calculations using the Gogny-D1M[58] EDF.[59] Spectroscopic data are scarce on neutron-deficient nuclei with $N \approx 54$, since these nuclei are close to the proton drip line and have not been accessible experimentally. The calculations are here performed in that region also, so as to explore

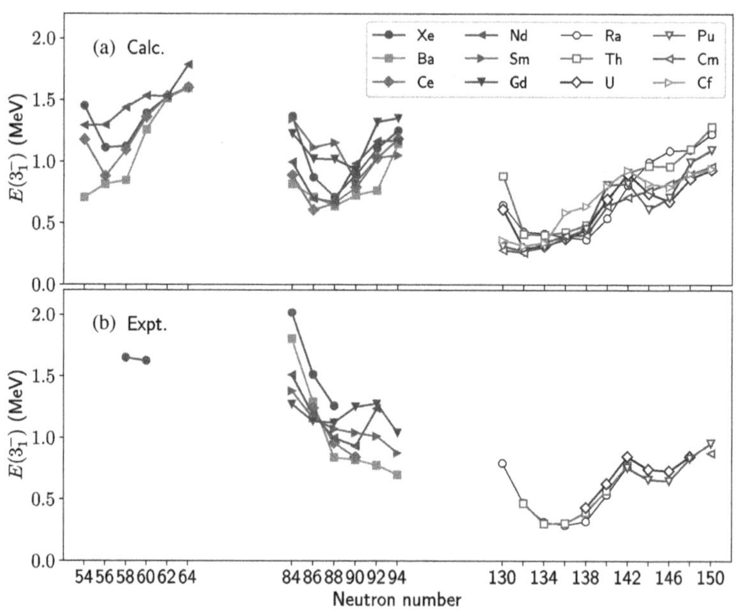

Fig. 5. Evolution of (a) calculated and (b) experimental[54] 3_1^- excitation energies as functions of the neutron number along those isotopic chains characteristic of the octupole deformation.

theoretically the robustness of the octupole deformation when one approaches the proton drip line.

In general, the predicted 3_1^- energy shows a parabolic behavior that is centered around $N = 56$, 88 and 134. This tendency is consistent with what is suggested experimentally, apart from the Ba and Sm isotopes. In many of the nuclei in the light actinides close to $N = 134$, the 3^- state appears at substantially low excitation energy $E_x < 0.5\,\text{MeV}$. The corresponding Gogny-HFB SCMF calculations indeed give an octupole-deformed ($\beta_3 \neq 0$) equilibrium minimum for these actinide nuclei.[51,55] One notices in Fig. 5 another parabolic variation of the 3^- level around $N = 146$. This local behavior implies the effect of dynamical octupole correlations that could be important even away from the neutron number $N = 134$. The 3^- energy is calculated to be rather high for the neutron-deficient Ba region with $54 \leq N \leq 64$. In this region, therefore, the octupole correlations appear to be relatively weak. It is also noted that for the nuclei with $N \approx Z$, some additional correlations such as the neutron–proton pairs may play a role, which are, however, not taken into account in the present calculations.

The electric octupole ($E3$) transition probability $B(E3; 3_1^- \to 0_1^+)$ is another indicator of the enhanced octupole collectivity. Figure 6 shows the

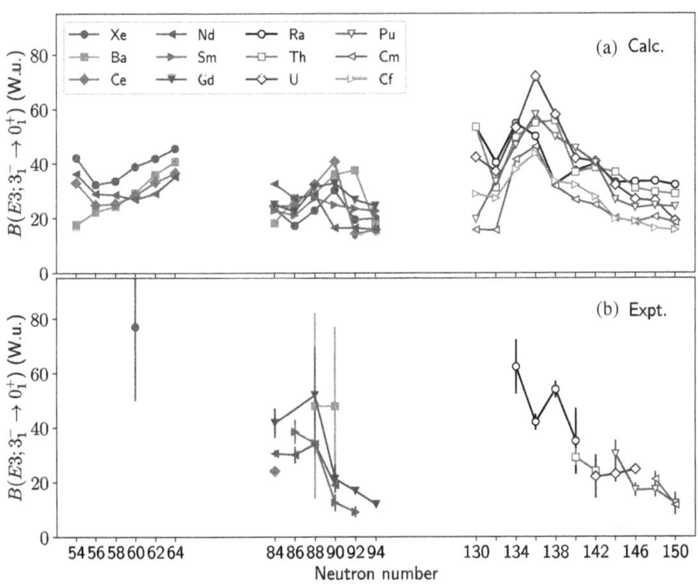

Fig. 6. (a) Calculated and (b) experimental[6,7,54,60,61] $B(E3; 3_1^- \to 0_1^+)$ transition probabilities for the same nuclei as those considered in Fig. 5.

$B(E3; 3_1^- \to 0_1^+)$ values for the same nuclei as those shown in Fig. 5, computed by the sdf-IBM with the microscopic input from the same Gogny-HFB calculations.[33,51,55–57] For the $E3$ transition operator, the standard boson $E3$ effective charges that are often used in realistic sdf-IBM calculations are chosen.[33,51,55–57] The calculated $B(E3; 3_1^- \to 0_1^+)$ rate exhibits an inverse parabolic behavior exhibiting a maximum value around $N = 88$ and 134, This behavior corroborates the one obtained for the 3^- energy level (see Fig. 5), and is consistent with the data.

5. Coupling with Additional Degrees of Freedom

5.1. *Particle-boson coupling in odd-mass systems*

Octupole correlations are expected to be relevant to the low-lying states of odd-mass nuclei, as well as to the even–even ones. To treat odd-mass nuclear systems, coupling between a single-particle (or unpaired nucleon) degree of freedom and collective even–even boson space is considered within the interacting boson–fermion model (IBFM).[62,63] The IBFM Hamiltonian consists of the IBM Hamiltonian $\hat{H}_{\mathrm{B}}$, single-nucleon Hamiltonian $\hat{H}_{\mathrm{F}}$, and a term $\hat{V}_{\mathrm{BF}}$ representing the boson–fermion coupling:

$$\hat{H}_{\mathrm{IBFM}} = \hat{H}_{\mathrm{B}} + \hat{H}_{\mathrm{F}} + \hat{V}_{\mathrm{BF}}. \tag{6}$$

Here, $\hat{H}_{\mathrm{B}}$ was defined in Eq. (1), and $\hat{H}_{\mathrm{F}} = \sum_j \epsilon_j a_j^\dagger \cdot \tilde{a}_j$, with $a_j^\dagger$ (a_j) and ϵ_j being the creation (annihilation) operator of a nucleon in the orbital j and the single-particle energy, respectively. The interaction $\hat{V}_{\mathrm{BF}}$ consists of the monopole, dynamical quadrupole and octupole, and exchange terms.[64] By using the generalized seniority scheme,[64,65] coefficients of the boson–fermion interaction terms are shown to be given as functions of particle occupation numbers (v_j^2).

As an illustrative example, let us consider the nucleus ^{145}Ba, a system of a single neutron coupled to the even–even nucleus ^{144}Ba. Experimentally, a static octupole deformation has been identified in ^{144}Ba,[6] while the octupole correlations have been suggested to be present in the odd-mass neighbor ^{145}Ba.[66] For the single-particle space of the odd neutron we include the entire $N = 82$–126 neutron major oscillator shell consisting of the $3p_{1/2}$, $3p_{3/2}$, $2f_{5/2}$, $2f_{7/2}$, $1h_{9/2}$ and $1i_{13/2}$ orbits.

The procedure to fix the IBFM Hamiltonian in Eq. (6) goes as follows.[34,64] First, the sdf-IBM Hamiltonian $\hat{H}_{\mathrm{B}}$ is determined by the $\beta_2 - \beta_3$ energy-surface mapping as described in Sec. 3.2. Second, the spherical single-particle energies ϵ_j and occupation probabilities v_j^2, which enter $\hat{H}_{\mathrm{F}}$ and $\hat{V}_{\mathrm{BF}}$, respectively, are provided by the SCMF calculation constrained to zero quadrupole and octupole deformations. Finally, only the strength parameters for the boson–fermion interactions are treated as free parameters, and are fitted to reproduce the empirical low-energy spectra of the odd-mass nucleus.

Figure 7 shows SCMF axially-symmetric $\beta_2 - \beta_3$ energy surface for the even–even core ^{144}Ba, obtained from the RHB method that employs the functional DD-PC1 and the separable pairing force. One clearly sees an octupole deformed minimum at $(\beta_2, \beta_3) \approx (0.22, 0.14)$, suggesting a pronounced octupole collectivity. The intrinsic nucleon density projected on the x–z plane, also shown in Fig. 7, indeed resembles a reflection-asymmetric nuclear shape.

In Fig. 8, low-energy excitation spectra for the positive- and negative-parity yrast states of ^{144}Ba resulting from the sdf-IBM are compared with the experimental data.[6] For states belonging to the negative-parity yrast band, the matrix element of the f-boson number operator is calculated to be $\langle \hat{n}_f \rangle \approx 1$, that is, the band is mainly constructed by one-f boson ($n_f = 1$) components coupled with the sd boson space. The positive-parity band is, however, solely made of the sd bosons, as $\langle n_f \rangle \approx 0$. The calculated $B(E3; 3_1^- \to 0_1^+)$ transition probability of 23 W.u. is comparable to the measured value, 48^{+25}_{-34} W.u.

The low-energy excitation spectra for the neighboring odd-mass nucleus ^{145}Ba computed by the sdf-IBFM are given in Fig. 9. The sdf-IBFM wave functions for the lowest two $5/2^-$ and $7/2^-$ negative-parity band are mainly composed of the configuration of the odd neutron in the $1h_{9/2}$ orbit coupled to the sd boson space, with the matrix element $\langle n_f \rangle \approx 0$. The lowest positive-parity band with the bandhead $9/2_1^+$ is mostly made of the odd

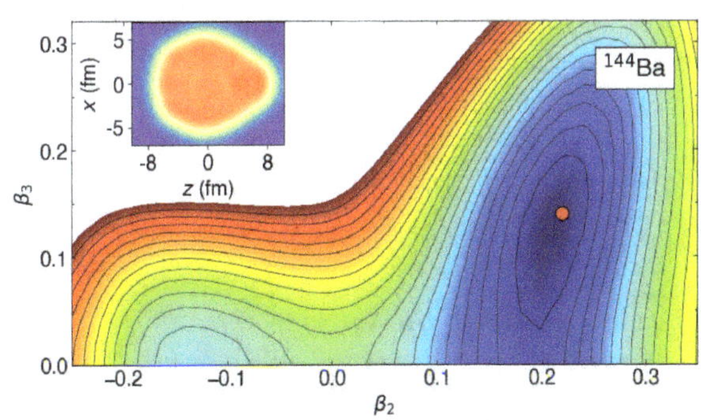

Fig. 7. Same as Fig. 1, but for ^{144}Ba.

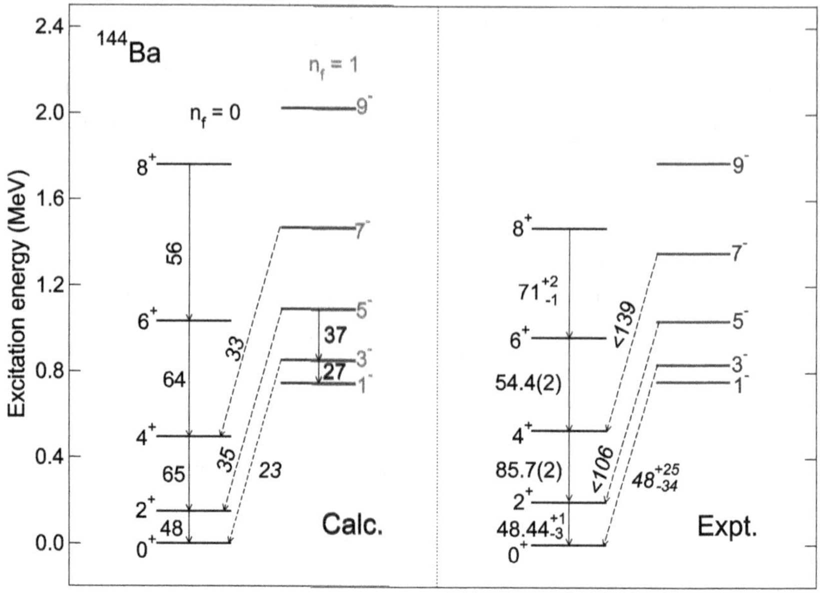

Fig. 8. Calculated and experimental[6] low-energy excitation spectra for positive- and negative-parity yrast states of ^{144}Ba. Numbers along solid and dotted lines are $B(E2)$ and $B(E3)$ values in W.u., respectively. The calculated positive- and negative-parity states are made of zero ($n_f = 0$) and one ($n_f = 1$) f boson configurations, respectively. The DD-PC1 EDF is employed for the microscopic input to the sdf-IBM.

neutron in the $1i_{13/2}$ unique-parity orbit coupled to the sd boson space. Experimentally[66] the observed $\Delta I = 2$ negative-parity band built on the $15/2^-$ state at 1226 keV, and positive-parity band built on the $11/2^+$ state at 670 keV are interpreted as octupole bands. Corresponding theoretical bands are the ones based on the $15/2^-_4$ and $7/2^+_1$ states. These are, respectively, determined by the configurations of the odd neutron in the $1i_{13/2}$ and $1h_{9/2}$ orbits coupled to the boson space containing one f boson ($\langle n_f \rangle \approx 1$). The predicted octupole bands have rather large $E3$ transitions to nonoctupole ($\langle n_f \rangle \approx 0$) states: $B(E3; 15/2^-_4 \to 9/2^+_1) = 25$, $B(E3; 19/2^-_2 \to 13/2^+_1) = 31$, $B(E3; 11/2^+_1 \to 5/2^-_1) = 4.7$ and $B(E3; 15/2^+_1 \to 9/2^-_1) = 15$ (all in W.u.)

5.2. *Shape coexistence and low-lying* 0^+ *states*

Shape coexistence is an intriguing feature of nuclear structure, in which different intrinsic structures appear near the ground state of a single nucleus.[11,12] Manifestation of the shape coexistence is the existence of

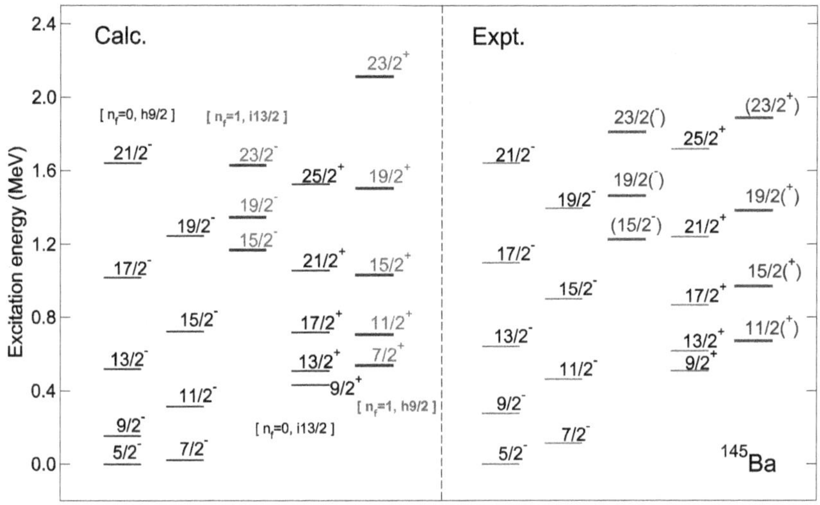

Fig. 9. Calculated and experimental[66] energy spectra for the low-lying negative- and positive-parity states for the odd-A nucleus ^{145}Ba. Those bands that are suggested to be of octupole nature are highlighted in thick lines. The calculations are performed within the sdf-IBFM that is based on the universal energy functional DD-PC1.

low-lying 0^+ excited states close in energy to the ground state. The extra low-lying 0^+ states are interpreted as intruder states arising from the cross-shell multiparticle–multihole excitations. With the particle-hole excitations the residual neutron–proton correlations are increased to such a degree as to lower the 0^+ energies.[67,68] The different intruder 0^+ states and subsequent bands can also be associated with different intrinsic deformations in the mean-field potential energy surface.[69,70]

To incorporate the intruder states in the IBM, one can adopt the prescription proposed by Duval and Barrett,[71] in which the usual IBM Hilbert space is extended to be a direct sum of the shell-model-like 0p-0h normal, and 2p-2h, 4p-4h, ... intruder configuration spaces. Several independent IBM Hamiltonians, corresponding to different configurations, are introduced and are allowed to be mixed. Assuming that the particle and hole states are not distinguished, these Hamiltonians differ in boson number by two. The total IBM Hamiltonian that is required to carry out the configuration mixing of the $2k$p-$2k$h $(k = 0, 1, 2, \ldots)$ states is then given by

$$\hat{H} = \hat{P}_0 \hat{H}_0 \hat{P}_0 + \sum_{k=1} \hat{P}_k (\hat{H}_k + \Delta_k) \hat{P}_k + \sum_{k=0} \hat{P}_{k+1} \hat{V}_{k,k+1} \hat{P}_k + (H.c.). \quad (7)$$

$\hat{H}_k$ is the unperturbed Hamiltonian for the $2k$p-$2k$h configuration, the operator $\hat{P}_k$ projects states onto the $2k$p-$2k$h space, and the constant Δ_k $(k \geq 1)$ represents the energy required to promote $2k$ nucleons (or k bosons) from one major shell to another. The interaction $\hat{V}_{k,k+1}$ admixes the $2k$p-$2k$h and $2(k+1)$p-$2(k+1)$h configurations, hence does not conserve boson number.

Since the configuration mixing IBM framework significantly alleviates the computational burden often accompanying massive computations as in the large-scale shell model, it has been extensively employed as an alternative approach to shape coexistence. A number of phenomenological studies using the configuration-mixing IBM have been carried out, e.g., for the neutron-deficient Hg,[72] neutron-rich Zr,[73] and even–even Cd.[74] On the other hand, since several independent IBM Hamiltonians and mixing terms are considered, this framework naturally involves a large number of adjustable parameters.

The method of deriving the configuration-mixing IBM Hamiltonian from the SCMF calculation has been developed[30] for the description of the quadrupole collective states in the Pb region. More recently, this method was extended to include the f boson degrees of freedom so as to deal with the shape coexistence that involves both quadrupole and octupole states.[75] The following discussion concerns this new development, taking the nucleus ^{76}Kr as a representative case. Note that there are a few instances in which the configuration mixing *sdf* (or *spdf*) IBM has been used in a purely phenomenological way for describing structure of even–even Cd isotopes.[76,77]

To begin with, two sets of SCMF calculations are performed, with constraints on the triaxial quadrupole (β_2, γ) deformations with $\beta_3 = 0$, and on the axially-symmetric quadrupole–octupole (β_2, β_3) deformations. The angle variable γ corresponds to the triaxial deformation, and is defined within the range $0^\circ \leq \gamma \leq 60^\circ$. Figure 10 shows the potential energy surfaces within the (β_2, γ) and (β_2, β_3) planes, computed by the RHB method using the functional DD-PC1 and the separable pairing force. In the $\beta_2 - \gamma$ energy surface, a spherical global minimum is obtained and, in addition, a weakly oblate deformed and a strongly prolate deformed local minima can be seen at $(\beta_2, \gamma) \approx (0.2, 60^\circ)$ and $(0.45, 0^\circ)$, respectively. The spherical ground state reflects the effect of the neutron $N = 40$ subshell closure. The $\beta_2 - \beta_3$ energy surface has the global minimum at $(\beta_2, \beta_3) \approx (0.0, 0.05)$ with nonzero octupole deformation, while the potential is soft along the β_3 deformation.

Since the energy surface exhibits three minima, the IBM space consists of the three configurations corresponding to the 0p-0h, 2p-2h and 4p-4h states

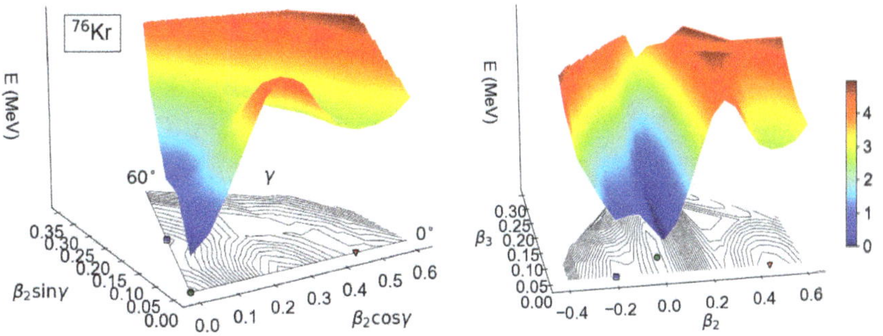

(a) (b)

Fig. 10. Potential energy surfaces for ^{76}Kr in terms of the (a) triaxial quadrupole (β_2, γ) and (b) axially-symmetric quadrupole β_2 and octupole β_3 deformations, calculated within the constrained RHB method based on the DD-PC1 functional and separable pairing force. Two-dimensional contour plots are also shown, with the energy difference between neighboring contours being 0.2 MeV. The symbols circle, square and triangle represent the spherical global minimum, and oblate and prolate local minima, respectively.

that comprise n, $n+2$ and $n+4$ bosons. The normal 0p-0h space here corresponds to the $N/Z = 28$–50 major shell, hence the boson number $n = 9$. In addition, the unperturbed 0p-0h Hamiltonian is associated with the spherical global minimum on the energy surfaces, while the 2p-2h and 4p-4h Hamiltonians are assigned to the oblate and prolate local minima, respectively. The full configuration mixing sdf IBM Hamiltonian is then determined by the following procedure.[75] Each unperturbed Hamiltonian $\hat{H}_k$ is determined to reproduce topology of the SCMF energy surface around the minimum with which the Hamiltonian is associated, the energy offset Δ_k is determined so that the energy difference between two neighboring minima is reproduced, and the mixing interaction $\hat{V}_{k,k+1}$ is fixed so as to reproduce as close as possible the overall topology of the barrier separating the two minima. The corresponding Hamiltonian is diagonalized in the boson space

$$[(sdf)^n] \oplus [(sdf)^{n+2}] \oplus [(sdf)^{n+4}]. \tag{8}$$

In the middle part of Fig. 11 the low-energy excitation spectra for ^{76}Kr, resulting from the configuration mixing sdf-IBM, are compared with the experimental data,[54,78] and with the usual sdf-IBM result obtained by using the single configuration associated with the spherical ground state. In both sets of the sdf-IBM, all the negative-parity states are based on the one-f boson configuration with the expectation value $\langle \hat{n}_f \rangle \approx 1$, while $\langle \hat{n}_f \rangle \approx 0$ for the positive-parity states.

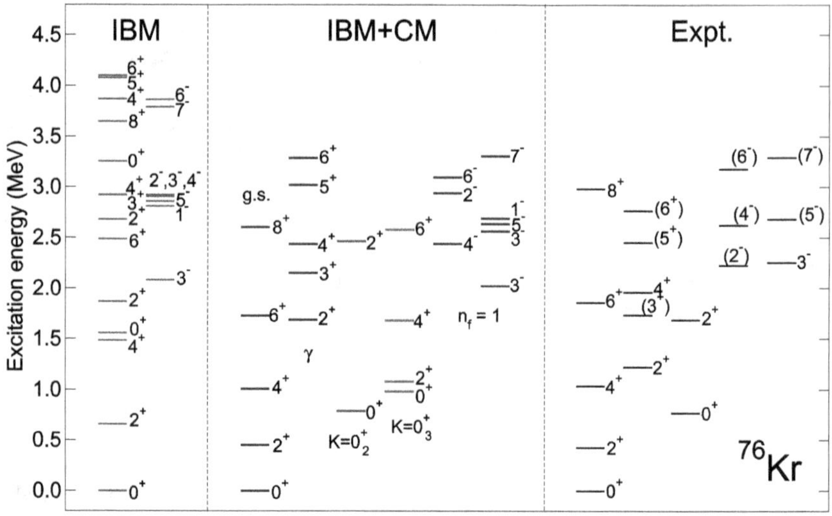

Fig. 11. Excitation spectra for positive- and negative-parity states for ^{76}Kr calculated by using a single configuration ("IBM") and taking into account the configuration mixing ("IBM + CM"). The IBM calculations are performed with the microscopic basis on the universal functional DD-PC1. Experimental data[54,78] are shown on the right.

A notable effect of taking into account the configuration mixing is the lowering in energy of the excited 0^+ states. In the usual *sdf*-IBM with the single configuration, the 0^+_2 and 0^+_3 excitation energies are calculated to be $E_x \approx 1.5\,\text{MeV}$ and $3.25\,\text{MeV}$, respectively, whereas they are both $E_x < 1\,\text{MeV}$ in the configuration mixing IBM. The energy levels for the positive-parity yrast states with spin $I > 0$ and also for the negative-parity states are generally lowered as a result of the mixing. A near degeneracy of the 1^-_1, 2^-_1, 3^-_2, 4^-_1 and 5^-_1 levels, obtained in the single-configuration calculation, can be interpreted as a quintet arising from the quadrupole–octupole phonon coupling $2^+ \otimes 3^-$. The degeneracy is, however, removed in the configuration mixing calculation as a consequence of the level repulsion.

Regarding the transition properties, the $B(E3; 3^-_1 \to 0^+_1)$ rates are calculated to be $\approx 30\,\text{W.u.}$ in this mass region.[75] An effect of the configuration mixing is, for instance, increase of the ratio $B(E3; 3^-_1 \to 2^+_1)/B(E3; 3^-_1 \to 0^+_1)$ by an order of magnitude. However, experimental data for the $E3$ and $E1$ transitions in this region are scarce, and a more thorough assessment of the model remains to be done.

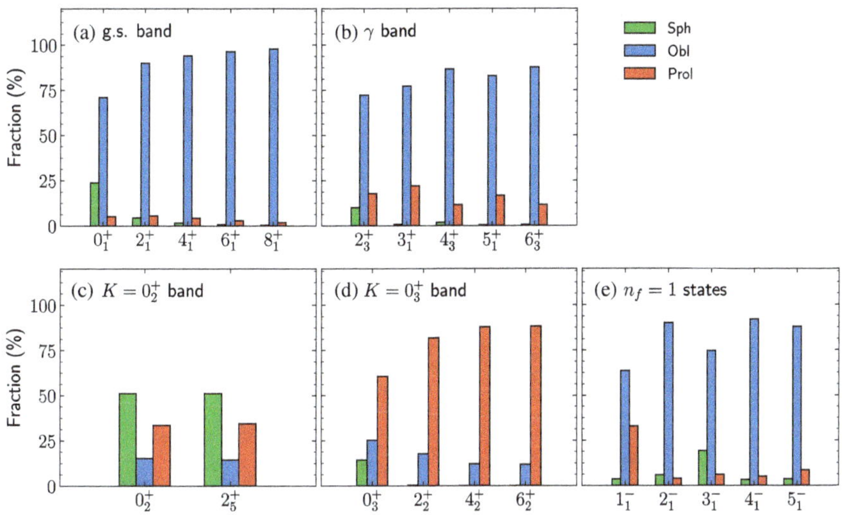

Fig. 12. Compositions of the spherical, oblate, and prolate configurations in the wave functions of the positive-parity states belonging to the (a) ground state, (b) γ-, (c) $K^\pi = 0_2^+$ and (d) $K^\pi = 0_3^+$ bands, and of the (e) low-spin negative-parity (one f boson, $n_f = 1$) states for ^{76}Kr.

To analyze in more detail the nature of the proposed low-spin bands, Fig. 12 shows compositions of the spherical (0p-0h), oblate (2p-2h) and prolate (4p-4h) configurations in the wave functions of the members of these bands. Even though the spherical ground state is predicted at the SCMF level, at the IBM level the oblate 2p-2h configuration constitutes most of the wave functions for the ground state and γ bands, and all the negative-parity states shown in Fig. 11. The $K^\pi = 0_2^+$ band with the bandhead 0_2^+ state is, in turn, spherical in nature. However, substantial amounts of the oblate and prolate components are admixed into the members of the proposed $K^\pi = 0_2^+$ band. The $K^\pi = 0_3^+$ rotational-like band based upon the 0_3^+ state is dominated by the strongly deformed prolate configuration.

5.3. *Double octupole phonon states*

A number of low-lying excited 0^+ states are also observed in those nuclei in rare-earth region.[79] Microscopic origin of these 0^+ states is under active debate, and has been attributed, e.g., to the pairing vibrations,[80,81] intruder excitations,[82] and coupling of the double octupole phonons.[83] To explain the occurrence of these low-energy 0^+ states the sdf- or $spdf$-IBM has often been used. In particular, calculation within the sdf-IBM that is based on the Gogny

EDF has shown[33] that the 0_2^+ excited states in many of the Sm and Gd nuclei are formed mainly by the coupling of two f bosons. The double octupole phonon nature of the 0_2^+ state has been suggested in light actinides as well within the sdf-IBM calculation based on the relativistic EDF.[32]

5.4. *Coupling with p boson*

A straightforward extension of the sdf-boson model would be to include the dipole p boson. Inclusion of this boson degree of freedom would indeed improve the description of the negative-parity levels and their $E1$ transition properties,[84] and has also been considered in the calculations on the α-clustering phenomena.[85–88] At the microscopic level, it was shown,[89] by using the Nilsson model, that large amount of the dipole pair, i.e., p boson, component is contained in the wave functions of negative-parity states, and that the dipole boson is as essential a degree of freedom as the octupole one. The physical meaning of the p boson has been attributed to the center of mass motion or giant dipole resonance,[89,90] but is not as entirely clear as that of the f boson, since the dipole mode is more or less of single-particle nature.

6. Conclusions

We have reviewed recent theoretical investigations of the octupole shapes and collective excitations in nuclei within the sdf-IBM that is formulated microscopically using the nuclear EDF framework. The illustrative application to Th isotopes has shown the occurrence of the stable octupole state around $N = 134$ as well as the shape phase transition that involves both quadrupole and octupole modes. A global spectroscopic study has confirmed the enhancement of the octupole collectivity in a wide mass region. The coupling between the octupole and additional degrees of freedom has been shown to play a role in the descriptions of a variety of related nuclear structure phenomena, including the shape coexistence.

This model's descriptions are based on the assumptions that the nuclear EDFs, upon which the IBM Hamiltonian is built, provide for all nuclei the correct mean-field properties (potential energy surfaces, single-particle energies, etc.), and that the EDF-to-IBM mapping procedure is valid. In this respect, for a more refined description of the octupole-related properties, further improvements of the method would be made, e.g., through (i) the assessment of the capability of a given EDF of predicting complex nuclear shape phenomena as well as spectroscopy, (ii) the inclusions of additional terms in the boson Hamiltonian and/or degrees of freedom, such as the p

boson, in the boson space, and (iii) the microscopic derivation of the boson effective charges for the $E1$ and $E3$ transition operators.

Furthermore, the theoretical method discussed throughout this paper has a potential to explore problems in interdisciplinary fields. For instance, the semi-microscopic sdf-IBFM, described in Sec. 5.1, can be used to study the proposed parity doublet and EDM for the odd-mass Ra and Th isotopes, which serve as a testing ground for the CP symmetry violation.[91]

Acknowledgments

The author thanks D. Vretenar, T. Nikšić, L. M. Robledo and R. Rodríguez-Guzmán for their contributions to the works discussed in this paper. The author has been supported by the Tenure Track Pilot Programme of the Croatian Science Foundation and the École Polytechnique Fédérale de Lausanne, and the Project TTP-2018-07-3554 Exotic Nuclear Structure and Dynamics, with funds of the Croatian-Swiss Research Programme.

References

1. P. A. Butler and W. Nazarewicz, *Rev. Mod. Phys.* **68** (1996) 349.
2. P. A. Butler, *J. Phys. G, Nucl. Part. Phys.* **43** (2016) 073002.
3. L. P. Gaffney *et al.*, *Nature (London)* **497** (2013) 199.
4. M. M. R. Chishti *et al.*, *Nat. Phys.* **16** (2020) 853.
5. P. A. Butler *et al.*, *Phys. Rev. Lett.* **124** (2020) 042503.
6. B. Bucher *et al.*, *Phys. Rev. Lett.* **116** (2016) 112503.
7. B. Bucher *et al.*, *Phys. Rev. Lett.* **118** (2017) 152504.
8. J. Engel, M. J. Ramsey-Musolf and U. van Kolck, *Prog. Part. Nucl. Phys.* **71** (2013) 21.
9. F. Iachello and A. Arima, *The Interacting Boson Model* (Cambridge University Press, Cambridge, UK, 1987).
10. P. Cejnar, J. Jolie and R. F. Casten, *Rev. Mod. Phys.* **82** (2010) 2155.
11. J. L. Wood, K. Heyde, W. Nazarewicz, M. Huyse and P. van Duppen, *Phys. Rep.* **215** (1992) 101.
12. K. Heyde and J. L. Wood, *Rev. Mod. Phys.* **83** (2011) 1467.
13. A. Arima and F. Iachello, *Phys. Rev. Lett.* **35** (1975) 1069.
14. T. Otsuka, A. Arima, F. Iachello and I. Talmi, *Phys. Lett. B* **76** (1978) 139.
15. T. Otsuka, A. Arima and F. Iachello, *Nucl. Phys. A* **309** (1978) 1.
16. A. Arima and F. Iachello, *Ann. Phys. (NY)* **111** (1978) 201.
17. J. Engel and F. Iachello, *Phys. Rev. Lett.* **54** (1985) 1126.
18. J. Engel and F. Iachello, *Nucl. Phys. A* **472** (1987) 61.
19. A. F. Barfield *et al.*, *Ann. Phys.* **182** (1988) 344.
20. N. V. Zamfir and D. Kusnezov, *Phys. Rev. C* **63** (2001) 054306.
21. P. D. Cottle and N. V. Zamfir, *Phys. Rev. C* **54** (1996) 176.

22. D. Kusnezov and F. Iachello, *Phys. Lett. B* **209** (1988) 420.
23. D. Kusnezov, *J. Phys. A, Math. Gen.* **22** (1989) 4271.
24. D. Kusnezov, *J. Phys. A, Math. Gen.* **23** (1990) 5673.
25. T. Mizusaki and T. Otsuka, *Prog. Theor. Phys. Suppl.* **125** (1996) 97.
26. K. Nomura, N. Shimizu and T. Otsuka, *Phys. Rev. Lett.* **101** (2008) 142501.
27. K. Nomura, N. Shimizu and T. Otsuka, *Phys. Rev. C* **81** (2010) 044307.
28. K. Nomura, T. Otsuka, N. Shimizu and L. Guo, *Phys. Rev. C* **83** (2011) 041302.
29. K. Nomura, N. Shimizu, D. Vretenar, T. Nikšić and T. Otsuka, *Phys. Rev. Lett.* **108** (2012) 132501.
30. K. Nomura, R. Rodríguez-Guzmán, L. M. Robledo and N. Shimizu, *Phys. Rev. C* **86** (2012) 034322.
31. K. Nomura, D. Vretenar and B.-N. Lu, *Phys. Rev. C* **88** (2013) 021303.
32. K. Nomura, D. Vretenar, T. Nikšić and B.-N. Lu, *Phys. Rev. C* **89** (2014) 024312.
33. K. Nomura, R. Rodríguez-Guzmán and L. M. Robledo, *Phys. Rev. C* **92** (2015) 014312.
34. K. Nomura, T. Nikšić and D. Vretenar, *Phys. Rev. C* **93** (2016) 054305.
35. K. Nomura, R. Rodríguez-Guzmán and L. M. Robledo, *Phys. Rev. C* **101** (2020) 044318.
36. K. Nomura, R. Rodríguez-Guzmán and L. M. Robledo, *Phys. Rev. C* **101** (2020) 024311.
37. M. Bender, P.-H. Heenen and P.-G. Reinhard, *Rev. Mod. Phys.* **75** (2003) 121.
38. D. Vretenar, A. V. Afanasjev, G. A. Lalazissis and P. Ring, *Phys. Rep.* **409** (2005) 101.
39. T. Nikšić, D. Vretenar and P. Ring, *Prog. Part. Nucl. Phys.* **66** (2011) 519.
40. J. Meng (ed.), *Relativistic Density Functional for Nuclear Structure* (World Scientific, 2016).
41. L. M. Robledo, T. R. Rodríguez and R. R. Rodríguez-Guzmán, *J. Phys. G, Nucl. Part. Phys.* **46** (2019) 013001.
42. N. Schunck (ed.), *Energy Density Functional Methods for Atomic Nuclei* (IOP Publishing, 2019).
43. P. Ring and P. Schuck, *The Nuclear Many-Body Problem* (Springer, Berlin, 1980).
44. T. Nikšić, D. Vretenar and P. Ring, *Phys. Rev. C* **78** (2008) 034318.
45. Y. Tian, Z. Y. Ma and P. Ring, *Phys. Lett. B* **676** (2009) 44.
46. Z. P. Li, T. Nikšić and D. Vretenar, *J. Phys. G. Nucl. Part. Phys.* **43** (2016) 024005.
47. J. N. Ginocchio and M. W. Kirson, *Nucl. Phys. A* **350** (1980) 31.
48. A. E. L. Dieperink, O. Scholten and F. Iachello, *Phys. Rev. Lett.* **44** (1980) 1747.
49. A. Bohr and B. R. Mottelson, *Phys. Scr.* **22** (1980) 468.
50. K. Nomura, L. Lotina, T. Nikšić and D. Vretenar, *Phys. Rev. C* **103** (2021) 054301.
51. K. Nomura *et al.*, *Phys. Rev. C* **102** (2020) 064326.
52. Z. P. Li *et al.*, *Phys. Lett. B* **726** (2013) 866.
53. S. Y. Xia *et al.*, *Phys. Rev. C* **96** (2017) 054303.
54. Brookhaven National Nuclear Data Center, http://www.nndc.bnl.gov.

55. K. Nomura *et al.*, *Phys. Rev. C* **103** (2021) 044311.
56. K. Nomura *et al.*, *Phys. Rev. C* **104** (2021) 044324.
57. K. Nomura *et al.*, *Phys. Rev. C* **104** (2021) 054320.
58. S. Goriely, S. Hilaire, M. Girod and S. Péru, *Phys. Rev. Lett.* **102** (2009) 242501.
59. J. Decharge and M. Girod and D. Gogny, *Phys. Lett. B* **55** (1975) 361.
60. T. Kibédi and R. H. Spear, *At. Data Nucl. Data Tables* **80** (2002) 35.
61. G. de Angelis *et al.*, *Phys. Lett. B* **535** (2002) 93.
62. F. Iachello and P. Van Isacker, *The Interacting Boson–Fermion Model* (Cambridge University Press, Cambridge, UK, 1991).
63. F. Iachello and O. Scholten, *Phys. Rev. Lett.* **43** (1979) 679.
64. K. Nomura, T. Nikšić and D. Vretenar, *Phys. Rev. C* **97** (2018) 024317.
65. O. Scholten, *Prog. Part. Nucl. Phys.* **14** (1985) 189.
66. T. Rzaca-Urban *et al.*, *Phys. Rev. C* **86** (2012) 044324.
67. P. Federman and S. Pittel, *Phys. Lett. B* **69** (1977) 385.
68. K. Heyde *et al.*, *Nucl. Phys. A* **466** (1987) 189.
69. R. Bengtsson *et al.*, *Phys. Lett. B* **183** (1987) 1.
70. W. Nazarewicz, *Phys. Lett. B* **305** (1993) 195.
71. P. D. Duval and B. R. Barrett, *Phys. Lett. B* **100** (1981) 223.
72. J. E. García-Ramos and K. Heyde, *Phys. Rev. C* **89** (2014) 014306.
73. N. Gavrielov, A. Leviatan and F. Iachello, *Phys. Rev. C* **99** (2019) 064324.
74. A. Leviatan *et al.*, *Phys. Rev. C* **98** (2018) 031302.
75. K. Nomura, *Phys. Rev. C* **106** (2022) 024330.
76. P. E. Garrett *et al.*, *Phys. Rev. C* **59** (1999) 2455.
77. D. Bandyopadhyay *et al.*, *Phys. Rev. C* **68** (2003) 014324.
78. E. Clément *et al.*, *Phys. Rev. C* **75** (2007) 054313.
79. A. Aprahamian *et al.*, *Phys. Rev. C* **98** (2018) 034303.
80. J. Xiang *et al.*, *Phys. Rev. C* **101** (2020) 064301.
81. K. Nomura, D. Vretenar, Z. P. Li and J. Xiang, *Phys. Rev. C* **102** (2020) 054313.
82. P. Van Isacker *et al.*, *Nucl. Phys. A* **380** (1982) 383.
83. N. V. Zamfir, J.-Y. Zhang and R. F. Casten, *Phys. Rev. C* **66** (2002) 057303.
84. M. Sugita, T. Otsuka and P. von Brentano, *Phys. Lett. B* **389** (1996) 642.
85. F. Iachello and A. Jackson, *Phys. Lett. B* **108** (1982) 151.
86. H. Daley and F. Iachello, *Phys. Lett. B* **131** (1983) 281.
87. M. Spieker, S. Pascu, A. Zilges and F. Iachello, *Phys. Rev. Lett.* **114** (2015) 192504.
88. R. Bijker and F. Iachello, *Nucl. Phys. A* **957** (2017) 154.
89. T. Otsuka, *Phys. Lett. B* **182** (1986) 256.
90. T. Otsuka and M. Sugita, *Phys. Lett. B* **209** (1988) 140.
91. J. Dobaczewski and J. Engel, *Phys. Rev. Lett.* **94** (2005) 232502.

Experimental review of octupole correlations in the $A \approx 80$ mass region*

X. C. Han, C. Liu† and S. Y. Wang‡

Shandong Provincial Key Laboratory of Optical Astronomy and Solar-Terrestrial Environment, School of Space Science and Physics, Shandong University at Weihai, Weihai 264209, P. R. China

WeiHai Research Institute of Industrial Technology of Shandong University, Weihai 264209, P. R. China

†*cliu@sdu.edu.cn*

‡*sywang@sdu.edu.cn*

This review presents the nuclei with octupole correlations in the $A \approx 80$ mass region. The status of experimental data on the behavior of the energy displacement, the $E1$ transition probability and intrinsic electric dipole moment have been reviewed. The evolutions of the octupole strength with the numbers of protons and neutrons are discussed.

Keywords: Octupole correlations; 80 mass region; systematic studies.

1. Introduction

The atomic nucleus is a many-body quantum system. The number of nucleons and the interactions between them determine the shape of the nucleus.[1] For the nuclear quantum system, nuclei with certain numbers of protons and neutrons (called magic numbers) are very stable and have a spherical shape in their ground states. When sufficient nucleons are added beyond the magic numbers, or the nucleonic- configuration is excited, the long-range correlations between nucleons cause the spontaneous symmetry

*To cite this article, please refer to its earlier version published in the *International Journal of Modern Physics E*, Volume 32(10), (7 pages), 2340003 (2023), DOI: 10.1142/S0218301323400037.

†Corresponding author.

breaking and distort the nuclear shape from spherical symmetry in the lowest energy states. For certain combinations of protons and neutrons, the nucleus can lower its energy through octupole–octupole interactions, resulting in the deviation from reflection symmetry in the intrinsic frame. Studies of the octupole-deformed nuclei facilitate our understanding for the origin of simple patterns in such complex quantum many-body systems and might help to reveal extensions to the standard model.

Following the mean-field description of the nucleus, octupole deformation depends on the strong octupole interactions between single particle states with orbital (ℓ) and total (j) angular momenta differing by $3\hbar$.[2] This condition is met when the particle numbers lie close to 34 ($g_{9/2} \leftrightarrow p_{3/2}$ coupling), 56 ($h_{11/2} \leftrightarrow d_{5/2}$ coupling), 88 ($i_{13/2} \leftrightarrow f_{7/2}$ coupling) and 134 ($j_{15/2} \leftrightarrow g_{9/2}$ coupling), which results in the experimental observation of the strong $E1$ transitions between the alternating-parity rotational bands. Octupole correlations have been experimentally identified in many nuclei, especially in the $A \approx 150$ and 220 mass regions where octupole correlations are strong enough to form stable octupole deformation. The $A \approx 80$ mass region is a relatively new octupole correlation region. Very recently, octupole correlations have been reported in ^{73,76,78}Br,[3–5] ^{74}As,[6] ^{72,79}Se[7,8] and ^{81}Kr,[9] which makes octupole correlation a common experimental character in the $A \approx 80$ mass region. But the corresponding systematic studies are still rather limited.[10–12] Therefore, a systematic study of octupole correlations in the $A \approx 80$ mass region is necessary, which stimulates the present investigation.

2. Experimental Findings and Systematic Studies of Octupole Correlations in the $A \approx 80$ Mass Region

In the $A \approx 80$ mass region, ^{71,72}Ge,[13,14] ^{69,74}As,[6,15] ^{72,74,78,79}Se,[7,8,16–18] ^{73,76,78}Br[3–5] and ^{80}Kr[19] are reported to have octupole correlations. Besides, the $E1$ transitions between the alternating-parity rotational bands are also observed in ^{70,74}Ge,[20–22] ^{71,73,75}As,[23–26] ^{75,76,77}Se,[27–29] ^{75,77,80}Br,[30–32] ^{74,76,77,78,79,81}Kr[9,33–37] and ^{78,79,81,83}Rb,[38–42] which can be regarded as candidates of octupole correlations. These nuclei are illustrated in Fig. 1. Majority of the octupole correlations in these nuclei arise from the valence proton occupying $p_{3/2}$ and $g_{9/2}$ orbitals, which results in the $I^- \to (I-1)^+ E1$ transitions. It should be noted that octupole correlations of some odd-neutron nuclei (^{75,77,79}Se and ^{79}Kr) result from the valence neutron occupying $p_{3/2}$ and $g_{9/2}$ orbitals. Besides the observed

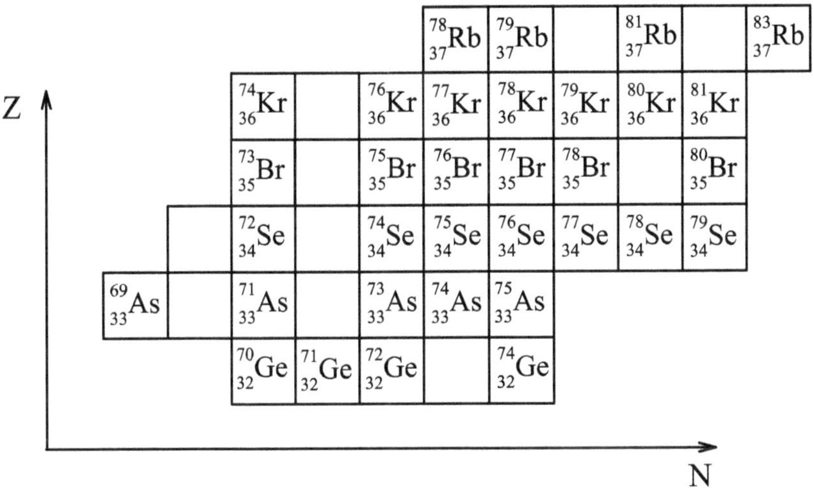

Fig. 1. Nuclei with the reported octupole correlations or the observed $E1$ transitions between the alternating-parity rotational bands in the $A \approx 80$ mass region.

$I^- \to (I-1)^+E1$ transitions, several $I^- \to I^+E1$ transitions have also been reported in these nuclei.

In order to perform systematic study for octupole correlations in the above mentioned nuclei, the energy displacement $\delta E = E(I) - [(I+1)E(I-1) + IE(I+1)]/(2I+1)$ between the positive- and the negative-parity bands is presented in Fig. 2. The energy displacement should be close to zero in the limit of stable octupole deformation and substantially different from zero of octupole correlation.[10,43] As shown in Fig. 2, the octupole correlations do exist in the Ge, As, Se, Kr and Rb isotopes in the $A \approx 80$ mass region. Meanwhile, the δE of all these nuclei is substantially different from zero, indicating that the octupole correlations in the $A \approx 80$ mass region are not strong enough to lead to a stable octupole deformation. According to the experimental features of the δE values, we suggest that octupole correlations also exist in the ^{70,74}Ge, ^{71,73,75}As, ^{75,76,77}Se, ^{75,77,80}Br, $^{74,76-79,81}$Kr and ^{78,79,81,83}Rb nuclei.

Electromagnetic transition probabilities like the $B(E1)$ and intrinsic electric dipole moment D_0 values are effective experimental signatures to probe the strength of octupole correlations.[2] To study the octupole strength in these nuclei in the $A \approx 80$ mass region, available experimental $B(E1)$ as well as D_0 values are given in Table 1. The $B(E1)$ values were extracted

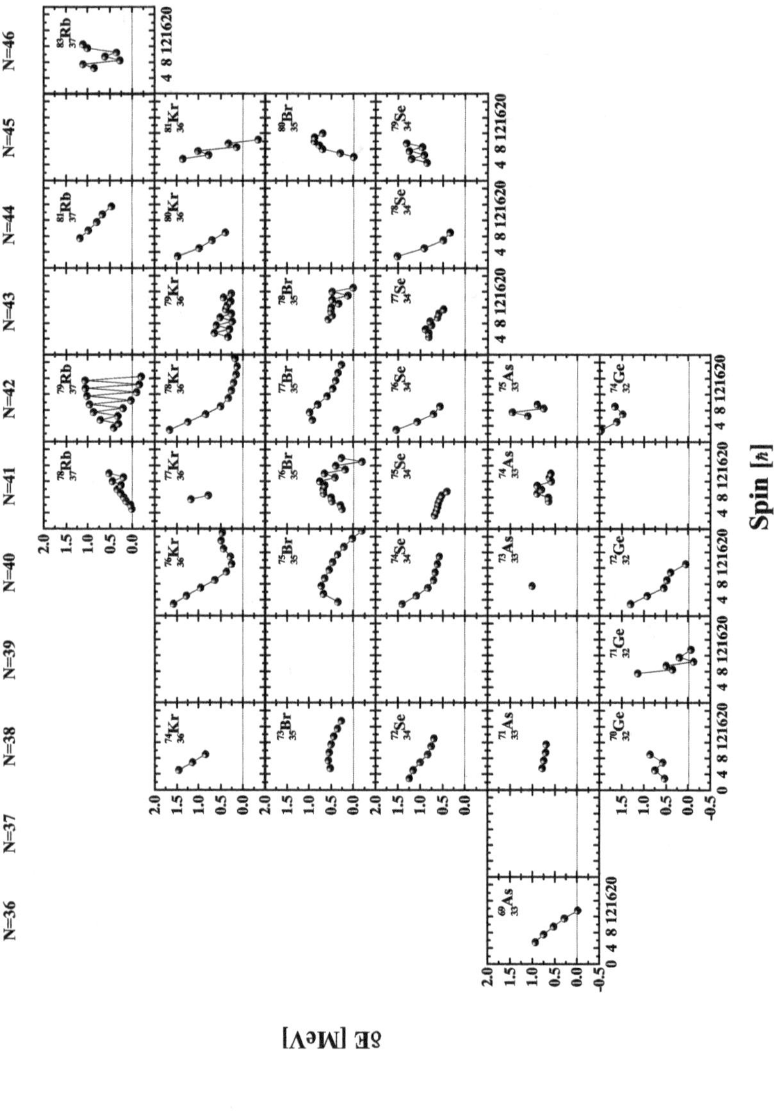

Fig. 2. The energy displacement δE between the alternating-parity bands as a function of spin in the nuclei shown in Fig. 1.

Table 1. Experimental $B(E1)$ and intrinsic electric dipole moment D_0 values in the nuclei in Fig. 1.

	$I_i^\pi \to I_f^\pi$	$B(E1)$ (10^{-5}W.u.)	D_0 (e fm)
^{70}Ge[20,21]	$3^{(-)} \to 2^+$	28.000 (7.000)	0.052 (0.009)
^{71}As[23,24]	$15/2^- \to 13/2^+$	6.817 (3.035)	0.028 (0.009)
	$19/2^- \to 17/2^+$	4.069 (7.305)	0.021 (0.026)
^{74}Se[16,17]	$3^- \to 2^+$	0.110 (0.020)	0.003 (0.001)
	$3^- \to 2_2^+$	0.320 (0.060)	0.006 (0.001)
	$3^- \to 4^+$	0.590 (1.000)	0.008 (0.009)
^{78}Se[18]	$3^- \to 2^+$	17.000 (6.00)	0.041 (0.010)
	$5^- \to 4^+$	51.000 (15.000)	0.086 (0.018)
	$3^- \to 2_2^+$	190.000 (60.000)	0.078 (0.018)
^{79}Se[8]	$9/2^- \to 7/2^+$	> 70.000	> 0.097
	$11/2^- \to 9/2^+$	90.000 (20.000)	0.102 (0.016)
	$15/2^- \to 13/2^+$	> 60.000	> 0.078
^{73}Br[3]	$11/2^- \to 9/2^+$	4.600 (0.900)	0.021 (0.003)
	$15/2^- \to 13/2^+$	6.400 (0.600)	0.024 (0.002)
^{76}Br[4]	$9^- \to 8^+$	4.295 (1.005)	0.022 (0.005)
	$10^- \to 9^+$	5.520 (1.383)	0.026 (0.005)
^{80}Br[32]	$9^- \to 8^+$	7.100 (2.500)	0.028 (0.007)
	$11^- \to 10^+$	10.500 (3.400)	0.033 (0.007)
^{78}Kr[35,36]	$3^- \to 2^+$	8.100 (20.000)	0.028 (0.049)
^{80}Kr[19]	$5^- \to 4^+$	6.945 (0.335)	0.031 (0.001)
^{79}Rb[39,40]	$9/2^- \to 7/2^+$	5.238 (1.635)	0.023 (0.005)
	$13/2^- \to 11/2^+$	5.678 (1.791)	0.023 (0.005)
^{81}Rb[41]	$23/2^- \to 21/2^+$	5.000 (1.000)	0.031 (0.004)
^{83}Rb[42]	$17/2^- \to 15/2^+$	21.092 (10.530)	0.043 (0.015)

directly from the lifetimes of the related excited states, and the D_0 values were extracted from the $E1$ transition probabilities. The $B(E1)$ values of the nuclei with octupole correlations should be greater than 10^{-5} W.u.,[2] while the $B(E1)$ values increase with enhanced octupole-octupole interaction and reach $10^{-4} \sim 10^{-2}$ W.u. when the nuclei form the stable octupole deformation.[2] The D_0 values of the nuclei with stable octupole deformation should be greater than 0.1 e fm.[44] The $B(E1)$ values of ^{69,70}Ge, ^{71}As, ^{78,79}Se, ^{73,76,80}Br, ^{78}Kr and ^{79,81,83}Rb displayed in Table 1 are greater than 10^{-5} W.u., even reach 1.9×10^{-3} W.u. in the 3^- state in ^{78}Se,[18] which represents unambiguous determination of the octupole collectivities in these nuclei. The D_0 values of the relevant states listed in Table 1 are one order of magnitude smaller than 0.1 e fm, which indicates octupole correlations rather than stable octupole deformation in these nuclei.

Among the available experimental data, there is a clear maximum of $B(E1)$ value in ^{78}Se[18] with 34 protons and 44 neutrons. This observation is consistent with the theoretical predictions of the "magic" octupole numbers of $Z = 34$.[2] As shown in Table 1, an accurate description of the evolution of the strength of octupole correlations with different neutron numbers in the $A \approx 80$ mass region could not be concluded due to the absence of the experimental data. However, the δE values of the alternating-parity bands display a trend closer to zero around the neutron $N = 41$, as shown in Fig. 2. Therefore, we tentatively consider the best candidates for the maximal octupole collectivity occurs around the $N = 41$ isotones, which is consistent with the results through the systematic investigations on the behavior of the 3^- octupole states by Cottle.[10]

3. Conclusion and Outlook

The nuclei with octupole correlations in the $A \approx 80$ mass region are reviewed. In particular, the experimental δE, $B(E1)$ and D_0 values, as well as the criteria for octupole correlations are discussed. Octupole correlations in the $A \approx 80$ mass region relating to the reflection symmetry breaking of the nuclear shape are mainly from the interaction between the $\pi p_{3/2}$ and the $\pi g_{9/2}$ orbitals in the configurations. Additionally, a few odd-neutron nuclei (^{75,77,79}Se and ^{79}Kr) were also reported to have octupole correlations, which might have originated from the interaction between their valence neutrons. In this work, we suggest that octupole correlations also exist in the ^{70,74}Ge, ^{71,73,75}As, ^{75,76,77}Se, ^{75,77,80}Br, $^{74,76-79,81}$Kr and ^{78,79,81,83}Rb. The trend observed in $B(E1)$ values indicates a significant enhanced strength of octupole correlations in the protons $Z = 34$, which is exactly consistent with the predicted "magic" octupole number. Meanwhile, the octupole strength is enhanced with neutron number around $N = 41$ according to the δE values. The systematic studies indicate that there might be a broader region with octupole correlations in the $A \approx 80$ mass region. Therefore, more experimental results, especially the lifetime measurements of nuclei states, are desirable in this mass region.

Acknowledgments

The authors would like to thank A. Rohilla for his efforts in reading and revising the English writing of the manuscript. This work is partly supported by the Major Program of Natural Science Foundation of Shandong Province (No. ZR2020ZD30), the Outstanding Youth Fund of Natural Science Foundation of Shandong Province (No. ZR2020YQ07), the National

Natural Science Foundation of China (Nos. 12075137 and 12075138), the Young Scholars Program of Shandong University, Weihai, and Young Foreign Expert Program (No. QN2022150001L).

References

1. L. P. Gaffney *et al.*, *Nature (London)* **497** (2013) 199.
2. P. A. Butler and W. Nazarewicz, *Rev. Mod. Phys.* **68** (1996) 349.
3. S. Bhattacharya *et al.*, *Phys. Rev. C* **100** (2019) 014315.
4. W. Z. Xu *et al.*, *Phys. Lett. B* **833** (2022) 17287.
5. C. Liu *et al.*, *Phys. Rev. Lett.* **116** (2016) 112501.
6. X. Xiao *et al.*, *Phys. Rev. C* **106** (2022) 064302.
7. A. Mukherjee *et al.*, *Phys. Rev. C* **105** (2022) 014322.
8. S. Rajbanshi *et al.*, *Phys. Rev. C* **104** (2021) 064316.
9. L. Mu *et al.*, *Phys. Lett. B* **827** (2022) 137006.
10. P. D. Cottle, *Phys. Rev. C* **42** (1990) 1264.
11. P. D. Cottle, *Phys. Rev. C* **47** (1993) 1529.
12. M. Spieker, *Phys. Rev. C* **106** (2022) 054305.
13. C. G. Wang *et al.*, *Phys. Rev. C* **106** (2022) L011303.
14. D. G. Roux *et al.*, *Eur. Phys. J. A* **48** (2012) 99.
15. I. Stefanescu *et al.*, *Phys. Rev. C* **70** (2004) 044304.
16. J. Döring *et al.*, *Phys. Rev. C* **57** (1998) 2912.
17. B. Singh, *Nucl. Data Sheets* **51** (1987) 225.
18. R. Schwengner *et al.*, *Z. Phys. A* **326** (1987) 287.
19. J. Döring *et al.*, *Phys. Rev. C* **52** (1995) 76.
20. M. Sugawara, *Phys. Rev. C* **81** (2010) 024309.
21. M. R. Bhat, *Nucl. Data Sheets* **51** (1987) 95.
22. J. J. Sun *et al.*, *Phys. Lett. B* **734** (2014) 308.
23. R. S. Zighelboim *et al.*, *Phys. Rev. C* **50** (1994) 716.
24. R. A. Kaye *et al.*, *Phys. Rev. C* **83** (2011) 044316.
25. M. Kumar Raju *et al.*, *Phys. Rev. C* **92** (2015) 064324.
26. X. Xiao *et al.*, This paper is still in prepare.
27. T. D. Johnson *et al.*, *Phys. Rev. C* **46** (1992) 516.
28. J. C. Wells *et al.*, *Phys. Rev. C* **22** (1980) 1126.
29. G. D. Johns *et al.*, *Phys. Rev. C* **55** (1997) 660.
30. G. Z. Solomon *et al.*, *Phys. Rev. C* **59** (1999) 1339.
31. G. N. Sylvan *et al.*, *Phys. Rev. C* **48** (1993) 2252.
32. R. J. Guo *et al.*, *Phys. Lett. B* **833** (2022) 137344.
33. J. J. Valiente-Dobón *et al.*, *Phys. Rev. Lett.* **95** (2005) 232501.
34. J. J. Valiente-Dobón *et al.*, *Phys. Rev. C* **71** (2005) 034311.
35. H. Sun *et al.*, *Phys. Rev. C* **59** (1999) 655.
36. S. Rab, *Nucl. Data Sheets* **63** (1991) 1.
37. G. D. Johns *et al.*, *Phys. Rev. C* **50** (1994) 2786.
38. R. A. Kaye *et al.*, *Phys. Rev. C* **54** (1996) 1038.
39. R. K. Sinha *et al.*, *Eur. Phys. J. A* **28** (2006) 277.

40. J. W. Holcomb *et al.*, *Phys. Rev. C* **48** (1993) 1020.
41. J. Döring *et al.*, *Phys. Rev. C* **50** (1994) 1845.
42. S. Ganguly *et al.*, *Nucl. Phys. A* **768** (2006) 43.
43. W. Nazarewicz and P. Olanders, *Nucl. Phys. A* **441** (1985) 420.
44. I. Ahmad and P. A. Butler, *Annu. Rev. Nucl. Part. Sci.* **43** (1993) 71.

Rotational spectroscopies of reflection-asymmetric nuclei within particle rotor model*

Yuanyuan Wang†

Mathematics and Physics Department,
North China Electric Power University,
Beijing 102206, P. R. China

Hebei Key Laboratory of Physics and Energy Technology,
North China Electric Power University, Baoding 071000, P. R. China
yywang1021@ncepu.edu.cn

Shuangquan Zhang†

State Key Laboratory of Nuclear Physics
and Technology, School of Physics, Peking
University, Beijing 100871, P. R. China
sqzhang@pku.edu.cn

The rotational spectroscopic properties of reflection-asymmetric nuclei investigated by the reflection-asymmetric axial or triaxial particle rotor model are briefly reviewed. The status of observations of the alternated parity bands in even–even nuclei and the parity doublet bands in odd-A and odd–odd nuclei for the $A \sim 220$ mass region are summarized. The reflection-asymmetric axial particle rotor model has been used to systematically describe the parity doublet bands of the odd-A nuclei in this region. With the triaxial degree of freedom included, the reflection-asymmetric triaxial particle rotor model shows the ability to study the multiple chiral doublet bands with octupole correlations and explore the possible chirality-parity quartet bands.

Keywords: Reflection-asymmetric nuclei; rotational spectra; electromagnetic transitions; chirality-parity violation; particle rotor model.

1. Introduction

The atomic nucleus is a microscopic quantum many-body system, and its shape provides an intuitive understanding of nuclear spatial density

*To cite this article, please refer to its earlier version published in the *International Journal of Modern Physics E*, Volume 32(10), (19 pages), 2340010 (2023), DOI: 10.1142/S0218301323400104.

†Corresponding authors.

distributions which is usually manifested in the observations of such as the energy spectra and the electromagnetic transitions.

The nuclei were assumed to be spherical in the 1950s.[1] The observations of the large quadrupole moments in some nuclei led to the suggestions that these nuclei might have axially deformed shapes.[2] The existence of stable quadrupole-deformed nuclear shapes with axial and reflection symmetry in the intrinsic frame and the corresponding rotational spectra established according to the energy formula $E_I \approx I(I+1)$, have been known for a long time.[3] Considering that such shape is symmetric under the space reflection operation, all members of the observed rotational bands in quadrupole-deformed nuclei have the same parity. However, with the first observation of the low-lying negative-parity 1^- states in the even–even radium isotopes by Berkeley group,[4,5] the possibility arose that nuclei might have octupole-deformed shapes with reflection-asymmetry in the intrinsic frame such as the so-called pear shape.[6]

The study of the structure of pear-shaped nuclei has been at the frontiers of both nuclear physics and particle physics, since it provides one probe to test the charge-parity (CP) violation beyond the standard model.[7] From a microscopic view, the pear shape of a nucleus can be arisen from the strong octupole correlations of the nucleons near the Fermi surface occupying states of opposite parity with orbital and total angular momentum differing by $3\hbar$, i.e., $\Delta l = \Delta j = 3\hbar$. Empirically, this condition occurs for proton or neutron particle numbers around 34 ($g_{9/2} \leftrightarrow p_{3/2}$), 56 ($h_{11/2} \leftrightarrow d_{5/2}$), 88 ($i_{13/2} \leftrightarrow f_{7/2}$) and 134 ($j_{15/2} \leftrightarrow g_{9/2}$), i.e., when the Fermi surface locates between the normal-parity shell and an intruder orbital with opposite parity.[8] So far, the octupole correlations and the pear-shaped structures have been studied extensively in $A \sim 150$ mass region with $Z \approx 56$ and $N \approx 88$ and in $A \sim 220$ mass region with $Z \approx 88$ and $N \approx 134$, see, for reviews, Refs. 8–10. Evidence for octupole correlations in $A \sim 80$ mass region was also discussed,[11–19] which opens a new territory of the observation of octupole correlations.

For the even–even nuclei, an important signal of the reflection asymmetry arising from strong octupole correlations is the presence of the low-lying negative-parity states connecting to the members of the rotational ground-state band via enhanced $E1$ transitions, i.e., interleaved positive- and negative-parity bands.[3] Further evidence is provided by the sizeable value of the electric octupole ($E3$) moment for the transition to the ground state, indicating unambiguous signature of octupole collectivity.[8] On account of an enhancement in the $E3$ moment, evidences have been presented that the lighter radium isotopes ^{222,224,226}Ra[7,20,21] and the neutron-rich barium isotopes ^{144,146}Ba[22,23] have static octupole deformation.[10]

For the odd-A nuclei, with the odd proton or neutron angular momentum j, the total angular momentum I has projection $K = j_z$ on the z-axis when axially deformed. In this case, the experimental signal for static octupole deformation is parity doublet bands with spin-parity $K^{\pm}, (K+1)^{\pm}, \ldots$, i.e., two nearly degenerate rotational bands with the same spin I but opposite parity.[9] Parity doublet bands can be identified by their similar magnetic moments and enhanced $E1$ transitions between these bands. It is noted that octupole-deformed nuclei with an odd nucleon have attracted special interest since they will have enhanced nuclear Schiff moments owing to the presence of the large octupole collectivity and nearly degenerate parity doublets.[10]

Figure 1 summarizes the present status of observations of the alternating parity bands in even–even nuclei and the parity doublet bands in odd-A and odd–odd nuclei for the $A \sim 220$ mass region, which highlights the wealth of the information concerning octupole correlations. For the

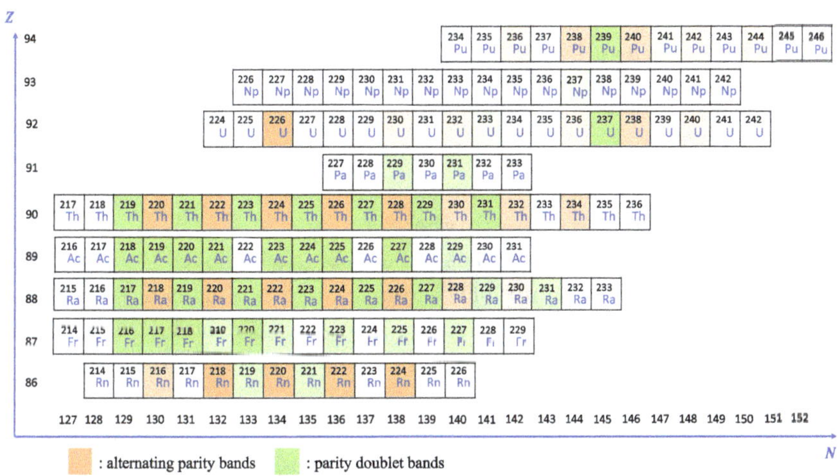

Fig. 1. (Color online) Nuclei with the observation of the alternating parity bands and parity doublet bands in $A \sim 220$ mass region. The even–even nuclei colored orange represent the negative-parity bands which are observed and alternate with the ground state positive-parity band in low spin region (dark orange), in high spin region (medium orange) and without alternating (light orange). The odd nuclei colored in dark green represent the nuclei in which the parity doublet bands, i.e., two nearly degenerate rotational bands with the same spins but opposite parities are observed, while those in medium green demonstrate only a few parity doublet states are observed. The nuclei colored in light green represent the observed parity partner bands which are interpreted as being formed by coupling of the ground state to the $K = 0^-$ octupole band.

even–even nuclei, the low-lying interleaved positive- and negative-parity bands have been observed in $^{218-224}$Rn,[7,24–26] $^{218-226}$Ra,[7,25,27,28] $^{220-228}$Th[25,28,29] and ^{226}U[30] nuclei, indicating the strong octupole correlations or even static octupole deformation in these nuclei. Meanwhile, the alternating parity bands have only been observed at high-spin regions ($> 10\hbar$) in nuclei like ^{216}Rn,[31] ^{228}Ra,[25] $^{230-234}$Th,[25] ^{238}U[32] and ^{238,240}Pu,[33,34] which show the manifestation of the octupole collectivity with the increase of spin. Apart from the cases of the low- and high-lying alternating parity bands, there are some nuclei in which the negative-parity states along with the positive-parity ground state band have been observed but no alternating parity structure was formed, for example ^{230}Ra,[25] ^{230,232,236,240}U[35–37] and ^{236,242,244}Pu,[38,39] indicating the lack of a good octupole collectivity. For the odd-A and odd–odd nuclei, the so-called parity doublet bands have been observed extensively in Fr, Ra, Ac and Th isotopes, for example in ^{216,217,218,220}Fr,[40–43] $^{217-227}$Ra,[44–49] ^{218,219,220,221,223,224,225,227}Ac,[50–55] $^{219-231}$Th,[25,56–61] as well as in the $N = 145$ isotones ^{237}U[62] and ^{239}Pu.[62] Meanwhile in some nuclei neighboring, a few parity doublet states but not forming doublet bands were observed. In addition, there are nuclei in which the parity partner bands formed by the coupling of the ground state to the $K = 0^-$ octupole band were observed, i.e., ^{233}U,[63] ^{237}Np[64] and ^{241}Pu.[65] Generally speaking, these different features of the spectroscopic properties reflect the strengthness of the octupole correlations in odd-A and odd–odd nuclei. In view of the abundant information of octupole collectivities in $A \sim 220$ mass region, more efforts are highly expected to other mass regions.

Theoretically, various approaches have been developed to study the observed experimental features of the even–even, odd-A and odd–odd nuclei characterized by static and dynamic octupole deformation, such as the quadrupole–octupole collective model,[66–76] the interacting boson (or interacting boson-fermion) model,[77,78] the reflection asymmetric shell model,[79,80] the microscopic core-quasiparticle coupling model,[81] the cluster model,[82,83] and the reflection-asymmetric particle rotor model,[84–87] etc.

As a quantum-mechanical model consisting of the collective rotation and the intrinsic single-particle motions, the particle rotor model (PRM) describes the nuclear system in the laboratory reference frame with total angular momentum and total parity as good quantum numbers.[3] In the PRM, the energies and transition probabilities can be treated in a fully quantal way. It has been extensively applied to investigate various exotic nuclear phenomena such as backbending,[88] signature inversion,[89,90] chiral rotation,[91–98] wobbling motion[3,99–102] and parity doublet bands.[84]

To investigate the parity doublet bands observed in pear-shaped nuclei, a reflection-asymmetric PRM with the axial rotor has been developed in Ref. 84, and applied to extensively describe the low-lying spectra properties in nuclei with $A = 219$–229.[85,86] Further including the triaxial degree of freedom, a reflection-asymmetric triaxial particle potor model (RAT-PRM) has been developed in Ref. 87. This RAT-PRM has been successfully applied to investigate the multiple chiral doublet (MχD) bands with octupole correlations in ^{78}Br,[87] ^{131}Ba[103] and ^{124}Cs,[104] the octupole collectivities in ^{143}Ba,[105] ^{223}Th,[106] ^{73}Br[107] and ^{79}Se,[18] and to explore the novel structure for an ideal chirality-parity (ChP) violation system.[108]

In Sec. 2, we will briefly introduce the reflection-asymmetric particle rotor model with the axial or triaxial rotor. Some investigation of the spectroscopies of reflection-asymmetric nuclei within reflection-asymmetric axial or triaxial PRM will be discussed in Secs. 3 and 4, respectively. A short summary is given in Sec. 5.

2. Reflection-Asymmetric Particle Rotor Model

In order to describe and explain the spectroscopic properties of the reflection-asymmetric nuclei, there have been several efforts with the spirit of particle rotor model that was originally proposed by Bohr and Mottelson.[109] Based on the PRM, one considers only a few valence nucleons and couples them to the inert core as a collective rotor. Generally, the PRM Hamiltonian includes two parts that can be written as follows[3]:

$$\hat{H} = \hat{H}_{\text{intr}} + \hat{H}_{\text{coll}}, \tag{1}$$

where $\hat{H}_{\text{intr}}$ is the intrinsic Hamiltonian for valence nucleons in the deformed potential of the core and the collective Hamiltonian $\hat{H}_{\text{coll}}$ describes the rotations of the core. For a general introduction to the PRM, see Refs. 3, 110. Here, we briefly outline the reflection-asymmetric particle rotor model developed with the axial symmetry and the one in the triaxial case.

2.1. *Axial symmetry case*

In 1984, Leander and Sheline formulated the theoretical description for odd-A nuclei whose intrinsic shape is axially elongated and statically or dynamically reflection asymmetric.[84] For the statically reflection-asymmetric case, in the strong-coupling scheme, they introduced a phenomenological core Hamiltonian that accounts for the parity splitting in the octupole deformed core. For the dynamically reflection-asymmetric case, they formulated in the intermediate coupling scheme where a basis which is strong-coupled with respect to elongation but weak-coupled with respect to reflection asymmetry was employed. Here, we focus on the static case.

Based on the strong-coupling scheme, the two parts in PRM Hamiltonian (1) are the intrinsic part

$$\hat{H}_{\text{intr}} = \hat{H}_{\text{s.p.}} + \hat{H}_{\text{pair}}, \tag{2}$$

where $\hat{H}_{\text{s.p.}}$ is the intrinsic single particle Hamiltonian with a reflection-asymmetric folded Yukawa form and $\hat{H}_{\text{pair}}$ is the standard Bardeen–Cooper–Schrieffer (BCS) paring Hamiltonian, and the core part

$$\hat{H}_{\text{coll}} = \frac{\hat{R}_1^2 + \hat{R}_2^2}{2\mathcal{J}} + \frac{1}{2}E(0^-)(1 - \hat{P}_c), \tag{3}$$

where $\hat{R}_{1,2}$ denote the core angular momentum components perpendicular to the symmetry axis, $E(0^-)$ is the phenomenological parameter to consider the parity splitting of the core resulting from the reflection asymmetry and $\hat{P}_c$ is the parity operator of the core. In the PRM, the core angular momentum operator can be written as $\hat{\boldsymbol{R}} = \hat{\boldsymbol{I}} - \hat{\boldsymbol{j}}$ with $\hat{\boldsymbol{I}}$ and $\hat{\boldsymbol{j}}$ the angular momentum operators for the nucleus and the valence particle, respectively.

The space of the PRM Hamiltonian is spanned by the symmetrized basis with good parity and angular momentum[84,86]

$$\psi^{\nu}_{IM\Omega p} = \frac{1}{2N}(1 + \hat{R}_1)D^I_{M\Omega}(1 + p\hat{P}_c\hat{\pi})\Phi_a\tilde{\chi}^{\nu}_{\Omega}, \tag{4}$$

where N is a normalization constant, $D^I_{M\Omega}$ is the rotation matrix, $p = \pm 1$ is the total parity, $\hat{\pi}$ is the valence (single particle) parity operator, Φ_a denotes the core with the same orientation in space as the single particle potential and $\tilde{\chi}^{\nu}_{\Omega}$ is the BCS state with the odd nucleon in the orbital χ^{ν}_{Ω}.

This model was subsequently developed by Leander and Chen with $\hat{H}_{\text{s.p.}}$ of a reflection-asymmetric Woods–Saxon form.[86]

2.2. *Axial asymmetry case*

For the triaxial case, based on the strong coupling scheme, the reflection-asymmetric triaxial particle rotor model has been developed in Ref. 87. The collective Hamiltonian for the reflection-asymmetric triaxial core is generalized straightforwardly from the reflection-asymmetric axial rotor and the intrinsic Hamiltonian for valence protons (neutrons) is in a reflection-asymmetric triaxially deformed potential.

The collective Hamiltonian reads

$$\hat{H}_{\text{coll}} = \sum_{k=1}^{3} \frac{\hat{R}_k^2}{2\mathcal{J}_k} + \frac{1}{2} E(0^-)(1 - \hat{P}_c), \tag{5}$$

where $\hat{R}_k$ denote the core angular momentum components along the intrinsic principal axes, and $\hat{R}_k = \hat{I}_k - \hat{j}_{pk} - \hat{j}_{nk}$. Here, $\hat{I}_k$, $\hat{j}_{pk}$ and $\hat{j}_{nk}$ are the angular momentum operators for the nucleus, the valence protons and the valence neutrons, respectively. For the moments of inertia (MoIs), the irrotational flow type $\mathcal{J}_k = \mathcal{J}_0 \sin^2(\gamma - 2k\pi/3)$ can be adopted.

The intrinsic Hamiltonian for valence protons (neutrons) is[89,94,111]

$$\hat{H}_{\text{intr}}^{p(n)} = \sum_{\nu>0} (\varepsilon_\nu^{p(n)} - \lambda)(a_\nu^\dagger a_\nu + a_{\bar{\nu}}^\dagger a_{\bar{\nu}}) - \frac{\Delta}{2} \sum_{\nu>0} (a_\nu^\dagger a_{\bar{\nu}}^\dagger + a_{\bar{\nu}} a_\nu), \tag{6}$$

where λ denotes the Fermi energy, Δ the pairing gap parameter and $|\bar{\nu}\rangle$ the time-reversal state of $|\nu\rangle$. The single-particle energy $\varepsilon_\nu^{p(n)}$ is obtained by diagonalizing a single-particle Hamiltonian that has the form of a Nilsson Hamiltonian[112]

$$\hat{H}_{\text{s.p.}}^{p(n)} = -\frac{1}{2}\hbar\omega_0 \nabla^2 + V(r; \theta, \varphi) + C\boldsymbol{l} \cdot \boldsymbol{s} + D[\boldsymbol{l}^2 - \langle \boldsymbol{l}^2 \rangle_N], \tag{7}$$

with the four terms being kinetic energy, the reflection-asymmetric triaxially deformed potential,[87] the spin-orbit term and the shifting term,[110] respectively. The parameters C and D are related to the standard Nilsson parameters κ and μ with $C = -2\kappa$ and $D = -\kappa\mu$, respectively.

The total RAT-PRM Hamiltonian is diagonalized numerically in the symmetrized strong-coupled basis with good parity and angular momentum, which gives rise to the eigenvalues and eigen wave functions. Then the reduced electromagnetic transition probabilities can be calculated.[87]

3. Rotational Properties of Reflection-Asymmetric Axial Nuclei

In many odd-A nuclei, there are experimentally observed parity doublet states which are closely spaced in energy and with the same spin but the opposite parity. These parity doublets might originate from a single parity-mixed orbital in an octupole-deformed mean field (strong coupling), or alternatively from two distinct single-particle orbitals in a reflection symmetric field which are partly mixed by the octupole vibrational mode of the core (weak or intermediate coupling).

To distinguish between the above two alternatives, with the assumption of axial quadrupole deformation and considering reflection asymmetric degree of freedom either statically or dynamically, Leander and Sheline applied their model to the odd-A Ra, Ac, Th and Pa isotopes.[84] They have evaluated several items of experimental information, including energy levels, decoupling factors, α-transitions and electromagnetic $M1$, $E1$ and $E3$ matrix elements. Their results show, relative to the experimental data on all these aspects, strong coupling does give a uniform improvement over weak coupling. Nevertheless, the strong-coupling limit seems to be reached only in a few cases. In particular, this study provided empirical evidence for a shape transition from octupole vibrations above $A \sim 229$ to more or less stable octupole deformation in lighter isotopes.[84]

By using the intrinsic single particle Hamiltonian of a reflection-asymmetric Woods–Saxon form, Leander and Chen performed the reflection-asymmetric particle rotor model descriptions to the parity doublet bands in a number of odd-A nuclei with $A \sim 219 - 229$.[85,86] The first realistic calculation was applied to the nucleus ^{227}Ac.[85] It was found that the low-energy structure observables of ^{227}Ac, such as the spectra of parity doublet bands, are well accounted for with a static octupole deformation; the deformation parameters adopted in the corresponding calculations are $\beta_2 = 0.168$, $\beta_3 = 0.1$, $\beta_4 = 0.094$, $\beta_5 = 0.01$ and $\beta_6 = 0.0052$. Systematic comparisons between theory and experiment were performed in Ref. 86 for the nuclei with available data at that time: ^{219,221,223,225}Rn, ^{221,223,225,227}Fr, ^{219,221,223,225,227}Ra, ^{219,223,225,227,229}Ac, ^{221,223,225,227,229}Th and ^{229}Pa. Overall agreement has been achieved by the reflection-asymmetric rotor model with a static octupole deformation $\beta_3 \sim 0.1$. Combined with newly experimental observations on parity doublet structure, more theoretical applications have been devoted to ^{219}Ra,[113] ^{221}Ra,[46] ^{223}Ra,[47,115] ^{227}Fr and ^{223,225,227}Ac.[54]

Very recently, the parity doublet structures in ^{223}Th have been restudied with the reflection-asymmetric PRM,[106] owing that a new fusion-evaporation experiment in 2017 had largely extended the structures to the positive-parity state $49/2^+$ and the negative-parity state $47/2^-$.[58] A further study is interesting to investigate the new spectroscopy information and to explore the parity doublet structures in the medium- and high-spin regions. The new reflection-asymmetric PRM calculations were performed with the quadrupole and octupole deformation parameters taken from Ref. 86, i.e., $\beta_2 = 0.12$, $\gamma = 0°$ and $\beta_3 = 0.10$, and an intrinsic Hamiltonian of the reflection-asymmetric Nilsson form.

The comparison between the theoretical and experimental results for ^{223}Th is shown in Fig. 2. As seen in Fig. 2(a), the calculated excitation energies are in good agreement with the data for both positive-parity band A and

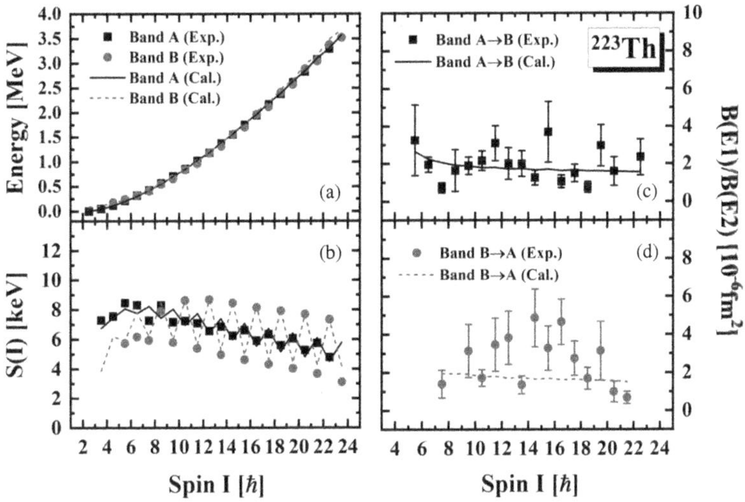

Fig. 2. The calculated energies $E(I)$ (a), the energy staggering parameters $S(I) = [E(I) - E(I-1)]/2I$ (b), the $B(E1)/B(E2)$ ratios between the interband $E1$ transitions and the intraband $E2$ transitions for (c) band A→B and (d) band B→ A in ^{223}Th by RAT-PRM (lines), in comparison with the experimental data[59] (symbols). Reproduced from Ref. 106.

negative-parity band B. For the observed spin range of $9/2\hbar \leq I \leq 45/2\hbar$, the average energy difference between the parity doublets is only 37 keV and the calculated one is 56 keV; both values indicate strong octupole correlations in ^{223}Th. The energy staggering parameters $S(I) = [E(I) - E(I-1)]/2I$ depicted in Fig. 2(b) exhibit a pronounced signature splitting for band B in contrast to the considerably smaller splitting for band A. The calculated $S(I)$ values excellently reproduce the experimental data in both staggering phase and amplitude except for $I \leq 21/2h$ region in band A. In addition to the degeneracy in energy spectra, the enhanced $E1$ transitions connecting the parity doublet bands also serve as one of important experimental evidences for octupole correlations. As seen in Figs. 2(c) and 2(d), the experimental $B(E1)/B(E2)$ ratios with $E1$ transitions starting from the positive parity and the negative parity states are similar in general, and can be reasonably reproduced by the calculated results. The main components of the intrinsic wave functions which account for the parity doublet bands were also analyzed. It was found that the parity doublet bands in ^{223}Th are mainly based on a single neutron reflection-asymmetric configuration with its largest component being $\nu(g_{9/2}, j_{15/2})$ $[\Omega = 5/2]$ and the second largest being $\nu(g_{9/2}, j_{15/2})[\Omega = 3/2]$. With increasing spin, the amplitude of the largest component decreases whereas second largest increases, and through the variation of these components the signature splitting behaviors in the parity doublet bands can be understood.[106]

4. Rotational Properties of Reflection-Asymmetric Triaxial Nuclei

The existence of nuclear shape without axial symmetry, i.e., the triaxial deformation has attracted extensive interests and become one of the hot topics in current nuclear physics frontiers. Among a variety of nuclear phenomena that may manifest the triaxial degree of freedom, of particular interest is chiral doublet bands, which originate from the spontaneous breaking of chiral symmetry in the intrinsic frame of reference and serve as a direct evidence for the existence of triaxial nuclear shape.[91] Moreover, the diversity of nuclear chirality and its experimental signal multiple chiral doublets, which were predicted theoretically in 2006[116] and confirmed experimentally in 2013,[117] provide the evidence of the triaxial shape coexistence and make it possible to explore the interplay between the spontaneous breakings of chiral symmetry and other intrinsic symmetries. For reviews on nuclear chirality, see Refs. 118–125.

It is therefore interesting to ask whether the spontaneous breakings of chiral symmetry and spatial reflection symmetry can happen simultaneously in atomic nucleus and what are feature signals of this exotic phenomenon.

In 2016, the observation of the MχD bands with octupole correlations in nucleus ^{78}Br provides the first evidence of chiral geometry in octupole soft nuclei and indicates the spontaneous breakings of chiral and spatial reflection symmetries can happen simultaneously.[16] In ^{78}Br, the octupole deformation and the reflection symmetry breaking is dynamic. If a nucleus has both static triaxial and octupole deformations, the nuclear ChP violation, a strong breaking of both chiral and reflection symmetries may occur in the intrinsic frame. The symmetry restoration in the laboratory frame for the ChP violated nuclei would result in both chiral and parity splittings which give rise to four nearly degenerate states, i.e., ChP quartet states. The rotational excitation on the ChP quartet states then generates four nearly degenerate rotational bands, i.e., the so-called ChP quartet bands, as expected in Refs. 16, 118. Up to now, the ChP quartet bands have not been observed in experiment, while the MχD bands with octupole correlations have been suggested in ^{78}Br,[16] ^{124}Cs[104,126] and ^{131}Ba.[127]

In this section, we will briefly introduce the applications of RAT-PRM to the MχD bands with octupole correlations and to the ChP quartet bands as well as the selection rules of electromagnetic transitions for an ideal ChP violation in atomic nuclei.

4.1. *MχD with octupole correlations*

The first evidence of the MχD bands with octupole correlations was observed in ^{78}Br.[16] The potential energy surface given by the microscopic multi-dimensionally-constrained covariant density functional theory (MDC-CDFT)[128–130] is very soft with respect to octupole deformation degree of freedom, which supports octupole correlations rather than static octupole deformation in ^{78}Br. As the octupole softness, two pairs of chiral bands built on different single particle configurations with octupole interactions in between will appear.

To investigate the MχD bands with octupole correlations in ^{78}Br, the RAT-PRM calculations with a quasi-proton and a quasi-neutron coupled with a reflection-asymmetric triaxial rotor have been performed. The quadrupole deformation parameters $\beta_2 = 0.28$ and $\gamma = 16.3°$ are obtained from the MDC-CDFT, and a small octupole deformation parameter $\beta_3 = 0.02$ is adopted to simulate the dynamically reflection asymmetric.[87]

Figure 3 shows the calculated excited energies, energy staggering parameters $S(I) = [E(I) - E(I-1)]/2I$, and $B(M1)/B(E2)$ ratios for the positive-parity doublet bands 1 and 2 as well as the negative-parity doublet bands 3 and 4 in ^{78}Br in comparison with experimental data. The calculated excited energies reproduce the data for both the positive- and negative-parity doublet bands satisfactorily. The average energy difference between the positive-parity doublet bands overestimates the data, which may be due to the small triaxial deformation ($\gamma = 16.3°$) adopted. By using a larger triaxial deformation, the RAT-PRM calculations could provide smaller average energy difference between the positive-parity doublet bands. The calculated $S(I)$ values exhibit an odd–even staggering behavior for the positive-parity doublet bands, while which are smooth till $14\hbar$ for the negative-parity doublet bands. The different $S(I)$ behaviors may be attributed to their corresponding configurations. While the proton configurations are both a $g_{9/2}$ particle, the neutron configurations are quite different, i.e., a neutron hole at the top of the $f_{5/2}$ shell for the negative-parity bands but a neutron at the middle of the $g_{9/2}$ shell for the positive-parity bands. The experimental $B(M1)/B(E2)$ ratios for both positive- and negative-parity doublet bands, including the odd–even staggering for the positive-parity, are well reproduced.

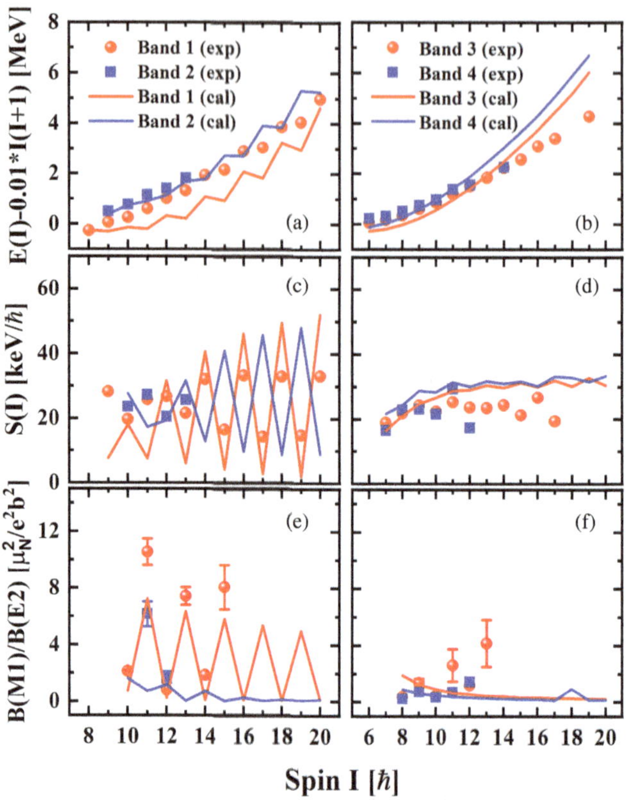

Fig. 3. The excitation energies $E(I)$ [panels (a) and (b)], the energy staggering parameters $S(I) = [E(I) - E(I-1)]/2I$ [panels (c) and (d)] and the $B(M1)/B(E2)$ ratios [panels (e) and (f)] for the positive-parity doublet bands 1 and 2 (left panels) as well as the negative-parity doublet bands 3 and 4 (right panels) in ^{78}Br by RAT-PRM (lines) in comparison with the data available[16] (symbols). The energies are relative to the bandhead of band 1 at $I = 8\hbar$, with a rotor reference subtracted. Taken from Ref. 87.

With the triaxial and octupole degrees of freedom included in the RAT-PRM, not only the positive- and negative-parity doublet bands but also the variety of the transition probabilities, including the interband $E1$ transitions, can be obtained simultaneously. The $E1$ transitions that reflect the relative strength of the octupole correlations deserve further attention. It was found that both the triaxial deformation γ and octupole deformation β_3 influence the calculated $B(E1)$ values. For $\beta_3 = 0.02$, by changing γ from 16° to 21° (given by cranked-shell-model calculations[131]), the $B(E1)$ values will be enhanced and better agreement with the $B(E1)/\ B(E2)$ data can be achieved. Increasing β_3 from 0.02 to 0.04 has no significant influence on the excited energies, staggering parameters and $B(M1)/B(E2)$ ratios except the $B(E1)$ values. The $B(E1)$ values are enhanced with $\beta_3 = 0.04$ and a better agreement with the $B(E1)/B(E2)$ data can be obtained. The interband $E1$ transitions from the negative-parity doublet bands 3 and 4 to the positive-parity doublet bands 1 and 2 with spin have been studied. It was shown that for both bands 3 and 4, the interband $E1$ transitions to band 1 and band 2 show similar variation patterns as expected for chiral structure.

Furthermore, the chiral geometry and its evolution for the positive- and negative-parity doublet bands were investigated by examining the components for the angular momenta of the core as well as the valence proton and neutron, and the effective angles between these angular momenta. For the positive-parity doublet bands, the angular momentum of the valence proton mainly aligns in the intermediate-short plane, while that of the valence neutron has nearly equal components on the three axes due to its mid-shell nature. Considering the fact that the angular momentum for the core mainly aligns along the intermediate axis, and grows rapidly, the total angular momentum lies close to the intermediate-short plane, which is consistent with the large energy difference between the doublet bands. For the negative-parity doublet bands, the calculated results showed that the chiral geometry is well constructed by the angular momenta of the valence proton along the intermediate-short plane, the valence neutron along the long axis and the core along the intermediate axis.

The above application to MχD bands observed in ^{78}Br shows the ability of the RAT-PRM to study the MχD with octupole correlations. It is interesting to find and describe more cases of this novel phenomenon by using the RAT-PRM.

By using the developed RAT-PRM, the possibility of the observed positive- and negative-parity bands in ^{124}Cs as MχD with octupole correlations has been investigated.[104] For the positive-parity bands, a pair of doublet bands with configuration $\pi h_{11/2} \otimes \nu h_{11/2}$ has been observed experimentally,[132] and further interpreted as chiral doublet bands theoretically.[134] For the negative-parity bands, one $\Delta I = 1$ and one $\Delta I = 2$ band with configuration $\pi h_{11/2} \otimes \nu(g_{7/2}d_{5/2})$ have been observed, indicating the possibility of a pair of chiral partners.[126,132] The evidences for the octupole correlations between the positive- and negative-parity bands have also been presented.[126,132]

The RAT-PRM calculation has been performed with the quadrupole deformation parameters $\beta_2 = 0.25$ and $\gamma = 33.8^\circ$ obtained from the configuration-fixed triaxial covariant density functional theory (CDFT) calculation[116] with density functional PC-PK1,[135] and a small octupole deformation parameter $\beta_3 = 0.03$ to consider the effect of octupole correlations. Figure 4 shows the calculated results in comparison with the data available.[126,132,133]

The calculated excitation energies and energy staggering parameters $S(I)$ well reproduce the experimental data for the positive-parity doublet bands and the negative-parity doublet bands. For the positive-parity doublet bands within $11\hbar \leq I \leq 17\hbar$ and negative-parity within $7\hbar \leq I \leq 18\hbar$, the calculated average energy differences are 0.49 MeV and 0.14 MeV, which reproduce data well. For both the calculated and experimental results, the quantity $S(I)$ shows a slight staggering behavior with spin for the positive-parity doublet bands, whereas the $S(I)$ is relatively independent of spin for the negative-parity ones. Therefore, from both the staggering parameter $S(I)$ and the energy differences between doublet bands, the negative-parity doublet bands coincide with the criteria for chirality better than the

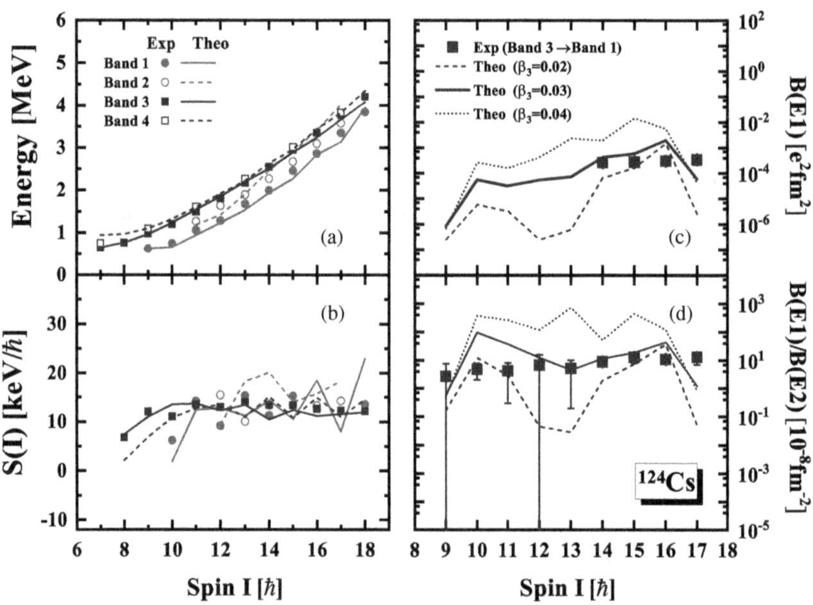

Fig. 4. The calculated excitation energies $E(I)$ [panel (a)], the energy staggering parameters $S(I) = [E(I) - E(I-1)]/2I$ [panel (b)], the calculated $B(E1)$ values of the interband $E1$ transitions (band 3 → band 1) [panel (c)] and the corresponding ratios over the $B(E2)$ values of the intraband $E2$ transitions (band 3) [panel (d)] in ^{124}Cs in comparison with the data available.[126,132,133] Reproduced from Ref. 104.

positive-parity doublet bands. The calculated $B(E1)$ and $B(E1)/B(E2)$ values with $\beta_3 = 0.03$ satisfactorily agree with the experimental data.

The RAT-PRM calculations with $\beta_3 = 0.02$ and 0.04 were performed to further investigate the influence of the octupole deformation. It is found that there is no significant influence on the excitation energies, staggering parameters, $B(M1)$ and $B(E2)$ values by changing β_3 from 0.02 to 0.04, while the $B(E1)$ and $B(E1)/B(E2)$ ratios sensitively depend on the octupole deformation parameter β_3. In general, the calculated $B(E1)$ and $B(E1)/B(E2)$ values can be enhanced up to three magnitudes of order by changing β_3 from 0.02 to 0.04.

To examine the chiral geometry of the two pairs of doublet bands in ^{124}Cs, the angular momentum geometry and its evolution with spin I have been investigated. For the positive-parity doublet bands, consistent results with Ref. 134 are obtained, which indicate that they mainly correspond to a typical chiral vibration pattern. For the negative-parity doublet bands, by examining the angular momentum components and the *azimuthal plot*, the chiral geometry and evolution have been presented which show a transient static chirality around $I = 12\hbar$. The two calculated pairs of positive- and negative-parity doublet bands in ^{124}Cs may be interpreted as a candidate of MχD with octupole correlations. So far, the yrare partner for the negative-parity chiral bands is not observed, and more efforts are still necessary to confirm the prediction of MχD with octupole correlations in this nucleus.

The observation of two pairs of positive-parity bands and one pair of negative-parity bands, as well as the $E1$ transitions among them in nucleus ^{131}Ba,[127] also provides the evidence of MχD candidates with octupole correlations. Especially, the four nearly degenerate positive-parity bands in this nucleus were suggested to be the pseudospin-chiral quartet bands. To investigate the observed MχD with octupole correlations and possible pseudospin-chiral quartet bands, a RAT-PRM with three quasiparticles and a reflection-asymmetric triaxial rotor has been developed.[103] The energy spectra, energy staggering parameters $S(I)$, $B(M1)/B(E2)$ ratios and $B(E1)/B(E2)$ ratios are well reproduced by the RAT-PRM.[127] To examine the chiral geometry of the three pairs of doublet bands in ^{131}Ba, the azimuthal plots, i.e., the probability distribution profiles for the orientation of the angular momentum on the (θ, φ) plane have been investigated.[103] It shows that the chiral vibration at low spins, static chirality at intermediate spins and a tendency to a principal axis rotation around the intermediate axis at high spins. Furthermore, by analyzing the intrinsic structure for the positive-parity MχD candidates, the possible pseudospin chiral quartet bands have been suggested.

4.2. *ChP quartet bands*

Note that the RAT-PRM with small β_3 deformation has been performed to investigate the above observed MχD candidates with octupole correlations established in octupole-soft nuclei. For the nuclei with both static octupole and triaxial deformation, it is of great interests to investigate the possible ChP quartet bands theoretically.

In a reflection-asymmetric triaxial nucleus, the possible simultaneous breaking of the chiral and reflection symmetries may occur. In such a nucleus, as shown in Fig. 5 for a schematic representation, it is expected to build four distinct sectors in the intrinsic frame, which can be regarded either as two pairs of chiral sectors[91] with opposite reflection orientations, or as two pairs of reflection partners with opposite chirality. In the laboratory frame, the quantum tunnelings among the four distinct sectors result in both chiral and parity splitting energies, and this gives rise to a composite of four nearly degenerate states, ChP quartet states. Therefore, it is possible to establish four nearly degenerate rotational bands in reflection-asymmetric triaxial nuclei,[118] i.e., the so-called ChP quartet bands.

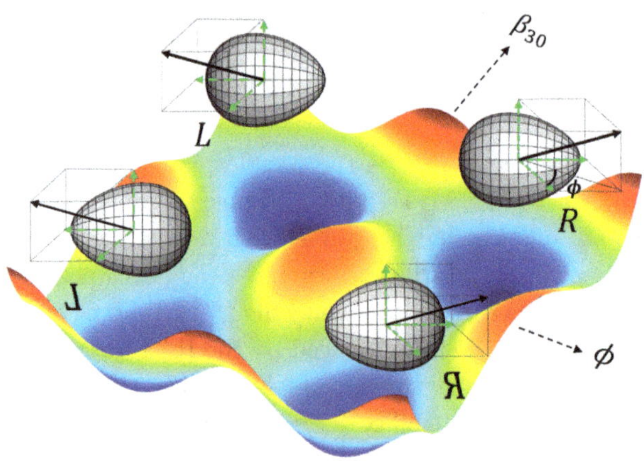

Fig. 5. (Color online) A schematic potential energy surface in the intrinsic (β_{30}, ϕ) plane for a ChP violation system, with β_{30} the octupole deformation parameter and ϕ the azimuthal orientation angle of the total angular momentum. The sign of β_{30} stands for the orientation of the nuclear distribution parallel or antiparallel with the intrinsic axis. The sign of ϕ stands for the right- and left-handed systems. Taken from Ref. 109.

Apart from energy spectra, the electromagnetic transitions also play a key role in understanding the mechanism of symmetry breaking in nuclei and the arisen collective phenomena. Therefore, to search for the experimental evidences for the exotic ChP quartet bands, it is highly desirable to find out theoretically the fingerprints of electromagnetic transitions in the expected ChP quartet bands.

In Ref. 108, the energy spectra and the electromagnetic transitions for ChP quartet bands were investigated in the RAT-PRM with an idealized configuration. The ideal ChP violation system corresponds to a maximum triaxial deformation, a pure particle–hole configuration and a static octupole deformation. The RAT-PRM Hamiltonian for the ideal ChP violation system has good total parity P, good angular momentum I and V_4 symmetry. A new symmetry, chiture $\hat{\mathcal{A}}$ or chiplex $\hat{\mathcal{B}} = \hat{\mathcal{A}}\hat{P}$, is derived for $\gamma = 90°$. The eigenstates of the RAT-PRM Hamiltonian can be characterized by the total parity P and the chiplex $\mathcal{B}$. This additional $\mathcal{B}$ symmetry gives rise to the selection rules for the electromagnetic transitions for the ChP quartet bands, including the parity conserved $E2$ and $M1$ transitions, as well as the parity reserved $E1$ and $E3$ transitions, i.e., $E2$, $M1$ and $E3$ transitions are found to link the states with different $\mathcal{B}$ only.

Furthermore, the ChP quartet bands and the corresponding selection rules are established numerically by using the RAT-PRM calculations. Figure 6 shows the consequence of the selection rules expected for the interbands and intrabands $E2$ and $M1$ transitions, as well as the interbands $E3$ transitions when ChP quartet geometry is realized in RAT-PRM.

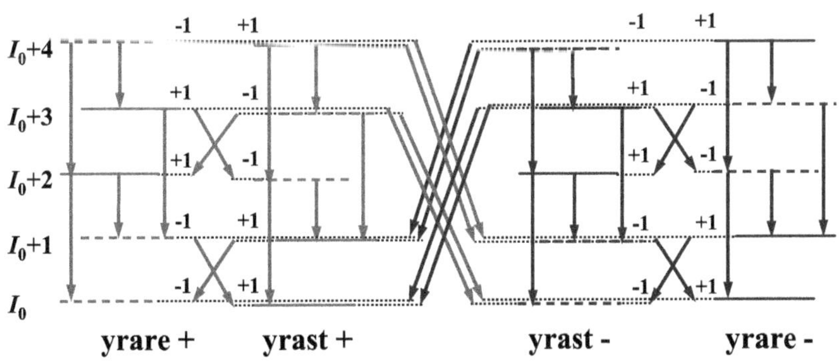

Fig. 6. (Color online) The ChP quartet bands organized by two pairs of chiral doublet bands with positive parity denoted by yrast+ and yrare+, and negative parity denoted by yrast− and yrare−. States with $\mathcal{B} = +1(-1)$ are denoted by solid (dashed) lines. Allowed $E2$, $M1$ and $E3$ transitions are denoted by arrows. Taken from Ref. 108.

The bands are arranged so that the $\Delta I = 2$ $E2$ transitions are always allowed within respective bands. The sign of $\mathcal{B}$ in a given band must change at every increase of spin I by $2\hbar$, and must be opposite to that in its chiral partner band for a given spin. It is clear that, in order to have allowed $E2$, $E3$ or strong $M1$ transitions, the quantum number $\mathcal{B}$ of the final state must have a different sign from that of the initial state. Note that for $\Delta I = 1$ transitions, the allowed $E2$ transitions occur together with stronger $M1$ transitions, since both types of transitions need a change of $\mathcal{B}$. This provides the fingerprints of electromagnetic transitions for the expected ChP quartet bands in reflection-asymmetric triaxial nuclei and trigger future experimental investigations.[136]

Note that the expected ChP quartet bands have not been observed experimentally, and it is of great interests to find out the ChP quartet bands satisfied the above fingerprints. The deviations of the actual situation in nuclei from the simple assumption made in the RAT-PRM may modify the selection rules described above. Nevertheless, the selection rules of the electromagnetic transitions for the expected ChP quartet bands presented in Ref. 108 can serve as a starting point for the study of more practical nuclear systems.

5. Summary

The pear-shaped nucleus has been at the frontiers of nuclear physics since 1950s. Its rotational spectroscopy can be characterized by the occurrence of, e.g., the interleaved positive and negative parity bands in even–even nuclei, the parity doublet bands in odd-mass nuclei, the enhanced electric dipole and octupole moments, etc. In this review, we focus on the rotational spectroscopies of reflection-asymmetric nuclei investigated by the reflection-asymmetric particle rotor model.

The status of observations of the alternated parity bands in even–even nuclei and the parity doublet bands in odd-A and odd–odd nuclei for the $A \sim 220$ mass region is summarized. In general, the strengthness of octupole correlations in nucleus could be reflected by different features of the summarized spectroscopic properties. In view of the existing abundant information in this mass region, more efforts are highly expected to other mass regions.

The formalism of the reflection-asymmetric particle rotor model with the axial or triaxial rotor is briefly introduced. By using the reflection-asymmetric axial particle rotor model, the parity doublet bands in a number of odd-A nuclei with $A \sim 219 - 229$ have been investigated. With the triaxial and octupole degrees of freedom included, the developed RAT-PRM shows the ability to study the MχD with octupole correlations in octupole soft nuclei and explore the possible ChP quartet bands in chirality-parity violation system. Nevertheless, to further search and identify the ChP, quartet bands in realistic nuclei are quite challenging for both experimental and theoretical sides and will be a subject of intense discussions.

Acknowledgments

We thank Y. K. Wang and Y. P. Wang for their careful reading of this paper and helpful discussions. This work was partly supported by the National Natural Science Foundation of China (NSFC) under Grant Nos. 12205097, 12070131001 and 11935003, the Fundamental Research Funds for the Central Universities (Grant No. 2022MS051) and the State Key Laboratory of Nuclear Physics and Technology, Peking University (Grant No. NPT2023ZX01).

ORCID

Yuanyuan Wang https://orcid.org/0000-0002-4340-6065
Shuangquan Zhang https://orcid.org/0000-0002-9590-1818

References

1. M. G. Mayer, *Phys. Rev.* **75** (1949) 1969.
2. J. Rainwater, *Phys. Rev.* **79** (1950) 432.
3. A. Bohr and B. R. Mottelson, *Nuclear Structure*, Vol. **II** (Benjamin, New York, 1975).
4. F. S. Stephens, F. Asaro and I. Perlman, *Phys. Rev.* **96** (1954) 1568.
5. F. S. Stephens, F. Asaro and I. Perlman, *Phys. Rev.* **100** (1955) 1543.
6. K. Lee and D. R. Inglis, *Phys. Rev.* **108** (1957) 774.
7. L. P. Gaffney *et al.*, *Nature* **497** (2013) 199.
8. P. A. Butler and W. Nazarewicz, *Rev. Mod. Phys.* **68** (1996) 349.
9. P. A. Butler, *J. Phys. G* **43** (2016) 073002.
10. P. A. Butler, *Proc. R. Soc. A* **476** (2020) 20200202.
11. L. Funke *et al.*, *Nucl. Phys. A* **355** (1981) 228.
12. R. Schwengner *et al.*, *Z Phys. A* **326** (1987) 287.
13. J. Adam *et al.*, *Z Phys. A* **332** (1989) 141.
14. F. Cristancho *et al.*, *Nucl. Phys. A* **540** (1992) 307.
15. J. Döring *et al.*, *Phys. Rev. C* **50** (1994) 1845.
16. C. Liu *et al.*, *Phys. Rev. Lett.* **116** (2016) 112501.

17. S. Bhattacharya *et al.*, *Phys. Rev. C* **100** (2019) 014315.
18. S. Rajbanshi *et al.*, *Phys. Rev. C* **104** (2021) 064316.
19. A. Mukherjee *et al.*, *Phys. Rev. C* **105** (2022) 014322.
20. P. A. Butler *et al.*, *Phys. Rev. Lett.* **124** (2020) 042503.
21. H. Wollersheim *et al.*, *Nucl. Phys. A* **556** (1993) 261.
22. B. Bucher *et al.*, *Phys. Rev. Lett.* **116** (2016) 112503.
23. B. Bucher *et al.*, *Phys. Rev. Lett.* **118** (2017) 152504.
24. J. F. C. Cocks *et al.*, *Phys. Rev. Lett.* **78** (1997) 2920.
25. J. Cocks *et al.*, *Nucl. Phys. A* **645** (1999) 61.
26. P. A. Butler *et al.*, *Nat. Commun.* **10** (2019) 2473.
27. M. Wieland *et al.*, *Phys. Rev. C* **46** (1992) 2628.
28. J. F. Smith *et al.*, *Phys. Rev. Lett.* **75** (1995) 1050.
29. W. Reviol *et al.*, *Phys. Rev. C* **74** (2006) 044305.
30. P. T. Greenlees *et al.*, *J. Phys. G* **24** (1998) L63.
31. M. E. Debray *et al.*, *Phys. Rev. C* **73** (2006) 024314.
32. S. Zhu *et al.*, *Phys. Rev. C* **81** (2010) 041306.
33. E. N. Shurshikov, *Nucl. Data Sheets* **53** (1988) 601.
34. X. Wang *et al.*, *Phys. Rev. Lett.* **102** (2009) 122501.
35. S. S. Ntshangase *et al.*, *Phys. Rev. C* **82** (2010) 041305.
36. P. Zeyen, B. Ackermann, U. Damrich, K. Euler and V. A. Grafen, *Z Phys. A* **332** (1989) 141.
37. T. Ishii *et al.*, *Phys. Rev. C* **72** (2005) 021301.
38. M. Asai *et al.*, *Eur. Phys. J. A* **23** (2005) 395.
39. I. Wiedenhöver *et al.*, *Phys. Rev. Lett.* **83** (1999) 2143.
40. Pragati *et al.*, *Phys. Rev. C* **97** (2018) 044309.
41. M. Aiche, A. Chevallier, J. Chevallier, S. Hulne, S. Khazrouni and J. C. Sens, *J. Phys. G: Nucl. Part. Phys.* **14** (1988) 1191.
42. M. E. Debray *et al.*, *Phys. Rev. C* **62** (2000) 024304.
43. C. F. Liang, P. Paris and R. K. Sheline, *Int. J. Mod. Phys. E* **1** (1992) 363.
44. N. Roy *et al.*, *Nucl. Phys. A* **426** (1984) 379.
45. T. C. Hensley *et al.*, *Phys. Rev. C* **96** (2017) 034325.
46. J. Fernandez-Niello, C. Mittag, F. Riess, E. Ruchowska and M. Stallknecht, *Nucl. Phys. A* **531** (1991) 164.
47. A. Abdul-Hadi, V. Barci, B. Weiss, H. Maria, G. Ardisson, M. Hussonnois and O. Constantinescu, *Phys. Rev. C* **47** (1993) 94.
48. J. Gasparro, G. Ardisson, V. Barci and R. K. Sheline, *Phys. Rev. C* **62** (2000) 064305.
49. A. Aas *et al.*, *Nucl. Phys. A* **611** (1996) 281.
50. M. E. Debray, A. J. Kreiner, M. Davidson, J. Davidson, D. Hojman and D. Santos, *Nucl. Phys. A* **568** (1994) 141.
51. F. Cristancho *et al.*, *Phys. Rev. C* **49** (1994) 663.
52. N. Schulz *et al.*, *Z Phys. A* **339** (1991) 325.
53. M. Ache *et al.*, *Nucl. Phys. A* **567** (1994) 685.
54. R. K. Sheline, C. F. Liang and P. Paris, *Int. J. Mod. Phys. E* **5** (1990) 2821.
55. I. Ahmad, J. Gindler, M. Carpenter, D. Henderson, E. Moore, R. Janssens, I. Bearden and C. Foster, *Nucl. Phys. A* **576** (1994) 246.

56. W. Reviol *et al.*, *Phys. Rev. C* **80** (2009) 011304.
57. W. Reviol *et al.*, *Phys. Rev. C* **90** (2014) 044318.
58. G. Maquart *et al.*, *Phys. Rev. C* **95** (2017) 034304.
59. J. Hughes *et al.*, *Nucl. Phys. A* **512** (1990) 275.
60. N. J. Hammond *et al.*, *Phys. Rev. C* **65** (2002) 064315.
61. K. Gulda *et al.*, *Nucl. Phys. A* **703** (2002) 45.
62. S. Zhu *et al.*, *Phys. Lett. B* **618** (2005) 51.
63. M. W. Johnson, R. C. Thompson and J. R. Huizenga, *Phys. Rev. C* **17** (1978) 927.
64. R. C. Thompson, J. R. Huizenga and T. W. Elze, *Phys. Rev. C* **13** (1976) 638.
65. D. H. White *et al.*, *Phys. Rev. C* **57** (1998) 1112.
66. P. G. Bizzeti and A. M. Bizzeti-Sona, *Phys. Rev. C* **70** (2004) 064319.
67. P. G. Bizzeti and A. M. Bizzeti-Sona, *Phys. Rev. C* **77** (2008) 024320.
68. P. G. Bizzeti and A. M. Bizzeti-Sona, *Phys. Rev. C* **81** (2010) 034320.
69. P. G. Bizzeti and A. M. Bizzeti-Sona, *Phys. Rev. C* **88** (2013) 011305.
70. D. Bonatsos, D. Lenis, N. Minkov, D. Petrellis and P. Yotov, *Phys. Rev. C* **71** (2005) 064309.
71. N. Minkov, P. Yotov, S. Drenska, W. Scheid, D. Bonatsos, D. Lenis and D. Petrellis, *Phys. Rev. C* **73** (2006) 044315.
72. N. Minkov, S. Drenska, M. Strecker and W. Scheid, *J. Phys. G* **36** (2009) 025108.
73. N. Minkov, S. Drenska, M. Strecker, W. Scheid and H. Lenske, *Phys. Rev. C* **85** (2012) 034306.
74. N. Minkov, *Phys. Scr.* **T154** (2013) 014017.
75. N. Minkov and A. Pálffy, *Phys. Rev. Lett.* **118** (2017) 212501.
76. A. Dobrowolski, K. Mazurek and A. Góźdź, *Phys. Rev. C* **94** (2016) 054322.
77. N. V. Zamfir and D. Kusnezov, *Phys. Rev. C* **63** (2001) 054306.
78. C. Alonso, J. Arias, A. Frank, H. Sofia, S. Lenzi and A. Vitturi, *Nucl. Phys. A* **586** (1995) 100.
79. Y. S. Chen and Z. C. Gao, *Phys. Rev. C* **63** (2000) 014314.
80. Y.-J. Chen, Z.-C. Gao, Y.-S. Chen and Y. Tu, *Phys. Rev. C* **91** (2015) 014317.
81. W. Sun, S. Quan, Z. P. Li, J. Zhao, T. Nikšić and D. Vretenar, *Phys. Rev. C* **100** (2019) 044319.
82. T. Shneidman, G. Adamian, N. Antonenko, R. Jolos and W. Scheid, *Phys. Lett. B* **526** (2002) 322.
83. G. G. Adamian, N. V. Antonenko, R. V. Jolos and T. M. Shneidman, *Phys. Rev. C* **70** (2004) 064318.
84. G. A. Leander and R. K. Sheline, *Nucl. Phys. A* **413** (1984) 375.
85. G. A. Leander and Y. S. Chen, *Phys. Rev. C* **35** (1987) 1145.
86. G. A. Leander and Y. S. Chen, *Phys. Rev. C* **37** (1988) 2744.
87. Y. Y. Wang, S. Q. Zhang, P. W. Zhao and J. Meng, *Phys. Lett. B* **792** (2019) 454.
88. F. Stephens and R. Simon, *Nucl. Phys. A* **183** (1972) 257.
89. I. Hamamoto and B. Mottelson, *Phys. Lett. B* **127** (1983) 281.
90. R. Bengtsson, H. Frisk, F. May and J. Pinston, *Nucl. Phys. A* **415** (1984) 189.

91. S. Frauendorf and J. Meng, *Nucl. Phys. A* **617** (1997) 131.
92. J. Peng, J. Meng and S. Q. Zhang, *Phys. Rev. C* **68** (2003) 044324.
93. T. Koike, K. Starosta and I. Hamamoto, *Phys. Rev. Lett.* **93** (2004) 172502.
94. S. Q. Zhang, B. Qi, S. Y. Wang and J. Meng, *Phys. Rev. C* **75** (2007) 044307.
95. B. Qi, S. Q. Zhang, J. Meng, S. Y. Wang and S. Frauendorf, *Phys. Lett. B* **675** (2009) 175.
96. H. Zhang and Q. B. Chen, *Chin. Phys. C* **40** (2016) 024102.
97. Q. B. Chen, B. F. Lv, C. M. Petrache and J. Meng, *Phys. Lett. B* **782** (2018) 744.
98. Q. B. Chen and J. Meng, *Phys. Rev. C* **98** (2018) 031303.
99. I. Hamamoto, *Phys. Rev. C* **65** (2002) 044305.
100. S. Frauendorf and F. Dönau, *Phys. Rev. C* **89** (2014) 014322.
101. E. Streck, Q. B. Chen, N. Kaiser and U.-G. Meißner, *Phys. Rev. C* **98** (2018) 044314.
102. Q. B. Chen and S. Frauendorf, *Eur. Phys. J. A* **58** (2022) 75.
103. Y. P. Wang, Y. Y. Wang and J. Meng, *Phys. Rev. C* **102** (2020) 024313.
104. Y. Y. Wang and S. Q. Zhang, *Phys. Rev. C* **102** (2020) 034303.
105. C. Morse *et al.*, *Phys. Rev. C* **102** (2020) 054328.
106. Y. Y. Wang, *Phys. Rev. C* **104** (2021) 014318.
107. Y. Y. Wang, Q. B. Chen and S. Q. Zhang, *Phys. Rev. C* **105** (2022) 044316.
108. Y. Y. Wang, X. H. Wu, S. Q. Zhang, P. W. Zhao and J. Meng, *Sci. Bull.* **65** (2020) 2001.
109. A. Bohr and B. R. Mottelson, *Mat. Fys. Medd. Dan. Vid. Selsk* **27** (1953) 1.
110. P. Ring and P. Schuck, *The Nuclear Many-body Problem* (Springer Science & Business Media, 2004).
111. I. Hamamoto, *Nucl. Phys. A* **271** (1976) 15.
112. S. G. Nilsson, *Mat. Fys. Medd. Dan. Vid. Selsk* **29** (1955) 1.
113. R. K. Sheline, C. F. Liang and P. Paris, *Czech J. Phys.* **43** (1993) 603.
114. R. K. Sheline, Y. S. Chen and G. A. Leander, *Nucl. Phys. A* **486** (1988) 306.
115. W. Kurcewicz *et al.*, *Nucl. Phys. A* **621** (1997) 827.
116. J. Meng, J. Peng, S. Q. Zhang and S.-G. Zhou, *Phys. Rev. C* **73** (2006) 037303.
117. A. D. Ayangeakaa *et al.*, *Phys. Rev. Lett.* **110** (2013) 172504.
118. S. Frauendorf, *Rev. Mod. Phys.* **73** (2001) 463.
119. J. Meng and S. Q. Zhang, *J. Phys. G* **37** (2010) 064025.
120. A. A. Raduta, *Prog. Part. Nucl. Phys.* **90** (2016) 241.
121. K. Starosta and T. Koike, *Phys. Scr.* **92** (2017) 093002.
122. S. Frauendorf, *Phys. Scr.* **93** (2018) 043003.
123. B. W. Xiong and Y. Y. Wang, *Atom. Data Nucl. Data Tabl.* **125** (2019) 193.
124. S. Y. Wang, *Chin. Phys. C* **44** (2020) 112001.
125. Q. B. Chen and J. Meng, *Nucl. Phys. News* **30** (2020) 11.
126. K. Selvakumar *et al.*, *Phys. Rev. C* **92** (2015) 064307.
127. S. Guo *et al.*, *Phys. Lett. B* **807** (2020) 135572.
128. B. N. Lu, E. G. Zhao and S. Zhou, *Phys. Rev. C* **85** (2012) 011301.
129. J. Zhao, B. N. Lu, E. G. Zhao and S. G. Zhou, *Phys. Rev. C* **86** (2012) 057304.
130. B. N. Lu, J. Zhao, E. G. Zhao and S. G. Zhou, *Phys. Rev. C* **89** (2014) 014323.

131. E. Landulfo *et al.*, *Phys. Rev. C* **54** (1996) 626.
132. A. Gizon *et al.*, *Nucl. Phys. A* **694** (2001) 63.
133. A. K. Singh *et al.*, *Phys. Rev. C* **97** (2018) 024323.
134. S. Y. Wang, B. Qi and D. P. Sun, *Phys. Rev. C* **82** (2010) 027303.
135. P. W. Zhao, Z. P. Li, J. M. Yao and J. Meng, *Phys. Rev. C* **82** (2010) 054319.
136. C. M. Petrache, *Sci. Bull.* **65** (2020) 1956.

Shape transition around ^{222}Ra based on finite-temperature covariant density functional theory*

Wei Zhang † and Jin-Ke Huang
School of Physics and Microelectronics, Zhengzhou University, Zhengzhou 450001, P. R. China
† *zw76@pku.org.cn*

Yi-Fei Niu
School of Nuclear Science and Technology, Lanzhou University, Lanzhou 730000, P. R. China
nyfster@gmail.com

The shape evolution and potential energy surfaces of even–even $^{220-240}$Ra are investigated in the (β_2, β_3) plane by the covariant density functional theory. For octupole-deformed ^{222}Ra, the free energy surfaces, the deformations, the pairing gaps, the excitation energy as well as the specific heat with increasing temperature are analyzed. Based on the specific heat, ^{222}Ra exhibits three distinct discontinuities as the temperature rises. The first pairing transition occurs at $T = 0.45$ MeV, while the second one happens at $T = 0.65$ MeV when octupole deformation disappears, and the third one takes place at $T = 0.80$ MeV as quadrupole deformation approaches zero. The gaps at $N = 132, 136$ and $Z = 88$ in the single-particle levels are responsible for the octupole global minima. The shape transitions not only occur between even–even $^{220-240}$Ra, but also occur with increasing temperature for ^{222}Ra.

Keywords: Shape transition; octupole deformation; temperature; covariant density functional theory.

1. Introduction

The octupole-deformed nuclei are of importance for nuclear structure theory as well as for physics beyond the Standard Model.[1] In recent years, the experimental evidence of the existence of octupole-deformed nuclei such as ^{222,224,226}Ra, ^{144,146}Ba and ^{228}Th has drawn much attention.[1–6] In 2020, by determining the electric octupole matrix elements, ^{222}Ra, very close to the

*To cite this article, please refer to its earlier version published in the *International Journal of Modern Physics E*, Volume 32(10), (11 pages), 2340008 (2023), DOI: 10.1142/S0218301323400086.

octupole-deformed ^{224}Ra experimentally confirmed in 2013,[1] is explained as a nucleus with stable octupole deformation.[6] It is known that the existence of low-energy negative parity bands and significant electric octupole moments are typical features of octupole-deformed nuclei.[7,8] And the introduction of octupole deformation can explain some phenomena, such as asymmetric mass distribution observed in ^{240}Pu fission.[9]

Theoretically, nuclei with strong octupole deformation are located near the nuclear regions where the proton or neutron number is close to 34 (coupling of $1g_{9/2}$ and $2p_{3/2}$ levels), 56 (coupling of $1h_{11/2}$ and $2d_{5/2}$ levels), 88 (coupling of $1i_{13/2}$ and $2f_{7/2}$ levels) and 134 (coupling of $1j_{15/2}$ and $2g_{9/2}$ levels).[7,10] By changing the nucleon numbers, shape evolution and phase transitions involving octupole deformations are studied in $^{146-156}$Sm and $^{218-246}$Th.[11–13] Most of the theoretical studies on first- and second-order nuclear quantum phase transitions have been conducted in even–even nuclei.[13–16] In Ref. 13, ^{224}Th is believed to be the critical nucleus in the possible quadrupole and octupole shape phase transitions. The theoretical frameworks employed in describing octupole shape evolution or transition are the covariant density functional theory (CDFT),[17,18] the macroscopic–microscopic models,[19,20] the interacting boson model,[21,22] etc.

Not only evolving with the nucleon numbers, the shape may evolve with the temperature. By establishing the giant dipole resonance (GDR) in excited states, experimental information about the shape changes of nuclei can be extracted.[23] For details, refer to Refs. 24–26. With increasing temperature, pairing transition as well as shape transitions may occur. It is well known that pairing correlation plays an important role and has a significant impact on the nuclear structure at low excitation energies.[27–29] The behavior of pairing transitions can be reflected in the specific heat and level density characteristics of the system.[30–32] And it is shown that the critical temperature T_c is proportional to the pairing gap at zero temperature as $T_c = \alpha\Delta(0)$ with the coefficient α between 0.50 and 0.60.[33–35] For octupole-deformed ^{224}Ra, two shape transitions may happen, one from quadrupole–octupole deformed to quadrupole-deformed, and the other from quadrupole-deformed to spherical.[36] For even–even $^{286-304}$Cm, a similar relation between the critical shape transition temperature and the deformation at zero temperature is found as $T_c = 6.6\beta(0)$.[37]

Prompted by recent experimental progress in ^{222}Ra,[6] this paper will focus on the octupole shape evolution around ^{222}Ra and with temperature using finite-temperature CDFT. The CDFT has achieved great success in describing ground-state properties of both spherical and deformed nuclei all over the nuclear chart.[38–43] Its successful applications include superheavy nuclei,[44–48] pseudospin symmetry,[43,49,50] single-particle resonances,[51–53] hypernuclei[54–59] and shell correction.[60–64]

This paper is organized as follows: in Sec. 2, a brief introduction of the finite-temperature CDFT theory with Bardeen–Cooper–Schrieffer (BCS) approach will be provided; in Sec. 3, the potential energy surfaces (PESs) of even–even $^{208-220}$Ra together with deformations, excitation energies, pairing gaps and the specific heat of ^{222}Ra will be discussed; and finally conclusions are drawn in Sec. 4.

2. Theoretical Framework

In order to study the properties of nuclei, this work employs a point-coupling model, which has an effective Lagrangian density read by the following equation:

$$\begin{aligned}\mathcal{L} = {} & \bar{\psi}(i\gamma_\mu\partial^\mu - m)\psi - \frac{1}{2}\alpha_S(\bar{\psi}\psi)(\bar{\psi}\psi) - \frac{1}{2}\alpha_V(\bar{\psi}\gamma_\mu\psi)(\bar{\psi}\gamma^\mu\psi) \\ & -\frac{1}{2}\alpha_{TV}(\bar{\psi}\vec{\tau}\gamma_\mu\psi)\cdot(\bar{\psi}\vec{\tau}\gamma^\mu\psi) - \frac{1}{3}\beta_S(\bar{\psi}\psi)^3 - \frac{1}{4}\gamma_S(\bar{\psi}\psi)^4 \\ & -\frac{1}{4}\gamma_V[(\bar{\psi}\gamma_\mu\psi)(\bar{\psi}\gamma^\mu\psi)]^2 - \frac{1}{2}\delta_S\partial_\nu(\bar{\psi}\psi)\partial^\nu(\bar{\psi}\psi) \\ & -\frac{1}{2}\delta_V\partial_\nu(\bar{\psi}\gamma_\mu\psi)\partial^\nu(\bar{\psi}\gamma^\mu\psi) - \frac{1}{2}\delta_{TV}\partial_\nu(\bar{\psi}\vec{\tau}\gamma_\mu\psi)\cdot\partial^\nu(\bar{\psi}\vec{\tau}\gamma^\mu\psi) \\ & -\frac{1}{4}F^{\mu\nu}F_{\mu\nu} - e\bar{\psi}\gamma^\mu\frac{1-\tau_3}{2}\psi A_\mu, \end{aligned} \tag{1}$$

where m is the nucleon mass, e is the charge unit, A_μ and $F_{\mu\nu}$ are the four-vector potential and field strength tensor of the electromagnetic field, respectively. α corresponding to the four-fermion term, β and γ, respectively, corresponding to the third- and fourth-order coupling terms, and δ corresponding to the derivative coupling. The subscripts S, V and T stand for scalar, vector and isovector, respectively. The Dirac equation for single nucleons reads[34]

$$[\gamma_\mu(i\partial^\mu - V^\mu) - (m + S)]\psi_k = 0, \tag{2}$$

where $\psi_k(\boldsymbol{r})$ represents the Dirac spinor field of a nucleon. Equation (3) represents the scalar $S(\boldsymbol{r})$ and Eq. (4) represents the vector $V^\mu(\boldsymbol{r})$

$$S(\boldsymbol{r}) = \alpha_S\rho_S + \beta_S\rho_S^2 + \gamma_S\rho_S^3 + \delta_S\Delta\rho_S, \tag{3}$$

$$V^{\mu}(\boldsymbol{r}) = \alpha_V j_V^{\mu} + \gamma_V (j_V^{\mu})^3 + \delta_V \Delta j_V^{\mu} + \tau_3 \alpha_{TV} \vec{j}_{TV}^{\mu} + \tau_3 \delta_{TV} \Delta \vec{j}_{TV}^{\mu} + eA^{\mu}. \quad (4)$$

The isoscalar density $\rho_S(\boldsymbol{r})$, isoscalar current $j_V^{\mu}(\boldsymbol{r})$ and isovector current $\vec{j}_{TV}^{\mu}(\boldsymbol{r})$ are represented by

$$\rho_S(\boldsymbol{r}) = \sum \bar{\psi}_k(\boldsymbol{r})\psi_k(\boldsymbol{r})[v_k^2(1-2f_k)+f_k], \quad (5)$$

$$j_V^{\mu}(\boldsymbol{r}) = \sum_k \bar{\psi}_k(\boldsymbol{r})\gamma^{\mu}\psi_k(\boldsymbol{r})[v_k^2(1-2f_k)+f_k], \quad (6)$$

$$\vec{j}_{TV}^{\mu}(\boldsymbol{r}) = \sum_k \bar{\psi}_k(\boldsymbol{r})\vec{\tau}\gamma^{\mu}\psi_k(\boldsymbol{r})[v_k^2(1-2f_k)+f_k]. \quad (7)$$

The thermal probability of quasiparticle states being occupied, which is represented by $f_k = 1/(1+e^{E_k/k_BT})$, has a directly correlation with the temperature. The quasiparticle energy for single-particle state k is given by $E_k = [(\epsilon_k - \lambda_q)^2 + (\Delta_k)^2]^{\frac{1}{2}}$ where Δ_k is the pairing gap parameter and λ_q represents the Fermi level. The occupancy probability v_k^2 in the BCS is obtained

$$v_k^2 = \frac{1}{2}\left(1 - \frac{\epsilon_k - \lambda_q}{E_k}\right). \quad (8)$$

The gap equation at finite temperature is satisfied here in the following equation:

$$\Delta_k = -\frac{1}{2}\sum_{k'>0} V_{k\bar{k}k'\bar{k}'}^{pp} \frac{\Delta_{k'}}{E_{k'}}(1-2f_{k'}). \quad (9)$$

The free energy for the system is $F = E - TS$ where the entropy S is evaluated by

$$S = -k_B \sum_k [f_k \ln f_k + (1-f_k)\ln(1-f_k)]. \quad (10)$$

The temperature is given in units of k_BT in units of MeV, and the entropy S/k_B is dimensionless where k_B is the Boltzmann constant. The specific heat is defined by the following equation:

$$C_{\text{v}} = \partial E^* / \partial T. \quad (11)$$

The excitation energy $E^*(T) = E(T) - E(T=0)$, where $E(T)$ is the binding energy of the nuclear system in its global state on the free energy surface at a given temperature.

3. Results and Discussion

In the CDFT calculations, a large major shell number $N_f = 20$ of the harmonic-oscillator basis is adopted to achieve the convergence for this region. The pairing strengths V_q of the δ pairing force $V(\boldsymbol{r}) = V_q\delta(\boldsymbol{r})$ are taken as -349.5 and $-330.0\,\text{MeV}\cdot\text{fm}^3$ for neutrons and protons, respectively.[65]

First, we explore the shape evolution around ^{222}Ra. The quadrupole deformation β_2, octupole deformation β_3 and hexadecapole deformation β_4 for the ground states of even–even $^{208-244}$Ra are shown in Fig. 1, together with available experimental data.[66] It is clear that the quadrupole deformation β_2 agrees well with the experiment. For octupole deformation, even–even $^{222-234}$Ra shows nonneglectable values with ^{226}Ra as a peak. With increasing neutron number, the isotopes ^{222}Ra to ^{226}Ra exhibit significant octupole deformations while the octupole deformations of $^{228-234}$Ra slowly decline. The curve of hexadecapole deformation is similar to that of the octupole deformation, and some part of the quadrupole deformation.

It is disadvantageous to see the octupole-softness in the ground state deformation figure like Fig. 1. For this purpose, the PESs of $^{220-240}$Ra in the (β_2, β_3) plane is plotted in Fig. 2. For ^{220}Ra, its ground state is quadrupole-deformed at $(\beta_2, \beta_3) \sim (0.08, 0.00)$, together with a octupole minimum at $(\beta_2, \beta_3) \sim (0.16, 0.12)$, 0.45 MeV higher. For ^{224}Ra and ^{228}Ra, there are deep

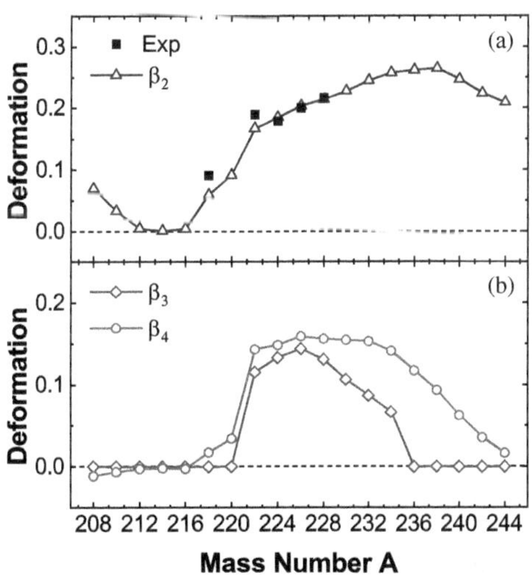

Fig. 1. The quadrupole deformation β_2, octupole deformation β_3 and hexadecapole deformation β_4 of even–even $^{208-244}$Ra isotopes as a function of mass number, obtained with the CDFT using the PC-PK1 parameter set.

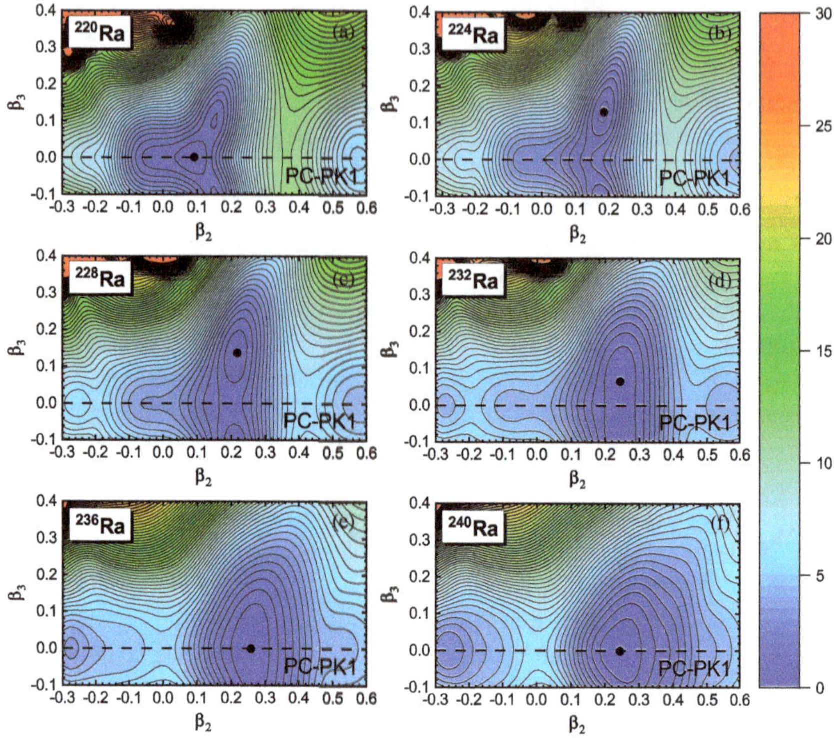

Fig. 2. The PESs of even–even ^{220}Ra (a), ^{224}Ra (b), ^{228}Ra (c), ^{232}Ra (d), ^{236}Ra (e), and ^{240}Ra (f) in the (β_2, β_3) plane, obtained with the CDFT using the PC-PK1 parameter set. The solid circles represent the global minima. The global minimum has been set as the energy zero point and the energy separation between contour lines is 0.5 MeV.

octupole minima, at least 0.5 MeV lower than the quadrupole saddle points. For ^{232}Ra, the octupole lowland becomes shallower, just 0.2 MeV lower than the quadrupole saddle point. For ^{236}Ra and heavier isotopes, the ground states are again quadrupole-deformed with certain octupole-softness. It should be noted that the pairing strength may alter the depth and stiffness of the octupole minimum. To this extent, the isotope around ^{232}Ra may exhibit octupole-softness while ^{224}Ra shows static octupole deformation. This shape evolution, from near-spherical quadruple deformation to octupole deformation, and back to quadrupole deformation, is consistent with previous study in even–even $^{218-246}$Th with NL3* parameter set.[12]

As an isotope between quadruple-deformed ^{220}Ra and octupole-deformed ^{224}Ra, ^{222}Ra is octupole-deformed, which is consistent with recent experiment.[6] Its shape evolution with temperature is of particular interest. Figure 3

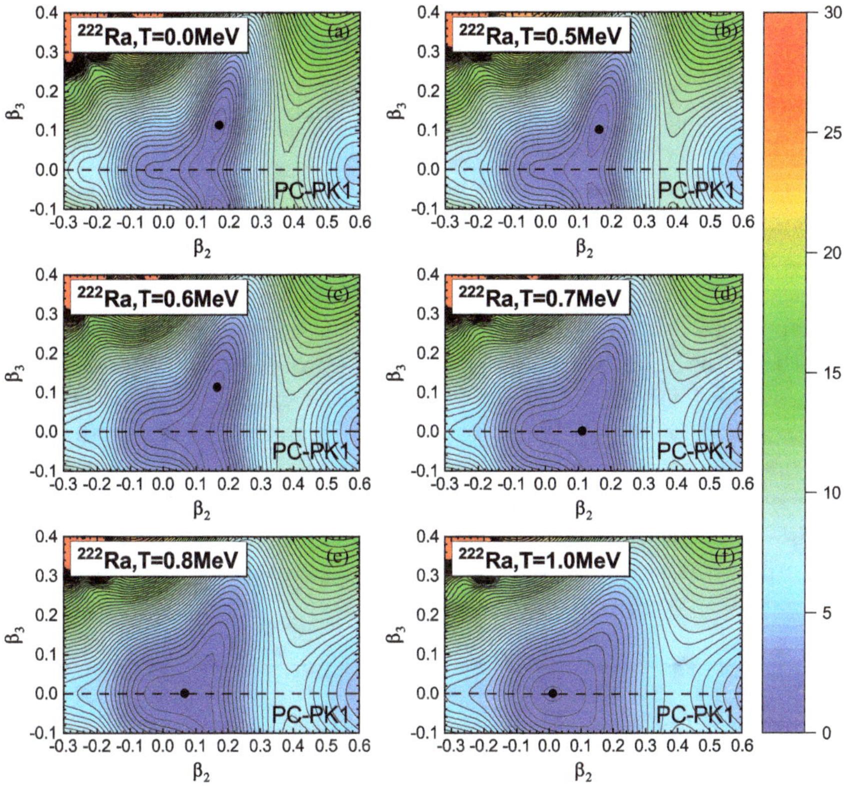

Fig. 3. The free energy surfaces of ^{222}Ra in the (β_2, β_3) plane at temperatures $T = 0.0$ MeV (a), $T = 0.5$ MeV (b), $T = 0.6$ MeV (c), $T = 0.7$ MeV (d), $T = 0.8$ MeV (e) and $T = 1.0$ MeV (f), obtained with the CDFT using the PC-PK1 parameter set. The solid circles represent the global minima. The global minimum has been set as the energy zero point and the energy separation between contour lines is 0.5 MeV.

shows the free energy surfaces of ^{222}Ra in the (β_2, β_3) plane at temperatures $T = 0.0$, 0.5, 0.6, 0.7, 0.8 and 1.0 MeV. The calculated binding energy of ^{222}Ra agrees with the experimental one within 0.13%. At $T = 0$ MeV in Fig. 3(a), the quadrupole-deformed saddle point is about 0.5 MeV higher than the octupole-deformed ground state at $(\beta_2, \beta_3) \sim (0.16, 0.12)$. As the temperature increases up to 0.7 MeV, like $T = 0.5$ and 0.6 MeV in Figs. 3(b) and 3(c), the deformation of the global minimum experience changes a little but the depth of the octupole minimum becomes shallower. When the temperature reaches 0.7 MeV in Fig. 3(d), the global minimum becomes quadrupole-deformed at $(\beta_2, \beta_3) \sim (0.11, 0.00)$ indicating the shape phase transition from octupole to quadrupole. For $0.7 < T < 0.8$ MeV, a relatively octupole-soft region around the global quadrupole minimum is developed. At $T = 0.8\,\mathrm{MeV}$ in Fig. 3(e), the free energy near the global minimum $(\beta_2, \beta_3) \sim (0.08, 0.00)$ is very close to the spherical free energy. When $T = 1.0\,\mathrm{MeV}$ in Fig. 3(f), the global minimum is spherical. For higher temperature, there is barely no change in the free energy surface.

To gain a detailed view of the bulk properties of ^{222}Ra, Fig. 4 shows the variations in deformations, excitation energies, pairing gaps and specific heat as functions of temperature. In Fig. 4(a), the deformation β_2, β_3 and β_4 of the global minimum are shown. At $T = 0.7\,\mathrm{MeV}$, the octupole deformation β_3 rapidly drops to zero; and at $T \sim 0.8\,\mathrm{MeV}$, the quadrupole deformation β_2 together with the hexadecapole deformation decreases quickly. Overall, as the temperature T increases, ^{222}Ra undergoes a shape transition from octupole to quadrupole-deformed, followed by a shape transition from quadrupole-deformed to spherical. In Fig. 4(b), the excitation energies E^* is generally a parabolic curve, displaying slight distortions near $0.65 \leq T \leq 0.85\,\mathrm{MeV}$. Its first derivative with respect to the temperature, the specific heat, is shown in Fig. 4(d), with two distinctive peaks at $T = 0.65$ and $0.80\,\mathrm{MeV}$, corresponding

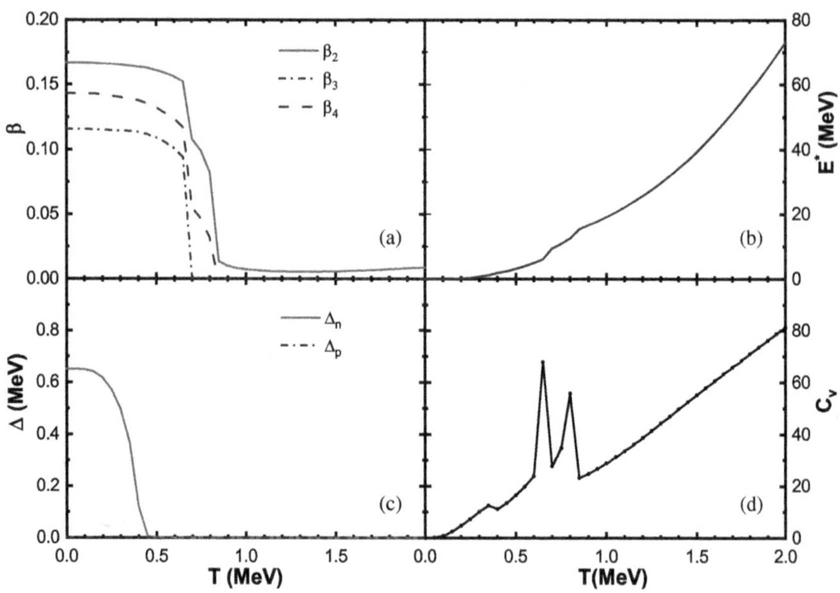

Fig. 4. The global minimum deformations β_2, β_3 and β_4 (a), excitation energies E^* (b), pairing gaps Δ_n and Δ_p (c) and specific heat C_v (d) of ^{222}Ra as functions of temperature T, obtained with the CDFT using the PC-PK1 parameter set.

to two shape transitions, and one discontinuity at $T_c = 0.45$ MeV, indicating the pairing phase transition. This pairing transition can be confirmed by pairing gap in Fig. 4(c). The neutron pairing gap $\Delta_{\rm n}$ at $T = 0$ MeV is 0.65 MeV, and it drops to 0 at the critical temperature $T_c = 0.45$ MeV. The well-known relation between the critical pairing transition temperature and the pairing gap at zero temperature $T_c = 0.6\Delta(0)$ is satisfied again for ^{222}Ra. Here, the proton pairing gap $\Delta_{\rm p}(0)$ collapses due to the large gap near the Fermi surface.

For a microscopic view of the shape phase transition, the single-particle (s.p.) levels of neutrons and protons at the global minima of ^{222}Ra are plotted as functions of temperature T in Fig. 5. Note the global minima have different deformations before and after the shape phase transition. Basically the single-particle levels are functions of both the deformation and the temperature, and they turn out to be sensitive to the deformation but not to the temperature. The quasi-particle probability that changes with the temperature alters the free energy at individual deformation. For neutrons in Fig. 4(a), there are large gaps at $N = 132$ and 136 for $T < 0.9$ MeV, while a large gap exists at $N = 126$ for $T > 0.9$ MeV. For protons in Fig. 4(b), the gaps for $T < 0.9$ MeV locates at $Z = 88$, while the gap for $T > 0.9$ MeV is at

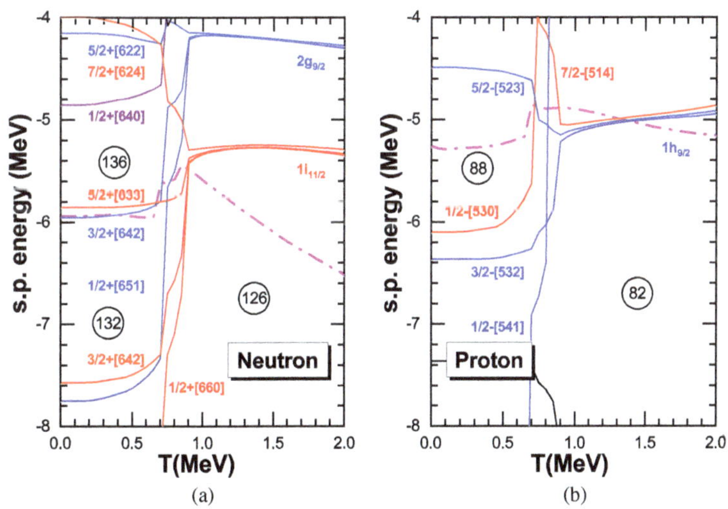

Fig. 5. The neutron (a) and proton (b) single-particle levels of ^{222}Ra as functions of temperature T obtained with the CDFT using the PC-PK1 parameter set. The dash-dot line represents the corresponding Fermi level, and the levels near the Fermi surface are labeled by Nilsson notations.

$Z = 82$. Due to the high sensitivity of single-particle levels to small deformations, when $T > 0.9\,\text{MeV}$, the single-particle levels constituting $\nu 2g_{9/2}$, $\nu 1i_{11/2}$ and $\pi 1h_{9/2}$ do not completely degenerate. This figure is very similar to Fig. 3 in Ref. 36 discussing the neighboring nucleus ^{224}Ra. The gaps at $N = 132$ and 136 and $Z = 88$ are responsible for stabilizing the octupole global minima while the well-known gaps at $N = 126$ and $Z = 82$ are responsible for maintaining the spherical global minima.

4. Summary

In summary, this paper employs the finite-temperature CDFT to study the properties of ^{222}Ra and its neighbor nuclei. It is found that even–even $^{208-244}$Ra undergoes a shape evolution from near-spherical to octupole-deformed, and then to quadrupole-deformed as the neutron number increases. For octupole-deformed ^{222}Ra, the free energy surfaces, the deformations, the pairing gaps, the excitation energy as well as the specific heat are studied. Two shape transitions, one at $T = 0.65\,\text{MeV}$ from octupole shape to quadrupole shape, and the other at $T = 0.80\,\text{MeV}$ from quadrupole shape to spherical shape, and one pairing phase transition at $T = 0.45\,\text{MeV}$ are shown in the specific heat curve. In addition, the single-particle levels of ^{222}Ra with temperature are analyzed. It is shown that at low temperatures, the large gaps at $N = 132$ and 136, and $Z = 88$ are responsible for the octupole global minima. The shape transitions not only occur with increasing neutron numbers for even–even $^{220-240}$Ra, but also occur with increasing temperature for ^{222}Ra.

Acknowledgments

This work was partly supported by the National Natural Science Foundation of China (Grant No. 12075104), the Young Scientist Scheme of the National Key R&D Program of China (Contract No. 2021YFA1601500), Natural Science Foundation of Henan Province (Grant No. 202300410480), Open Project of Lanzhou University, Open Project of Guangxi Key Laboratory of Nuclear Physics and Nuclear Technology (Grant No. NLK2021-01).

ORCID

Wei Zhang https://orcid.org/0000-0001-5450-6056

References

1. L. P. Gaffney *et al.*, *Nature* **497** (2013) 199.
2. H. Wollersheim *et al.*, *Nucl. Phys. A* **556** (1993) 261.
3. B. Bucher *et al.*, *Phys. Rev. Lett.* **116** (2016) 112503.
4. B. Bucher *et al.*, *Phys. Rev. Lett.* **118** (2017) 152504.
5. M. M. R. Chishti *et al.*, *Nat. Phys.* **16** (2020) 853.
6. P. A. Butler *et al.*, *Phys. Rev. Lett.* **124** (2020) 042503.
7. P. A. Butler and W. Nazarewicz, *Rev. Mod. Phys.* **68** (1996) 349.
8. I. Ahmad and P. A. Butler, *Annu. Rev. Nucl. Part. Sci.* **43** (1993) 71.
9. K. Rutz, J. Maruhn, P. G. Reinhard and W. Greiner, *Nucl. Phys. A* **590** (1995) 680.
10. J. Fernández Niello, H. Puchta, F. Riess and W. Trautmann, *Nucl. Phys. A* **391** (1982) 221.
11. W. M. Zhang, Z. P. Li, S. Q. Zhang and J. Meng, *Phys. Rev. C* **81** (2010) 034302.
12. J. Y. Guo, P. Jiao and X. Z. Fang, *Phys. Rev. C* **82** (2010) 047301.
13. Z. Li, B. Song, J. Yao, D. Vretenar and J. Meng, *Phys. Lett. B* **726** (2013) 866.
14. R. F. Casten and N. V. Zamfir, *Phys. Rev. Lett.* **87** (2001) 052503.
15. T. Nikšić, D. Vretenar, G. A. Lalazissis and P. Ring, *Phys. Rev. Lett.* **99** (2007) 092502.
16. Z. P. Li, T. Nikšić, D. Vretenar and J. Meng, *Phys. Rev. C* **81** (2010) 034316.
17. P. Möller, R. Bengtsson, B. Carlsson, P. Olivius, T. Ichikawa, H. Sagawa and A. Iwamoto, *At. Data Nucl. Data Tables* **94** (2008) 758.
18. W. Nazarewicz, P. Olanders, I. Ragnarsson, J. Dudek, G. Leander, P. Möller and E. Ruchowsa, *Nucl. Phys. A* **429** (1984) 269.
19. A. Dobrowolski, K. Mazurek and A. Góźdź, *Phys. Rev. C* **94** (2016) 054322.
20. P. G. Bizzeti and A. M. Bizzeti Sona, *Phys. Rev. C* **81** (2010) 034320.
21. N. V. Zamfir and D. Kusnezov, *Phys. Rev. C* **63** (2001) 054306.
22. T. Otsuka and M. Sugita, *Phys. Lett. B* **209** (1988) 140.
23. A. Schiller and M. Thoennessen, *At. Data Nucl. Data Tables* **93** (2007) 549.
24. P. Ring, L. Robledo, J. Egido and M. Faber, *Nucl. Phys. A* **419** (1984) 261.
25. Y. Niu, N. Paar, D. Vretenar and J. Meng, *Phys. Lett. B* **681** (2009) 315.
26. D. R. Chakrabarty, N. Dinh Dang and V. M. Datar, *Eur. Phys. J. A* **52** (2016) 143.
27. R. Balian, H. Flocard and M. Vénéroni, *Phys. Rep.* **317** (1999) 251.
28. K. Kaneko and M. Hasegawa, *Nucl. Phys. A* **740** (2004) 95.
29. M. Guttormsen, M. Hjorth Jensen, E. Melby, J. Rekstad, A. Schiller and S. Siem, *Phys. Rev. C* **64** (2001) 034319.
30. N. Sandulescu, *Phys. Rev. C* **70** (2004) 025801.
31. M. Fortin, F. Grill, J. Margueron, D. Page and N. Sandulescu, *Phys. Rev. C* **82** (2010) 065804.
32. A. Pastore, S. Baroni and C. Losa, *Phys. Rev. C* **84** (2011) 065807.
33. M. Sano and S. Yamasaki, *Prog. Part. Nucl. Phys.* **29** (1963) 397.
34. A. L. Goodman, *Nucl. Phys. A* **352** (1981) 30.
35. Z. M. Niu, Q. Liu, Y. F. Niu, W. H. Long and J. Y. Guo, *Phys. Rev. C* **87** (2013) 037301.

36. W. Zhang and Y. F. Niu, *Phys. Rev. C* **96** (2017) 054308.
37. W. Zhang and Y. F. Niu, *Phys. Rev. C* **97** (2018) 054302.
38. P. Ring, *Prog. Part. Nucl. Phys.* **37** (1996) 193.
39. S. G. Zhou, J. Meng and P. Ring, *Phys. Rev. Lett.* **91** (2003) 262501.
40. D. Vretenar, A. Afanasjev, G. Lalazissis and P. Ring, *Phys. Rep.* **409** (2005) 101.
41. J. Meng, J. Peng, S. Q. Zhang and S. G. Zhou, *Phys. Rev. C* **73** (2006) 037303.
42. J. Meng, H. Toki, S. Zhou, S. Zhang, W. Long and L. Geng, *Prog. Part. Nucl. Phys.* **57** (2006) 470.
43. H. Liang, J. Meng and S. G. Zhou, *Phys. Rep.* **570** (2015) 1.
44. W. Zhang, J. Meng, S. Zhang, L. Geng and H. Toki, *Nucl. Phys. A* **753** (2005) 106.
45. A. Sobiczewski and K. Pomorski, *Prog. Part. Nucl. Phys.* **58** (2007) 292.
46. N. Wang, E. G. Zhao, W. Scheid and S. G. Zhou, *Phys. Rev. C* **85** (2012) 041601.
47. W. Zhang, Z. P. Li and S. Q. Zhang, *Phys. Rev. C* **88** (2013) 054324.
48. B. N. Lu, J. Zhao, E. G. Zhao and S. G. Zhou, *Phys. Rev. C* **89** (2014) 014323.
49. T. T. Sun, W. L. Lu and S. S. Zhang, *Phys. Rev. C* **96** (2017) 044312.
50. T. T. Sun, W. L. Lu, L. Qian and Y. X. Li, *Phys. Rev. C* **99** (2019) 034310.
51. T. T. Sun, L. Qian, C. Chen, P. Ring and Z. P. Li, *Phys. Rev. C* **101** (2020) 014321.
52. T. H. Heng and Y. W. Chu, *Nucl. Sci. Tech.* **33** (2022) 117.
53. X. N. Cao *et al.*, *Nucl. Sci. Tech.* **34** (2023) 25.
54. B. N. Lu, E. G. Zhao and S. G. Zhou, *Phys. Rev. C* **84** (2011) 014328.
55. B. N. Lu, E. Hiyama, H. Sagawa and S. G. Zhou, *Phys. Rev. C* **89** (2014) 044307.
56. T. T. Sun, E. Hiyama, H. Sagawa, H. J. Schulze and J. Meng, *Phys. Rev. C* **94** (2016) 064319.
57. S. H. Ren, T. T. Sun and W. Zhang, *Phys. Rev. C* **95** (2017) 054318.
58. Z. X. Liu, C. J. Xia, W. L. Lu, Y. X. Li, J. N. Hu and T. T. Sun, *Phys. Rev. C* **98** (2018) 024316.
59. C. Chen, Q. K. Sun, Y. X. Li and T. T. Sun, *Sci. China Phys. Mech. Astron.* **64** (2021) 282011.
60. W. Zhang, S. S. Zhang, S. Q. Zhang and J. Meng, *Chin. Phys. Lett.* **20** (2003) 1694.
61. Y. F. Niu, H. Z. Liang and J. Meng, *Chin. Phys. Lett.* **26** (2009) 032103.
62. P. Jiang, Z. M. Niu, Y. F. Niu and W. H. Long, *Phys. Rev. C* **98** (2018) 064323.
63. W. Zhang, W. L. Lv and T. T. Sun, *Chin. Phys. C* **45** (2022) 024107.
64. W. Zhang, Z. Y. Li, W. Gao and T. T. Sun, *Chin. Phys. C* **46** (2022) 104105.
65. P. W. Zhao, Z. P. Li, J. M. Yao and J. Meng, *Phys. Rev. C* **82** (2010) 054319.
66. http://www.nndc.bnl.gov/nudat3/.

Octupole deformation and Ra puzzle with covariant density functional theory in three-dimensional lattice space*

D. D. Zhang

State Key Laboratory of Nuclear Physics and Technology,
School of Physics, Peking University, Beijing 100871, P. R. China
dandanzhang@pku.edu.cn

The anomaly of the residual proton–neutron interaction $\delta V_{\rm pn}$ for Ra isotopes around $N = 135$ deviating from the general trend, the so-called Ra puzzle, is investigated within the framework of covariant density functional theory in three-dimensional lattice space. The potential energy surfaces in (β_{20}, β_{30}) plane for Ra isotopes are given by the constrained calculations. The average values of proton–neutron interaction $\delta V_{\rm pn}$ extracted from the binding energies of Ra and Rn isotopes with axial, triaxial and reflection asymmetric calculations are compared with the data from AME2020 atomic mass evaluation. It is found that the octupole deformation provides a reasonable interpretation of the Ra puzzle.

Keywords: Covariant density functional theory; three-dimensional lattice space; proton–neutron interaction; Ra puzzle; octupole deformation.

1. Introduction

The understanding of proton–proton interaction, neutron–neutron interaction and proton–neutron interaction is critical to interpreting nuclear structure and its evolution with N and Z. In the past few decades, the proton–neutron interaction which affects many aspects of nuclear structure such as the single-particle energy levels, the onset of deformation and the nuclear shape transition has attracted much interest.[1–4]

In particular, the average interaction of the last proton(s) with the last neutron(s) can be empirically isolated and extracted from nuclear masses using appropriate mass differences. For an even–even nucleus, the average

*To cite this article, please refer to its earlier version published in the *International Journal of Modern Physics E*, Volume 32(10), (10 pages), 2340009 (2023), DOI: 10.1142/S0218301323400098.

interaction of the last two protons with the two last neutrons can be defined as[5,6]

$$\delta V_{\rm pn}^{\rm ee}(Z,N) = \frac{1}{4}\{[B(Z,N) - B(Z,N-2)] - [B(Z-2,N) - B(Z-2,N-2)]\}. \tag{1}$$

For an even-Z and odd-N nucleus, the average interaction of the last proton with the last neutron is given by[7–9]

$$\delta V_{\rm pn}^{\rm eo}(Z,N) = \frac{1}{2}\{[B(Z,N) - B(Z,N-1)] - [B(Z-2,N) - B(Z-2,N-1)]\}. \tag{2}$$

The strength of $\delta V_{\rm pn}$ has been systematically investigated throughout the mass surface based on the data from AME2003 atomic mass evaluation.[7,10–13] Besides, the $\delta V_{\rm pn}$ values are extensively studied both theoretically[14–16] and experimentally.[17–23]

One of the most attractive issues is called the Ra puzzle, i.e., the anomalous enhancement of the residual proton–neutron interaction $\delta V_{\rm pn}$ for Ra isotopes around $N = 135$.[7] A precise Penning-trap mass measurement on $^{223-229}$Rn has confirmed the existence of the Ra puzzle.[21] They find a well-developed peak of the $\delta V_{\rm pn}$ values around $N = 135$ which terminates at $N = 139$ for the odd-N Ra isotopes.

It has been pointed out that the Ra puzzle is related to the strong octupole correlations in regions of particle numbers near 88 ($i_{13/2} \leftrightarrow f_{7/2}$ coupling) and 134 ($j_{15/2} \leftrightarrow g_{9/2}$ coupling).[8,9,24,25] Therefore, it's essential to include the reflection asymmetric degree of freedom when one discusses the Ra puzzle.

The covariant density functional theory (CDFT) has been proved to be an effective theory to describe nuclear properties such as nuclear magnetic moments,[26–28] pseudospin symmetry,[29–34] low-lying excitations,[35–37] magnetic and antimagnetic rotations,[38–42] collective vibrations[43–47] and so on. The version of CDFT in three-dimensional (3D) lattice space was developed in Ref. 48, and had been successfully applied in describing nuclear alpha-cluster structure,[49,50] toroidal states[51] and so on. Considering that there is no symmetry assumption in 3D lattice CDFT, it's an appropriate approach to investigate how the deformation affects the proton–neutron interaction in Ra isotopes.

In this work, the shape evolution and the average proton–neutron interaction extracted from the binding energy of ground states in Ra and Rn isotopes are investigated using the 3D lattice CDFT, in which the variational collapse is avoided by the inverse Hamiltonian method[52] and the Fermion doubling problem is solved by spectral method.[48] This paper is organized as follows. In Sec. 2, the basic formalism of 3D lattice CDFT is briefly reviewed. Numerical details of the calculations are introduced in Sec. 3. The results of shape evolution and the average proton–neutron interaction in Ra isotopes are presented in Sec. 4. Finally, the conclusion is included in Sec. 5.

2. Theoretical Framework

The starting point of the point-coupling density functional theory is an effective nuclear Lagrangian density[53]

$$\begin{aligned}\mathcal{L} = {} & \bar{\psi}(i\gamma_\mu\partial^\mu - m)\psi - \frac{1}{2}\alpha_S(\bar{\psi}\psi)(\bar{\psi}\psi) - \frac{1}{2}\alpha_V(\bar{\psi}\gamma_\mu\psi)(\bar{\psi}\gamma^\mu\psi)\\ & -\frac{1}{2}\alpha_{TV}(\bar{\psi}\vec{\tau}\gamma_\mu\psi)(\bar{\psi}\vec{\tau}\gamma^\mu\psi) - \frac{1}{3}\beta_S(\bar{\psi}\psi)^3 - \frac{1}{4}\gamma_S(\bar{\psi}\psi)^4\\ & -\frac{1}{4}\gamma_V[(\bar{\psi}\gamma_\mu\psi)(\bar{\psi}\gamma^\mu\psi)]^2 - \frac{1}{2}\delta_S\partial_\nu(\bar{\psi}\psi)\partial^\nu(\bar{\psi}\psi)\\ & -\frac{1}{2}\delta_V\partial_\nu(\bar{\psi}\gamma_\mu\psi)\partial^\nu(\bar{\psi}\gamma^\mu\psi) - \frac{1}{2}\delta_{TV}\partial_\nu(\bar{\psi}\vec{\tau}\gamma_\mu\psi)\partial^\nu(\bar{\psi}\vec{\tau}\gamma_\mu\psi)\\ & -\frac{1}{4}F^{\mu\nu}F_{\mu\nu} - e\frac{1-\tau_3}{2}\bar{\psi}\gamma^\mu\psi A_\mu, \end{aligned} \tag{3}$$

where m is the nucleon mass, $\alpha_S, \alpha_V, \alpha_{TV}, \beta_S, \gamma_S, \gamma_V, \delta_S, \delta_V, \delta_{TV}$ are coupling constants, in which α refers to the four-fermion terms, β and γ to the third- and fourth-order terms, and δ the derivative couplings. The subscripts S, V and T stand for scalar, vector and isovector, respectively. A_μ and $F_{\mu\nu}$ are four-vector potential and field strength tensor of the electromagnetic field.

Based on the Lagrangian density (3), one can derive the Dirac equation,

$$[-i\boldsymbol{\alpha}\cdot\boldsymbol{\nabla} + \beta\gamma_\mu V^\mu + \beta(M+S)]\psi_k(\boldsymbol{r}) = \epsilon_k\psi_k(\boldsymbol{r}), \tag{4}$$

where

$$S(\boldsymbol{r}) = \alpha_S\rho_S + \beta_S\rho_S^2 + \gamma_S\rho_S^3 + \delta_S\Delta\rho_S, \tag{5}$$

$$V^\mu(\boldsymbol{r}) = \alpha_V j_V^\mu + \gamma_S(j_V^\mu)^3 + \delta_V\Delta j_V^\mu + \tau_3\alpha_{TV}j_{TV}^\mu + \tau_3\delta_{TV}\Delta j_{TV}^\mu + eA^\mu. \tag{6}$$

More details can be found in Ref. 54.

In this work, the Dirac equation (4) is solved in 3D lattice space. The inverse Hamiltonian method[52] and spectral method[48] are applied to solve the variational collapse and the Fermion doubling problem, respectively. For details, we refer the reader to Ref. 48.

The constrained calculations are done by minimizing

$$\langle H' \rangle = \langle H \rangle + \lambda(\langle \hat{\beta}_{\lambda\mu} \rangle - \beta^{c}_{\lambda\mu}) + \frac{1}{2} C(\langle \hat{\beta}_{\lambda\mu} \rangle - \beta^{c}_{\lambda\mu})^2, \tag{7}$$

where H is the Hamiltonian of nuclear system, λ and C are the coefficients, $\beta^{c}_{\lambda\mu}$ is the desired deformation parameter such as $\beta_{20}, \beta_{22}, \beta_{30}$. $\langle \hat{\beta}_{\lambda\mu} \rangle$ is the expectation value of the corresponding deformation operator.

3. Numerical Details

In this work, the density functional PC-PK1[54] is employed in self-consistent mean-field calculations. The pairing correlations are neglected. The number of grid points in 3D lattice space is 32 in the x, y and z directions. The step size along all three axes is 1 fm. The parameters used in the inverse Hamiltonian method are the same as in Ref. 50.

4. Results and Discussion

The constrained calculations are applied to investigate the shape evolution in Ra and Rn isotopes within the framework of CDFT in 3D lattice space. Figure 1 displays the potential energy surfaces in (β_{20}, β_{30}) plane for ^{212}Ra, ^{215}Ra, ^{218}Ra, ^{221}Ra, ^{224}Ra and ^{227}Ra. The energy difference between neighboring contour lines is 0.5 MeV. The ground states are labeled by solid stars.

It is shown that the ground states of ^{212}Ra, ^{215}Ra and ^{218}Ra are reflection symmetric. For ^{212}Ra, the ground state is oblate with $\beta_{20} = -0.06$. For ^{215}Ra and ^{218}Ra, the ground states are located in the prolate side with $\beta_{20} = 0.05$ and 0.09. Note that for ^{218}Ra, the potential energy surface around ground state becomes soft with respect to the direction of β_{30}. The octupole deformation appears in the ground states of ^{221}Ra, ^{224}Ra and ^{227}Ra. With the neutron number increasing, the values of deformation parameters (β_{20}, β_{30}) are $(0.16, 0.11), (0.18, 0.14), (0.22, 0.16)$, respectively. Meanwhile, one notices that the potential energy surfaces become very soft in the prolate side.

In order to extract the average proton–neutron interaction strength δV_{pn} for Ra isotopes, the ground states of $^{210-229}$Ra and $^{208-227}$Rn are obtained by self-consistent 3D lattice CDFT calculations with different initial potentials. For Ra isotopes, one can find that the ground states of $^{210-219}$Ra are slightly oblate or prolate with quadrupole deformation β_{20} changing from -0.06 to

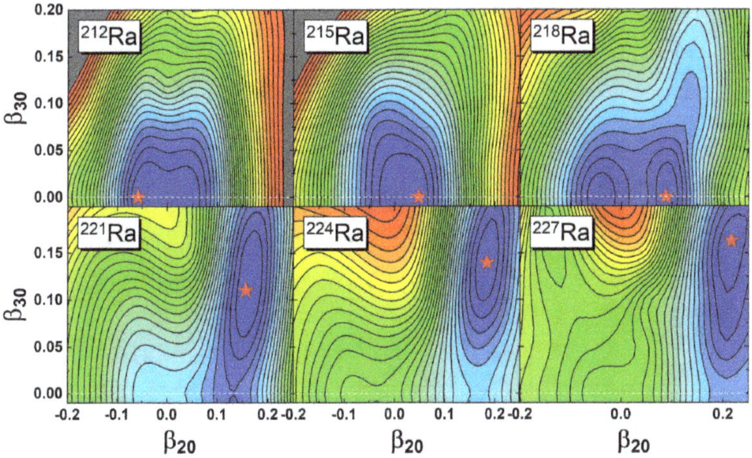

Fig. 1. The potential energy surfaces of ^{212}Ra, ^{215}Ra, ^{218}Ra, ^{221}Ra, ^{224}Ra and ^{227}Ra in (β_{20}, β_{30}) plane are given by the CDFT in 3D lattice space with PC-PK1. The energy difference between neighboring contour lines is 0.5 MeV. The ground states are labeled by solid stars.

0.09. For the ground states of $^{220-229}$Ra, β_{20} is remarkably increasing and the value of octupole deformation β_{30} becomes nonzero. With the neutron number increasing, β_{20} changes from 0.15 to 0.22 and β_{30} changes from 0.10 to 0.17.

For Rn isotopes, the shape evolution of the ground states is similar to Ra isotopes. For $^{208-217}$Rn, the ground states have only quadrupole deformation with β_{20} changing from -0.08 to 0.11. For the ground states of $^{218-227}$Rn, β_{20} continues to increase from 0.12 to 0.21 while the octupole deformation β_{30} arises in the ground states of $^{218-224}$Rn with $0.01 < \beta_{30} < 0.07$ and finally vanishes in the ground states of $^{225-227}$Rn. Note that the ground states of Rn isotopes have smaller values of octupole deformation compared to that of Ra isotopes.

According to Eqs. (1) and (2), the average proton–neutron interaction strength δV_{pn} can be calculated from the binding energy of the ground states of Ra and Rn isotopes. Figure 2 displays the values of δV_{pn} for odd-N Ra isotopes extracted from the AME2020 atomic mass evaluation[10] in comparison with the calculations given by the 3D lattice CDFT (Figs. 2(a)–2(c)), as well as HFB-17 mass table (Fig. 2(d)) taken from Ref. 55. In the self-consistent 3D lattice CDFT calculations, the ground states are obtained by searching for the minimum of energy with constrained conditions $\beta_{22} = 0$, $\beta_{30} = 0$ (Fig. 2(a)), $\beta_{30} = 0$ (Fig. 2(b)) and $\beta_{22} = 0$ (Fig. 2(c)).

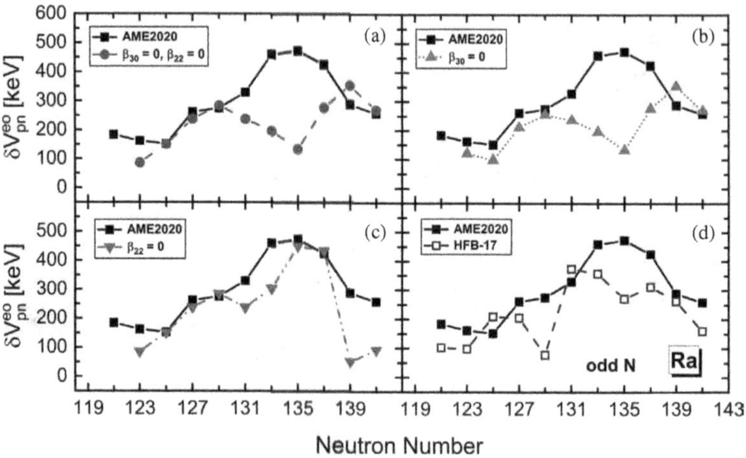

Fig. 2. The residual proton–neutron interaction δV_{pn} data for odd-N Ra isotopes extracted from the AME2020 atomic mass evaluation[10] in comparison with the calculations given by the CDFT in 3D lattice space with $\beta_{22}=0, \beta_{30}=0$ (a), $\beta_{30}=0$ (b) and $\beta_{22}=0$ (c), as well as HFB-17 mass table (d).[55]

In Fig. 2(a), the average interaction strength δV_{pn} extracted from axial calculations can fit well with the data from AME2020 except the peak around $N=135$. In Fig. 2(b), the peak around $N=135$ still cannot be reproduced after including the triaxial degree of freedom. It means that the axial deformation and the triaxial deformation are not responsible for the Ra puzzle. In Fig. 2(c), it's clear that the peak around $N=135$ is well reproduced after involving the octupole deformation. Note that δV_{pn} values of $N=139, 141$ for reflection asymmetric calculations are deviated from the data. It may be caused by the lack of pairing correlations in the present calculations. Figure 2(d) shows the result from HFB-17 mass table which is reflection symmetric. Although δV_{pn} has a better behavior than axial or triaxial calculations, it still fails to reproduce the peak around $N=135$. These results indicate that the octupole deformation can provide a reasonable interpretation of Ra puzzle.

Similar to Fig. 2, the residual proton–neutron interaction δV_{pn} data for even-N Ra isotopes are shown in Fig. 3. It can be seen from Figs. 3(a) and 3(b) that both axial calculations and triaxial calculations underestimate the values of δV_{pn} for $N=132, 134, 136$. The reflection asymmetric calculations improve the description of δV_{pn} for $N=132, 134$ and overestimate the value at $N=136$. The result from HFB-17 mass table fits well with the AME2020 data. It shows that the pairing correlations are

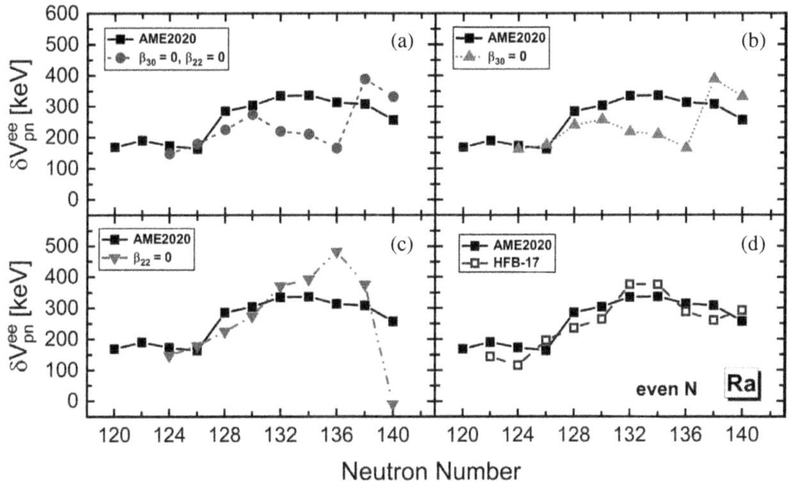

Fig. 3. Same as Fig. 2, but for the even-N Ra isotopes.

also an important factor to be considered in the description of $\delta V_{\rm pn}$ for even-N Ra isotopes.

The average proton–neutron interaction strength $\delta V_{\rm pn}$ of odd-N Ra isotopes calculated from Eq. (2) can be rewritten as

$$\delta V_{\rm pn}^{\rm eo}(Z,N) = \frac{1}{2}[S_n(Z,N) - S_n(Z-2,N)]. \tag{8}$$

Therefore, $\delta V_{\rm pn}$ can also be determined by the single-neutron separation energy of Ra and Rn isotopes with the same neutron number. As we all know, the octupole deformation would affect single-particle levels a lot and thus change single-neutron separation energy. As noted above, the values of octupole deformation are much larger for Ra isotopes than Rn isotopes. Analyzing the single-neutron levels with and without reflection symmetry for Ra isotopes contributes to understanding the behavior of $\delta V_{\rm pn}$. Figure 4 displays the single-neutron levels for $^{219-225}$Ra in 3D lattice CDFT calculations with and without octupole deformation. The solid circles denote the neutron levels occupied by the last neutron.

As shown in Fig. 4, the single-neutron levels of $^{220-225}$Ra with octupole deformation are quite different from the results obtained by constraining $\beta_{30} = 0$. For the axial calculations, the quadrupole deformation is increasing with the neutron number and thus the energy gap at magic number $N = 126$ gradually disappears. For the reflection asymmetry calculations, although the shell gap at magic number $N = 126$ also disappears with the neutron

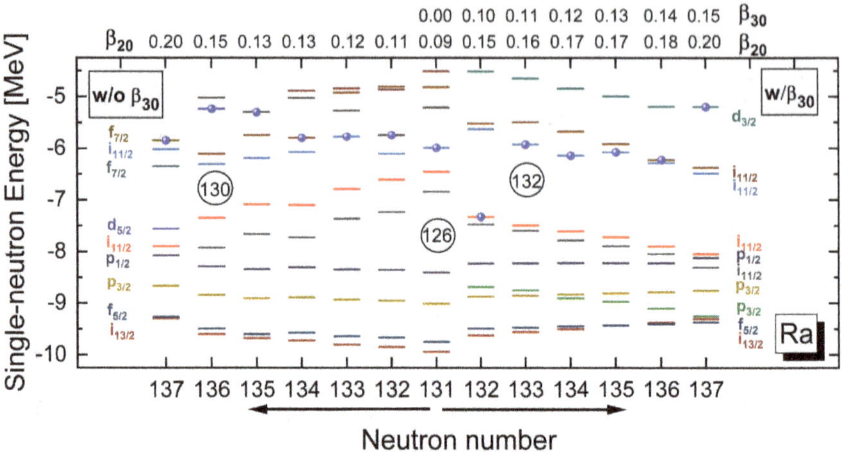

Fig. 4. Single-neutron levels for the ground states of $^{219-225}$Ra in the 3D lattice CDFT calculations with and without reflection symmetry. The quadrupole deformation β_{20} and octupole deformation β_{30} of ground states are shown on the top margin. The solid circles denote the neutron levels occupied by the last neutron. The spherical quantum numbers of each level are listed.

number increasing, the gap at $N = 132$ is formed. The neutron levels become deeply bound and provide larger single-neutron separation energy. That contributes a more reasonable description of the average proton–neutron interaction strength δV_{pn}.

5. Conclusion

The shape evolution in the ground states of Ra and Rn isotopes is investigated using the CDFT in 3D lattice space. The potential energy surfaces of ^{212,215,218,221,224,227}Ra in (β_{20}, β_{30}) plane are given by constrained calculations. The ground states of $^{210-229}$Ra and $^{208-217}$Rn are obtained in different constrained conditions corresponding to axial calculations, triaxial calculations and reflection asymmetric calculations. It is found that the ground states of $^{210-219}$Ra are only slightly oblate or prolate with β_{20} changing from -0.06 to 0.09. For $^{220-229}$Ra, the ground states have not only large quadrupole deformation with $0.15 < \beta_{20} < 0.22$ but also remarkable octupole deformation with $0.10 < \beta_{30} < 0.17$. The Rn isotopes exhibit similar behavior, but they have smaller octupole deformation than Ra isotopes.

Based on these 3D lattice CDFT calculations, the average proton–neutron interaction δV_{pn} for Ra isotopes is extracted from axial, triaxial and reflection asymmetric cases and compared with the data from AME2020. It is found that the peak around $N = 135$ cannot be reproduced unless the octupole deformation is included. Therefore, the octupole deformation provides a reasonable interpretation of the Ra puzzle.

The single-neutron levels with and without reflection symmetry are also shown. When the octupole degree of freedom is involved, the single-neutron levels are deeply bound and thus provide larger single-neutron separation energy as well as the energy gap at $N = 132$. It should be noted that the octupole deformation effects on Ra puzzle have been investigated with CDFT previously in Refs. 8 and 9 while this work has presented a more reliable calculation in a 3D lattice space without any symmetry assumption. It's found that the obtained ground states deformation are slightly different and the ground state energies are slightly slow for certain Ra and Rn isotopes. However, the Ra puzzle is still found to be highly related to the octupole deformation. Besides, the nuclei with particle number near 34 ($g_{9/2} \leftrightarrow p_{3/2}$ coupling), 56 ($h_{11/2} \leftrightarrow d_{5/2}$ coupling), 88 ($i_{13/2} \leftrightarrow f_{7/2}$ coupling) and 134 ($j_{15/2} \leftrightarrow g_{9/2}$ coupling) are considered to have strong octupole correlations. It will be interesting to investigate systematically such octupole deformation in these nuclei.

Acknowledgments

This work was partly supported by the National Natural Science Foundation of China (Grants Nos. 12141501, 12070131001, 11875075, 11935003 and 11975031), the National Key R&D Program of China (Contract Nos. 2017YFE0116700 and 2018YFA0404400) and the High-performance Computing Platform of Peking University.

ORCID

D. D. Zhang https://orcid.org/0000-0003-1328-3305

References

1. I. Talmi, *Rev. Mod. Phys.* **34** (1962) 704.
2. P. Federman and S. Pittel, *Phys. Lett. B* **64** (1977) 385.
3. K. Heyde, P. Van Isacker, R. F. Casten and J. L. Wood, *Phys. Lett. B* **155** (1985) 303.
4. W. Zhang, J. Meng, S. Q. Zhang, L. S. Geng and H. Toki, *Nucl. Phys. A* **753** (2005) 106.
5. J.-Y. Zhang, R. F. Casten and D. S. Brenner, *Phys. Lett. B* **227** (1989) 1.

6. D. S. Brenner, C. Wesselborg, R. F. Casten, D. D. Warner and J.-Y. Zhang, *Phys. Lett. B* **243** (1990) 1.
7. D. S. Brenner, R. B. Cakirli and R. F. Casten, *Phys. Rev. C* **73** (2006) 34315.
8. L. F. Yu, P. W. Zhao, S. Q. Zhang and J. Meng, arXiv:1211.0601 [nucl-th].
9. W. Zhang and S. Q. Zhang, *Phys. Rev. C* **100** (2019) 54303.
10. M. Wang, W. J. Huang, F. G. Kondev, G. Audi and S. Naimi, *Chin. Phys. C* **45** (2021) 30003.
11. R. B. Cakirli, D. S. Brenner, R. F. Casten and E. A. Millman, *Phys. Rev. Lett.* **94** (2005) 92501.
12. R. B. Cakirli and R. F. Casten, *Phys. Rev. Lett.* **96** (2006) 132501.
13. Y. Oktem, R. B. Cakirli, R. F. Casten, R. J. Casperson and D. S. Brenner, *Phys. Rev. C* **74** (2006) 27304.
14. M. Stoitsov, R. B. Cakirli, R. F. Casten, W. Nazarewicz and W. Satuła, *Phys. Rev. Lett.* **98** (2007) 132502.
15. Z. C. Gao and Y. S. Chen, *Phys. Rev. C* **59** (1999) 735.
16. G. J. Fu, H. Jiang, Y. M. Zhao, S. Pittel and A. Arima, *Phys. Rev. C* **82** (2010) 34304.
17. U. Hager, A. Jokinen, V.-V. Elomaa, T. Eronen, J. Hakala, A. Kankainen, S. Rahaman, J. Rissanen, I. D. Moore, S. Rinta-Antila, A. Saastamoinen, T. Sonoda and J. Äystö, *Nucl. Phys. A* **793** (2007) 20.
18. A. Kellerbauer, G. Audi, D. Beck, K. Blaum, G. Bollen, C. Guénaut, F. Herfurth, A. Herlert, H.-J. Kluge, D. Lunney, S. Schwarz, L. Schweikhard, C. Weber and C. Yazidjian, *Phys. Rev. C* **76** (2007) 45504.
19. M. B. Gómez-Hornillos, M. Chartier, W. Mittig, A. Lépine-Szily, L. Caballero, C. E. Demonchy, G. Georgiev, N. A. Orr, G. Politi, M. Rousseau, P. Roussel-Chomaz and A. C. C. Villari, *Phys. Rev. C* **78** (2008) 14311.
20. L. Chen, Y. A. Litvinov, W. R. Plaß, K. Beckert, P. Beller, F. Bosch, D. Boutin, L. Caceres, R. B. Cakirli, J. J. Carroll, R. F. Casten, R. S. Chakrawarthy, D. M. Cullen, I. J. Cullen, B. Franzke, H. Geissel, J. Gerl, M. Górska, G. A. Jones, A. Kishada, R. Knöbel, C. Kozhuharov, S. A. Litvinov, Z. Liu, S. Mandal, F. Montes, G. Münzenberg, F. Nolden, T. Ohtsubo, Z. Patyk, Z. Podolyák, R. Propri, S. Rigby, N. Saito, T. Saito, C. Scheidenberger, M. Shindo, M. Steck, P. Ugorowski, P. M. Walker, S. Williams, H. Weick, M. Winkler, H.-J. Wollersheim and T. Yamaguchi, *Phys. Rev. Lett.* **102** (2009) 122503.
21. D. Neidherr, G. Audi, D. Beck, K. Blaum, C. Böhm, M. Breitenfeldt, R. B. Cakirli, R. F. Casten, S. George, F. Herfurth, A. Herlert, A. Kellerbauer, M. Kowalska, D. Lunney, E. Minaya-Ramirez, S. Naimi, E. Noah, L. Penescu, M. Rosenbusch, S. Schwarz, L. Schweikhard and T. Stora, *Phys. Rev. Lett.* **102** (2009) 112501.
22. M. Breitenfeldt, C. Borgmann, G. Audi, S. Baruah, D. Beck, K. Blaum, C. Böhm, R. B. Cakirli, R. F. Casten, P. Delahaye, M. Dworschak, S. George, F. Herfurth, A. Herlert, A. Kellerbauer, M. Kowalska, D. Lunney, E. Minaya-Ramirez, S. Naimi, D. Neidherr, M. Rosenbusch, R. Savreux, S. Schwarz, L. Schweikhard and C. Yazidjian, *Phys. Rev. C* **81** (2010) 34313.
23. J. Ketelaer, G. Audi, T. Beyer, K. Blaum, M. Block, R. B. Cakirli, R. F. Casten, C. Droese, M. Dworschak, K. Eberhardt, M. Eibach, F. Herfurth, E. Minaya

Ramirez, S. Nagy, D. Neidherr, W. Nörtershäuser, C. Smorra and M. Wang, *Phys. Rev. C* **84** (2011) 14311.
24. P. A. Butler and W. Nazarewicz, *Rev. Mod. Phys.* **68** (1996) 349.
25. P. W. Zhao and Z. P. Li, *Int. J. Mod. Phys. E* **27** (2018) 1830007.
26. J. M. Yao, H. Chen and J. Meng, *Phys. Rev. C* **74** (2006) 24307.
27. J. Li, J. M. Yao, J. Meng and A. Arima, *Prog. Theor. Phys.* **125** (2011) 1185.
28. J. Li, J. Meng, R. Peter, J. M. Yao and A. Arima, *Sci. China Phys. Mech. Astron.* **54** (2011) 204.
29. S. G. Zhou, J. Meng and P. Ring, *Phys. Rev. Lett.* **91** (2003) 262501.
30. T. S. Chen, H. F. Lu, J. Meng, S. Q. Zhang and S. G. Zhou, *Chin. Phys. Lett.* **20** (2003) 358.
31. J. N. Ginocchio, *Phys. Rep.* **414** (2005) 165.
32. H. Z. Liang, P. W. Zhao, Y. Zhang, J. Meng and N. V. Giai, *Phys. Rev. C* **83** (2011) 41301.
33. J. Y. Guo, *Phys. Rev. C* **85** (2012) 21302.
34. B. N. Lu, E. G. Zhao and S. G. Zhou, *Phys. Rev. Lett.* **109** (2012) 72501.
35. T. Nikšić, D. Vretenar and P. Ring, *Prog. Part. Nucl. Phys.* **66** (2011) 519.
36. Z. P. Li, T. Nikšić, D. Vretenar, J. Meng, G. A. Lalazissis and P. Ring, *Phys. Rev. C* **79** (2009) 54301.
37. J. M. Yao, H. Mei, H. Chen, J. Meng, P. Ring and D. Vretenar, *Phys. Rev. C* **83** (2011) 14308.
38. J. Peng, J. Meng, P. Ring and S. Q. Zhang, *Phys. Rev. C* **78** (2008) 24313.
39. P. W. Zhao, S. Q. Zhang, J. Peng, H. Z. Liang, P. Ring and J. Meng, *Phys. Lett. B* **699** (2011) 181.
40. L. F. Yu, P. W. Zhao, S. Q. Zhang, P. Ring and J. Meng, *Phys. Rev. C* **85** (2012) 24318.
41. P. W. Zhao, J. Peng, H. Z. Liang, P. Ring and J. Meng, *Phys. Rev. Lett.* **107** (2011) 122501.
42. P. W. Zhao, J. Peng, H. Z. Liang, P. Ring and J. Meng, *Phys. Rev. C* **85** (2012) 54310.
43. N. Paar, D. Vretenar, E. Khan and G. Colò, *Rep. Prog. Phys.* **70** (2007) 691.
44. H. Z. Liang, N. V. Giai and J. Meng, *Phys. Rev. Lett.* **101** (2008) 122502.
45. N. Paar, Y. F. Niu, D. Vretenar and J. Meng, *Phys. Rev. Lett.* **103** (2009) 32502.
46. Y. F. Niu, N. Paar, D. Vretenar and J. Meng, *Phys. Lett. B* **681** (2009) 315.
47. H. Z. Liang, N. V. Giai and J. Meng, *Phys. Rev. C* **79** (2009) 64316.
48. Z. X. Ren, S. Q. Zhang and J. Meng, *Phys. Rev. C* **95** (2017) 24313.
49. Z. X. Ren, S. Q. Zhang, P. W. Zhao, N. Itagaki, J. A. Maruhn and J. Meng, *Sci. China Phys. Mech. Astron.* **62** (2019) 112062.
50. D. D. Zhang, Z. X. Ren, P. W. Zhao, D. Vretenar, T. Nikšić and J. Meng, *Phys. Rev. C* **105** (2022) 24322.
51. Z. X. Ren, P. W. Zhao, S. Q. Zhang and J. Meng, *Nucl. Phys. A* **996** (2020) 121696.
52. K. Hagino and Y. Tanimura, *Phys. Rev. C* **82** (2010) 57301.
53. J. Meng, *Int. Rev. Nucl. Phys.* **10** (2016) 21.
54. P. W. Zhao, Z. P. Li, J. M. Yao and J. Meng, *Phys. Rev. C* **82** (2010) 54319.
55. S. Goriely, N. Chamel and J. M. Pearson, *Phys. Rev. Lett.* **102** (2009) 152503.

Shell-model-like approach based on covariant density functional theory in 3D lattice space: Evolution of octupole shape in rotating ^{224}Th*

Fangfang Xu and Yiping Wang†

State Key Laboratory of Nuclear Physics and Technology,
School of Physics, Peking University, Beijing 100871, P. R. China
†*YipingWang@pku.edu.cn*

The evolution of the octupole shape with rotation in pear-shaped nuclei is a topic of broad interest. Based on the cranking covariant density functional theory in 3D lattice space, a shell-model-like approach is implemented to take into account the pairing correlations, and applied for the interleaved positive- and negative-parity bands in ^{224}Th. The experimental $I-\omega$ relations are well reproduced. It is found that the octupole deformation of yrast states in ^{224}Th rises slightly and then declines with increasing spin. After the band crossing, the octupole deformation suddenly disappears, i.e., a sharp transition from an octupole shape to a nearly spherical shape takes place at high spins. This is explained by the evolution of the coupling strength between the proton $i_{13/2}$ and $f_{7/2}$ orbitals with spin.

Keywords: Covariant density functional theory; shell-model-like approach; octupole shape; high spin.

1. Introduction

The existence of nuclei with stable deformed shapes was realized in the early stages of nuclear physics history.[1,2] The observation of large quadrupole moments revealed that many nuclei possess axially symmetric quadrupole deformed shapes, which was confirmed by the observation of rotational band structures and measurements of their properties.[3–6] All members of these observed rotational bands have the same parity since such shape is symmetric under the space inversion. However, with the first observation

*To cite this article, please refer to its earlier version published in the *International Journal of Modern Physics E*, Volume 32(10), (12 pages), 2340007 (2023), DOI: 10.1142/S0218301323400074.

†Corresponding author.

of the low-lying negative-parity states near the ground state in even–even Ra isotopes in the 1950s,[7–9] the possibility arose that some nuclei might have an asymmetric shape under space inversion, such as a pear shape. Since then, many effects have been devoted to study the pear-shaped nuclei[10–15] and their spectroscopic properties.[16–26] Moreover, the studies of fission process,[27–29] cluster structure,[30,31] and shape transition,[32–37] stimulate interest in octupole-deformed shapes.

The octupole shape of nuclei can be attributed to the strong octupole correlations, which arise from the coupling between single-particle orbitals of opposite parity near the Fermi surface with orbital and total angular momentum differing by $3\hbar$. This condition is predominantly met in nuclei with proton (neutron) numbers $Z(N) \approx 34$ ($g_{9/2} \leftrightarrow p_{3/2}$), 56 ($h_{11/2} \leftrightarrow d_{5/2}$), 88 ($i_{13/2} \leftrightarrow f_{7/2}$) and 134 ($j_{15/2} \leftrightarrow g_{9/2}$), characterized by the occurrence of low-lying negative-parity states, as well as pronounced $E1$ transition and $E3$ transition.[38–40]

The evolution of octupole deformation with rotation is a topic of broad interest in the study of pear-shaped nuclei.[41–48] There exists the possibility that the octupole deformation of the pear-shaped nuclei would increase, decrease, or even disappear with spin. This topic has been investigated with many theoretical approaches, e.g., the cranking model based on a Woods–Saxon[41–44] or Nilsson[45] potential combined with the Shell correction method, and the cranking nonrelativistic density functional theories (DFTs).[46–48] Among them, the DFTs provide a self-consistent mean field which depends entirely on a universal energy density functional based on the effective nucleon–nucleon interactions, and thus, would provide a microscopic understanding on the shape evolution of pear-shaped nuclei with spin. Recent decades, the covariant version of DFTs has been developed and achieved great successes.[49–52] The cranking covariant density functional theory (CDFT) offers a consistent treatment of nuclear currents, which is important in the description of nuclear rotations.[53–55] However, up to now, the cranking CDFT has not been applied to investigate the evolution of the octupole shape with spin.

For a full understanding of the evolution of the octupole shape with spin in the framework of the cranking CDFT, self-consistent inclusion of all deformation degrees of freedom is required. The cranking CDFT in three-dimensional (3D) lattice space provides a powerful way since the calculated single-particle wave functions have no symmetry limitation.[56] The cranking CDFT in 3D lattice space has been successfully applied to study the exotic shapes of nuclei and their excitation modes, such as the nuclear linear-chain[56,57] and toroidal structures.[58] In this work, the cranking CDFT in 3D lattice space will be used to investigate the evolution of the octupole shape with spin for the first time. Before that, considering that the pairing correlations are essential to describe the nuclear ground state[1,2,52,59] and excited-state properties,[59–64] a shell-model-like approach (SLAP)[64–72] is implemented in the cranking CDFT in 3D lattice space to take into account the pairing effects. Then, this framework is applied to study the pear-shaped nucleus ^{224}Th and its possible shape changes with rotation. A microscopic understanding on the evolution of octupole shape with spin will be provided.

2. Theoretical Framework

The cranking CDFT in 3D lattice space has been realized, and its detailed formalism can be found in Ref. 56. Here, a brief introduction is presented.

The starting point of the CDFT is a universal density functional.[49–51,73] In the cranking CDFT, the functional is transformed into a rotating frame with a constant rotational velocity $\boldsymbol{\omega}$ around the rotational axis. The corresponding Kohn–Sham equation for nucleons is a Dirac equation,

$$\hat{h}_0\psi_k = [\boldsymbol{\alpha}\cdot(-i\boldsymbol{\nabla} - \boldsymbol{V}(\boldsymbol{r})) + \beta(m + S(\boldsymbol{r})) + V(\boldsymbol{r}) - \boldsymbol{\omega}\cdot\widehat{\boldsymbol{J}}]\psi_k = \varepsilon_k\psi_k, \quad (1)$$

where the cranking single-particle states $\psi_k(\boldsymbol{r}) = \langle\boldsymbol{r}|\hat{b}_k^\dagger|0\rangle$, and ε_k is the corresponding single-particle Routhian. $\widehat{J}$ is the nuclear total angular momentum. The relativistic scalar field $S(\boldsymbol{r})$ and vector field $V^\mu(\boldsymbol{r})$ are connected in a self-consistent way to the nucleon densities and currents. The nucleon densities and currents are obtained from the cranking single-particle eigenstates ψ_k and the corresponding occupation probabilities v_k^2.[67,68]

Traditionally, the Dirac equation (1) is often solved in the harmonic oscillator (HO) bases.[74–78] In recent years, after overcoming the long-standing challenge of the variational collapse[79] and the Fermion doubling[80] problems with the inverse Hamiltonian[81] and the Fourier spectral methods,[82] respectively, the solution of the Dirac equation (1) in the 3D lattice space is available.[56]

The pairing correlations are taken into account by the SLAP.[64–72] The basic idea of the SLAP is to diagonalize the many-body Hamiltonian in a properly truncated many-particle configuration (MPC) space with exact particle number. The SLAP has many advantages, such as the conservation of the particle number, the strict treatment of the Pauli blocking effects, and the unified description for both the ground and excited states.

The cranking many-body Hamiltonian with pairing correlations reads

$$\hat{H} = \hat{H}_0 + \hat{H}_{\text{pair}}. \quad (2)$$

The one-body Hamiltonian, $\hat{H}_0 = \sum \hat{h}_0$, with $\hat{h}_0$ given in Eq. (1). The monopole pairing Hamiltonian $\hat{H}_{\text{pair}}$ is used for the two-body Hamiltonian,

$$\hat{H}_{\text{pair}} = -G \sum_{\xi,\eta>0} \hat{\beta}^{\dagger}_{\xi} \hat{\beta}^{\dagger}_{\bar{\xi}} \hat{\beta}_{\bar{\eta}} \hat{\beta}_{\eta}, \tag{3}$$

where G is the effective pairing strength, $\bar{\xi}(\bar{\eta})$ labels the time-reversal state of $\xi(\eta)$. Since the time-reversal symmetry is conserved in the nonrotational status, $\bar{\xi}(\bar{\eta})$ and $\xi(\eta)$ are chosen as the nonrotational single-particle bases obtained from Eq. (1) at $\hbar\omega = 0$ MeV. The cranking single-particle bases ψ_k and the nonrotational single-particle bases can transform each other, with the transformation coefficients $D_{\xi k} = \int \mathrm{d}^3 r \langle \boldsymbol{r} | \hat{\beta}^{\dagger}_{\xi} | 0 \rangle^* \langle \boldsymbol{r} | \hat{b}^{\dagger}_{k} | 0 \rangle$.

From the transformation coefficients $D_{\xi k}$, $\hat{H}_{\text{pair}}$ can be written in the cranking single-particle bases as

$$\hat{H}_{\text{pair}} = -G \sum_{k_1 k_2 k_3 k_4} \sum_{\xi,\eta>0} D^*_{\xi k_1} D^*_{\bar{\xi} k_2} D_{\bar{\eta} k_4} D_{\eta k_3} \hat{b}^{\dagger}_{k_1} \hat{b}^{\dagger}_{k_2} \hat{b}_{k_4} \hat{b}_{k_3}, \tag{4}$$

In the cranking single-particle bases, the MPC $|i\rangle$ for the A-particle system can be constructed as

$$|i\rangle \equiv |k_1 k_2 \cdots k_A\rangle = \hat{b}^{\dagger}_{k_1} \hat{b}^{\dagger}_{k_2} \cdots \hat{b}^{\dagger}_{k_A} |0\rangle. \tag{5}$$

Diagonalizing the cranking many-body Hamiltonian (2) in the MPC space, the eigenstates can be obtained,

$$\Psi = \sum_i C_i |i\rangle, \tag{6}$$

with the expanding coefficients C_i.

From the expanding coefficients C_i, the occupation probability v_k^2 for the cranking single-particle state ψ_k can be obtained,

$$v_k^2 = \sum_i |C_i|^2 P_i^k, \quad P_i^k = \begin{cases} 1, & \psi_k \text{ is occupied in } |i\rangle, \\ 0, & \text{otherwise.} \end{cases} \tag{7}$$

The occupation probabilities are used in the calculations for the nucleon densities and currents, which will be interacted back into Eq. (1) to achieve self-consistency.[67,68]

The physical observables, including the pairing energy, the angular momentum and the deformation parameters, can be calculated from the nuclear eigenstates Ψ.

The pairing energy is

$$E_{\text{pair}} = \langle \Psi | \hat{H}_{\text{pair}} | \Psi \rangle. \tag{8}$$

The expectation values of the angular momentum components are

$$J_q = \langle \Psi | \widehat{J}_q | \Psi \rangle, \quad q = x, y, z, \tag{9}$$

the corresponding spin I can be obtained through the semiclassical relation, $\langle \hat{\boldsymbol{J}} \rangle = \sqrt{I(I+1)}$.

The quadrupole deformation β_{20} and octupole deformation β_{30} are

$$\begin{aligned}
\beta_{20} &= \frac{4\pi}{3AR^2}\sqrt{\frac{5}{16\pi}}\langle 3z^2 - r^2 \rangle, \\
\beta_{30} &= \frac{4\pi}{3AR^3}\sqrt{\frac{7}{16\pi}}\langle (2z^2 - 3x^2 - 3y^2)z \rangle,
\end{aligned} \tag{10}$$

where $R = 1.2 \times A^{1/3}$ fm.

3. Numerical Details

In ^{224}Th, the evidence for the stable octupole deformation, i.e., the interleaved positive- and negative-parity bands connected by fast $E1$ transitions, has been observed.[83,84] Theoretically, ^{224}Th is predicted to have a stable octupole deformation by the nonrelativistic[15] and covariant[11,12,35] DFTs. The evolution of the octupole shape with spin in ^{224}Th has been investigated with the cranking Woods–Saxon–Bogoliubov–Strutinsky approach, and a sharp transition from octupole deformation to reflection-symmetric quadrupole deformation is predicted to take place at high spins.[42]

Here, the cranking CDFT-SLAP in 3D lattice space is applied for the interleaved positive- and negative-parity bands in ^{224}Th. The point-coupling density functional PC-PK1[85] is used in the particle-hole channel, and the monopole pairing interaction is used in the particle–particle channel. For the 3D lattice space, the step size and the grid number along the x, y and z axes are chosen as 1 fm and 30, respectively. The convergence of the 3D lattice space has been tested. By increasing the grid number from 30 to 32, or decreasing the step size from 1 fm to 0.8 fm, the change of the total Routhian at $\hbar\omega = 0.10$ MeV is within 0.1%. The dimensions of the MPC space are 1000 for both the neutron and proton. The effective neutron and proton pairing strengths are 0.45 MeV and 0.20 MeV by reproducing the experimental odd–even mass differences. By increasing the dimension of the MPC space and adjusting the effective pairing strengths, the results are essentially the same, indicating the convergence of the MPC space.

To check the validity of the cranking CDFT-SLAP in 3D lattice space, the nucleus ^{60}Fe is calculated as an example, and the results are compared with those from the cranking CDFT-SLAP in the HO bases with 10 major shells.[68] The effective pairing strengths are 0.8 MeV for both neutron and proton.[68] The size of the 3D lattice space and the MPC space are the same with those used in the calculations for ^{224}Th, which are sufficient to obtain convergent solutions. The single-neutron Routhians and the corresponding occupation probabilities in ^{60}Fe at $\hbar\omega = 0.10$ MeV are shown in Fig. 1. Satisfactory agreements are found with the differences less than 10^{-2} MeV for the single-neutron Routhians and 10^{-4} for the occupation probabilities, which confirms the correct implementation for the cranking CDFT-SLAP in 3D lattice space.

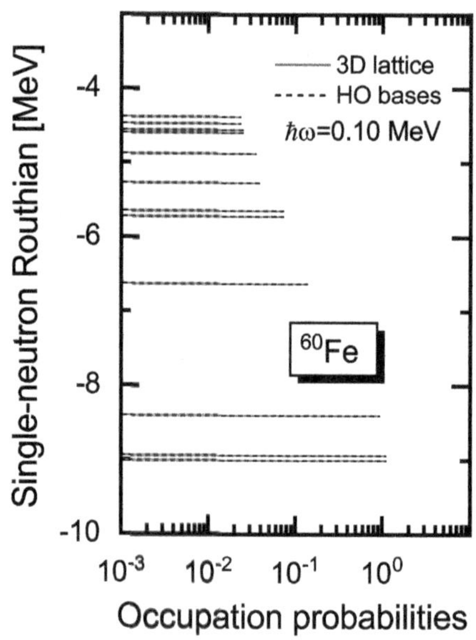

Fig. 1. (Color online) Single-neutron Routhians and the corresponding occupation probabilities in ^{60}Fe at $\hbar\omega = 0.10$ MeV calculated by the cranking CDFT-SLAP in 3D lattice space (red solid lines), in comparison with the results calculated by the cranking CDFT-SLAP in the harmonic oscillator bases (black dashed lines).

4. Results and Discussions

The spin as functions of the rotational frequency for the yrast band in ^{224}Th calculated by the cranking CDFT-SLAP in 3D lattice space with and without pairing correlations, is shown in Fig. 2, in comparison with the data available.[84] When the pairing correlations are not taken into account, the calculated spins are apparently too large compared with the data. After the pairing correlations are included, the spins are reduced significantly, and a good description for the data is achieved. This indicates that the pairing correlations would reduce the nuclear moments of inertia, and are important for the rotational excitation states of ^{224}Th. In the following, only the results including the pairing correlations are shown.

The cranking CDFT-SLAP calculations in 3D lattice space show that ^{224}Th has an axially symmetric and reflection asymmetric shape at the ground state, with $\beta_{20} \approx 0.17$ and $\beta_{30} \approx 0.12$. The evolutions of the quadrupole deformation β_{20} and octupole deformation β_{30} of the yrast states in ^{224}Th with the rotational frequency, are shown in Fig. 3. Below $\hbar\omega = 0.28$ MeV, β_{20} and β_{30} both have a slightly upward and then downward trend with spin. Within the region 0.28 MeV$< \hbar\omega < 0.4$ MeV, it's difficult to achieve convergent results because of the band crossing, as will be shown in Figs. 4 and 5. At $\hbar\omega = 0.4$ MeV, β_{30} drops to zero abruptly, and β_{20} reduces to 0.04. Thus there exists a sharp transition from an octupole shape to a

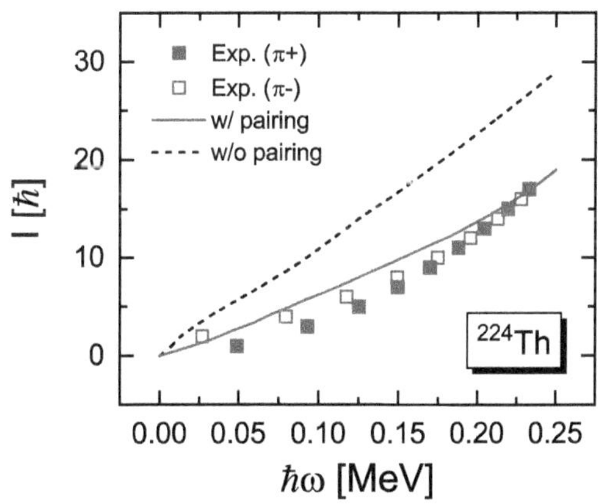

Fig. 2. (Color online) Spin as functions of the rotational frequency in ^{224}Th calculated by the cranking CDFT-SLAP in 3D lattice space with (red solid curve) and without (black dashed curve) pairing correlations, in comparison with the experimental interleaved positive-parity (solid symbols) and negative-parity bands (open symbols).

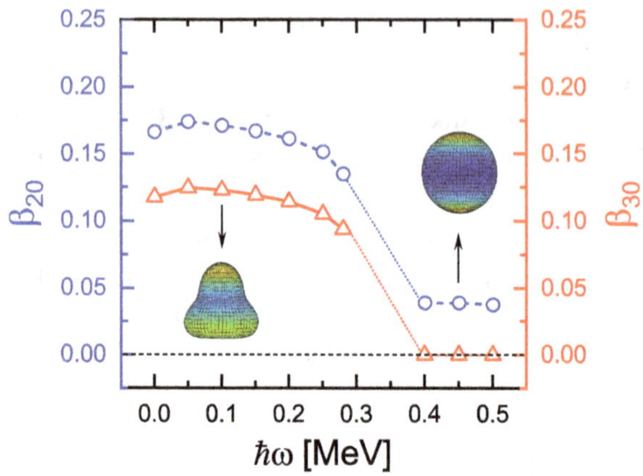

Fig. 3. (Color online) Quadrupole deformation β_{20} (blue) and octupole deformation β_{30} (red) of yrast states in ^{224}Th as functions of the rotational frequency. The 3D images inserted in the panel schematically show the shapes at $\hbar\omega = 0.10$ MeV and 0.45 MeV, respectively.

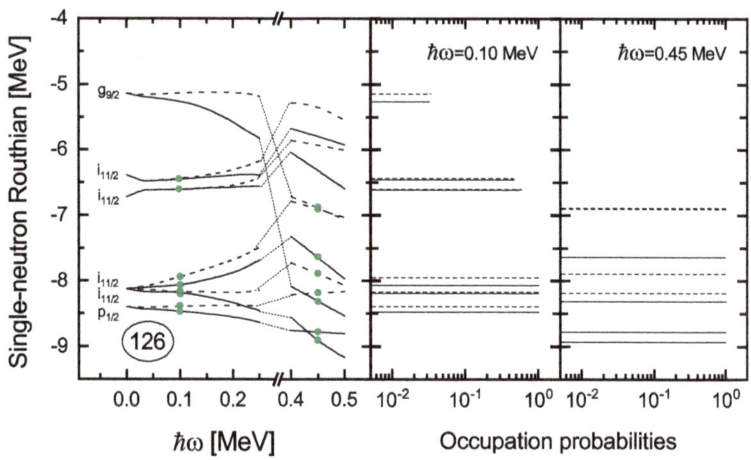

Fig. 4. Single-neutron Routhians of ^{224}Th as functions of the rotational frequency. Each orbital is labeled by the corresponding spherical quantum number of its main component. The orbitals with simplex $s = +i(-i)$ are denoted by solid (dashed) curves. The solid circles represent the occupied orbitals at $\hbar\omega = 0.10$ MeV and 0.45 MeV, and the corresponding occupation probabilities for each orbital are given in the right two columns, respectively.

nearly spherical shape in ^{224}Th at high spins. A similar sharp shape transition has been obtained in the cranking Woods–Saxon–Bogoliubov–Strutinsky approach, but the value of β_{20} at high spins is about 0.14, which is slightly larger than our result.[42]

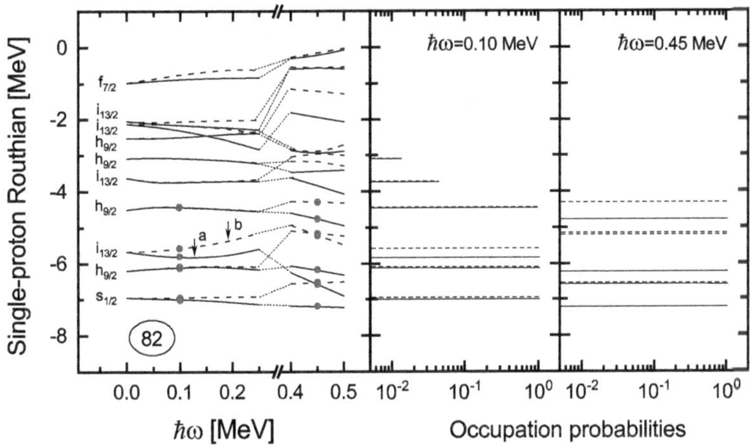

Fig. 5. The same as Fig. 4, but for single-proton Routhians. The proton $i_{13/2}$ orbitals are labeled by a and b.

In order to understand the evolution of the octupole shape of the yrast states in ^{224}Th with spin in Fig. 3, the single-neutron and -proton Routhians in ^{224}Th as functions of the rotational frequency are shown in Figs. 4 and 5, respectively.

In Fig. 4, it is found that the neutron configuration is $\nu(i_{11/2})^8$ at $\hbar\omega = 0.10\,\text{MeV}$. With the increase of the rotational frequency, the neutron $g_{9/2}$ simplex partners decrease rapidly. Thus, a neutron band crossing occurs at around $\hbar\omega = 0.28\,\text{MeV}$, resulting in a discontinuity in the single-neutron Routhians. After $\hbar\omega = 0.4\,\text{MeV}$, the neutron configuration changes to $\nu(i_{11/2})^6(g_{9/2})^2$, with the occupation probabilities of the $g_{9/2}$ simplex partners changing from less than 10^{-1} at $\hbar\omega = 0.10\,\text{MeV}$ to nearly 2 at $\hbar\omega = 0.45\,\text{MeV}$.

As shown in the single-proton Routhian diagram in Fig. 5, it is found that the proton configuration is $\pi(h_{9/2})^6(i_{13/2})^2$ at $\hbar\omega = 0.10\,\text{MeV}$. With the increase of the rotational frequency, the proton configuration remains unchanged and the occupation probabilities change smoothly. However, the single-proton Routhians and level sequence change obviously, resulting from the change of the mean field due to the neutron band crossing around $\hbar\omega = 0.28\,\text{MeV}$. Hence, a discontinuity also exists as in the single-neutron Routhian diagram.

For ^{224}Th in the $N \approx 134, Z \approx 88$ region, empirically, its octupole deformation originates from the $\nu(j_{15/2} \leftrightarrow g_{9/2})$ and $\pi(i_{13/2} \leftrightarrow f_{7/2})$ couplings. However, as shown in Fig. 4, the neutron $g_{9/2}$ orbitals are not

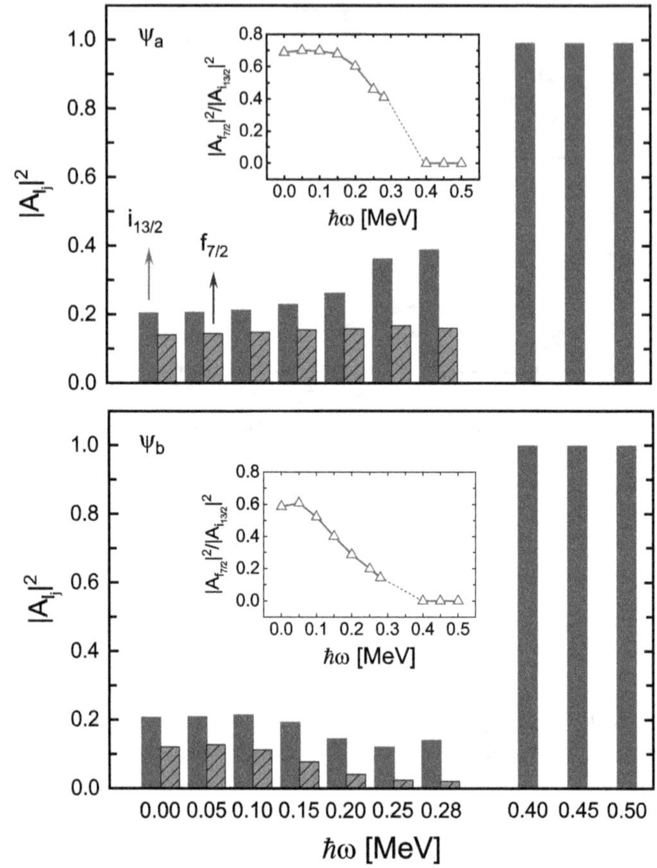

Fig. 6. The amplitudes of the top two components l_j of the single-proton levels ψ_ν with $\nu = a$ and b within $0\,\text{MeV} \leq \hbar\omega \leq 0.5\,\text{MeV}$. The inset is the ratio $|A_{f_{7/2}}|^2/|A_{i_{13/2}}|^2$ between the $f_{7/2}$ and $i_{13/2}$ components as functions of the rotational frequency.

occupied, with the occupation probabilities less than 10^{-1} at $\hbar\omega < 0.28\,\text{MeV}$, so the $\nu(j_{15/2} \leftrightarrow g_{9/2})$ coupling shall not be the main cause of the octupole deformation at low spins in ^{224}Th. Shown in Fig. 5, the proton $i_{13/2}$ orbitals labeled by a and b are occupied, and the coupling between the two orbitals and the $f_{7/2}$ orbitals would result in the octupole correlations. Thus, to understand the evolution of the octupole deformation with spin microscopically, the components of the single-proton orbitals ψ_a and ψ_b and their evolutions with spin are analyzed and shown in Fig. 6.

Figure 6 shows the top two components of the orbitals ψ_a and ψ_b expanded in the spherical Woods–Saxon bases l_j. It is found that the top two components are $i_{13/2}$ and $f_{7/2}$, and their amplitudes are comparable at low spins, which indicates the presence of strong $\pi(i_{13/2} \leftrightarrow f_{7/2})$ coupling. The ratios $|A_{f_{7/2}}|^2/|A_{i_{13/2}}|^2$, which reflect the coupling strength between the proton $i_{13/2}$ and $f_{7/2}$ orbitals, can be extracted from the amplitudes of the $f_{7/2}$ and $i_{13/2}$ components and are shown in the inset of Fig. 6. Below $\hbar\omega = 0.28$ MeV, the ratios $|A_{f_{7/2}}|^2/|A_{i_{13/2}}|^2$ of ψ_a and ψ_b rise slightly and then decline with the increase of the rotational frequency. This corresponds to the slightly upward and then downward trend of the octupole deformation β_{30} at low spins as shown in Fig. 3. Above $\hbar\omega = 0.4$ MeV, the $f_{7/2}$ component of the levels ψ_a and ψ_b drops to zero abruptly, and ψ_a and ψ_b almost become pure $i_{13/2}$ orbitals. This indicates that the coupling between the proton $i_{13/2}$ and $f_{7/2}$ orbitals no longer exists, which can explain the sudden disappearance of the octupole deformation at high spins in Fig. 3.

5. Summary

In summary, based on the cranking CDFT in 3D lattice space, the SLAP with exact particle number conservation is implemented to take into account the pairing correlations, and applied for the interleaved positive- and negative-parity bands in ^{224}Th. The microscopic understanding on the evolution of octupole shape of yrast states in ^{224}Th with spin is addressed.

The experimental $I - \omega$ relations are well reproduced by the cranking CDFT-SLAP calculations in 3D lattice space. It is found that the octupole deformation of yrast states in ^{224}Th rises slightly and then declines with increasing spin. After the band crossing, the octupole deformation suddenly disappears, i.e., a sharp transition from an octupole shape to a nearly spherical shape takes place at high spins. The mechanism is analyzed by investigating the evolution of the coupling strength between the proton $i_{13/2}$ and $f_{7/2}$ orbitals with spin. It is revealed that the coupling strength undergoes a slight increase followed by a decline below $\hbar\omega = 0.28$ MeV, and abruptly vanishes after the band crossing at $\hbar\omega = 0.40$ MeV, resulting in a consistent pattern in the evolution of the octupole shape.

Acknowledgments

The authors would like to express gratitude to P. W. Zhao for constructive guidance and valuable suggestions. Fruitful discussions with Y. K. Wang are acknowledged. This work was partly supported by the National Natural Science Foundation of China (Grants Nos. 12070131001, 11935003,

11975031 and 12141501), and the High-performance Computing Platform of Peking University.

References

1. A. Bohr and B. R. Mottelson, *Nuclear Structure*, Vol. **II** (Benjamin, New York, 1975).
2. P. Ring and P. Schuck, *The Nuclear Many-Body Problem* (Springer Science & Business Media, Berlin, 2004).
3. C. H. Townes, H. M. Foley and W. Low, *Phys. Rev.* **76** (1949) 1415.
4. W. Gordy, *Phys. Rev.* **76** (1949) 139.
5. A. N. Bohr and B. R. Mottelson, *Dan. Mat. Fys. Medd.* **27** (1953) 1.
6. S. G. Nilsson, *Dan. Mat. Fys. Medd.* **29** (1955) 1.
7. F. Asaro, F. S. Stephens and I. Perlman, *Phys. Rev.* **92** (1953) 1495.
8. F. S. Stephens, F. Asaro and I. Perlman, *Phys. Rev.* **96** (1954) 1568.
9. F. S. Stephens, F. Asaro and I. Perlman, *Phys. Rev.* **100** (1955) 1543.
10. C. Liu *et al.*, *Phys. Rev. Lett.* **116** (2016) 112501.
11. A. V. Afanasjev, H. Abusara and S. E. Agbemava, *Phys. Scr.* **93** (2018) 034002.
12. S. E. Agbemava, A. V. Afanasjev and P. Ring, *Phys. Rev. C* **93** (2016) 044304.
13. S. Ebata and T. Nakatsukasa, *Phys. Scr.* **92** (2017) 064005.
14. W. Nazarewicz, P. Olanders, I. Ragnarsson, J. Dudek, G. A. Leander, P. Möller and E. Ruchowsa, *Nucl. Phys. A* **429** (1984) 269.
15. J. L. Egido and L. M. Robledo, *Nucl. Phys. A* **494** (1989) 85.
16. L. P. Gaffney *et al.*, *Nature* **497** (2013) 199.
17. M. Scheck *et al.*, *J. Phys.: Conf. Ser.* **533** (2014) 012007.
18. B. Bucher *et al.*, *Phys. Rev. Lett.* **116** (2016) 112503.
19. B. Bucher *et al.*, *Phys. Rev. Lett.* **118** (2017) 152504.
20. Y. Y. Wang, *Phys. Rev. C* **104** (2021) 014318.
21. W. Sun, S. Quan, Z. P. Li, J. Zhao, T. Nikšić and D. Vretenar, *Phys. Rev. C* **100** (2019) 044319.
22. P. W. Zhao and Z. P. Li, *Int. J. Mod. Phys. E* **27** (2018) 1830007.
23. L. M. Robledo and P. A. Butler, *Phys. Rev. C* **88** (2013) 051302(R).
24. R. N. Bernard, L. M. Robledo and T. R. Rodríguez, *Phys. Rev. C* **93** (2016) 061302(R).
25. J. M. Yao, E. F. Zhou and Z. P. Li, *Phys. Rev. C* **92** (2015) 041304(R).
26. Y. S. Chen and Z. C. Gao, *Phys. Rev. C* **63** (2000) 014314.
27. H. Abusara, A. V. Afanasjev and P. Ring, *Phys. Rev. C* **82** (2010) 044303.
28. B. N. Lu, J. Zhao, E. G. Zhao and S. G. Zhou, *Phys. Rev. C* **89** (2014) 014323.
29. H. Tao, J. Zhao, Z. P. Li, T. Nikšić and D. Vretenar, *Phys. Rev. C* **96** (2017) 024319.
30. P. Marević, J.-P. Ebran, E. Khan, T. Nikšić and D. Vretenar, *Phys. Rev. C* **97** (2018) 024334.
31. E. F. Zhou, J. M. Yao, Z. P. Li, J. Meng and P. Ring, *Phys. Lett. B* **753** (2016) 227.
32. W. Zhang and Y. F. Niu, *Phys. Rev. C* **96** (2017) 054308.

33. J. Y. Guo, P. Jiao and X. Z. Fang, *Phys. Rev. C* **82** (2010) 047301.
34. K. Nomura, D. Vretenar and B. N. Lu, *Phys. Rev. C* **88** (2013) 021303(R).
35. Z. P. Li, B. Y. Song, J. M. Yao, D. Vretenar and J. Meng, *Phys. Lett. B* **726** (2013) 866.
36. Z. P. Li, T. Nikšić and D. Vretenar, *J. Phys. G: Nucl. Part. Phys.* **43** (2016) 024005.
37. H. L. Wang, J. Yang, M. L. Liu and F. R. Xu, *Phys. Rev. C* **92** (2015) 024303.
38. P. A. Butler and W. Nazarewicz, *Rev. Mod. Phys.* **68** (1996) 349.
39. P. A. Butler, *J. Phys. G* **43** (2016) 073002.
40. P. A. Butler, *Proc. R. Soc. A* **476** (2020) 20200202.
41. W. Nazarewicz, P. Olanders, I. Ragnarsson, J. Dudek and G. A. Leander, *Phys. Rev. Lett.* **52** (1984) 1272.
42. W. Nazarewicz, G. A. Leander and J. Dudek, *Nucl. Phys. A* **467** (1987) 437.
43. W. Nazarewicz and S. L. Tabor, *Phys. Rev. C* **45** (1992) 2226.
44. H. L. Wang, H. L. Liu and F. R. Xu, *Phys. Scr.* **86** (2012) 035201.
45. J. Skalski, S. Mizutori and W. Nazarewicz, *Nucl. Phys. A* **617** (1997) 282.
46. E. Garrote, J. L. Egido and L. M. Robledo, *Phys. Lett. B* **410** (1997) 86.
47. A. Tsvetkov, J. Kvasil and R.-G. Nazmitdinov, *J. Phys. G* **28** (2002) 2187.
48. W. Ryssens, M. Bender and P.-H. Heenen, *Acta. Phys. Pol. B* **49** (2018) 339.
49. P. Ring, *Prog. Part. Nucl. Phys.* **37** (1996) 193.
50. D. Vretenar, A. V. Afanasjev, G. A. Lalazissis and P. Ring, *Phys. Rep.* **409** (2005) 101.
51. J. Meng, H. Toki, S. G. Zhou, S. Q. Zhang, W. H. Long and L. S. Geng, *Prog. Part. Nucl. Phys.* **57** (2006) 470.
52. J. Meng (ed.), *Relativistic Density Functional for Nuclear Structure, International Review of Nuclear Physics*, Vol. **10** (World Scientific, Singapore, 2016).
53. J. König and P. Ring, *Phys. Rev. Lett.* **71** (1993) 3079.
54. A. V. Afanasjev and H. Abusara, *Phys. Rev. C* **88** (2010) 034329.
55. J. Meng, J. Peng, S. Q. Zhang and P. W. Zhao, *Front. Phys.* **8** (2013) 55.
56. Z. X. Ren, S. Q. Zhang, P. W. Zhao, N. Itagaki, J. A. Maruhn and J. Meng, *Sci. China: Phys. Mech. Astron.* **62** (2019) 112062.
57. D. D. Zhang, Z. X. Ren, P. W. Zhao, D. Vretenar, T. Nikšić and J. Meng, *Phys. Rev. C* **105** (2022) 024322.
58. Z. X. Ren, P. W. Zhao, S. Q. Zhang and J. Meng, *Nucl. Phys. A* **996** (2020) 121696.
59. R. A. Broglia and V. Zelevinsky (eds.), *Fifty Years of Nuclear BCS* (World Scientific, 2013).
60. A. V. Afanasjev, J. König and P. Ring, *Phys. Rev. C* **60** (1999) 051303.
61. Z. M. Niu, Y. F. Niu, H. Z. Liang, W. H. Long, T. Nikšić, D. Vretenar and J. Meng, *Phys. Lett. B* **723** (2013) 172.
62. P. W. Zhao, S. Q. Zhang and J. Meng, *Phys. Rev. C* **92** (2015) 034319.
63. Z. M. Niu, Y. F. Niu, H. Z. Liang, W. H. Long and J. Meng, *Phys. Rev. C* **95** (2017) 044301.
64. Y. P. Wang and J. Meng, *Phys. Lett. B* **841** (2023) 137923.
65. J. Y. Zeng and T. S. Cheng, *Nucl. Phys. A* **405** (1983) 1.

66. J. Y. Zeng, T. H. Jin and Z. J. Zhao, *Phys. Rev. C* **50** (1994) 1388.
67. J. Meng, J. Y. Guo, L. Liu and S. Q. Zhang, *Front. Phys. China* **1** (2006) 38.
68. Z. Shi, Z. H. Zhang, Q. B. Chen, S. Q. Zhang and J. Meng, *Phys. Rev. C* **97** (2018) 034317.
69. Y. L. Lin, T. Yan and L. Liu, *Int. J. Mod. Phys. E* **31** (2022) 2250035.
70. T. Yan, Y. L. Lin and L. Liu, *Phys. Rev. C* **104** (2021) 024303.
71. L. Liu, *Int. J. Mod. Phys. E* **30** (2021) 2150055.
72. L. Liu, *Phys. Rev. C* **99** (2019) 024317.
73. T. Nikšić, D. Vretenar and P. Ring, *Prog. Part. Nucl. Phys.* **66** (2011) 519.
74. P. W. Zhao, S. Q. Zhang, J. Peng, H. Z. Liang, P. Ring and J. Meng, *Phys. Lett. B* **699** (2011) 181.
75. W. Koepf and P. Ring, *Nucl. Phys. A* **493** (1989) 61.
76. J. Peng, J. Meng, P. Ring and S. Q. Zhang, *Phys. Rev. C* **78** (2008) 024313.
77. P. W. Zhao, J. Peng, H. Z. Liang, P. Ring and J. Meng, *Phys. Rev. Lett.* **107** (2011) 122501.
78. P. W. Zhao, *Phys. Lett. B* **773** (2017) 1.
79. Y. Zhang, H. Z. Liang and J. Meng, *Int. J. Mod. Phys. E* **19** (2010) 55.
80. Y. Tanimura, K. Hagino and H. Z. Liang, *Prog. Theor. Exp. Phys.* **2015** (2015) 073D01.
81. K. Hagino and Y. Tanimura, *Phys. Rev. C* **82** (2010) 057301.
82. Z. X. Ren, S. Q. Zhang and J. Meng, *Phys. Rev. C* **95** (2017) 024313.
83. B. Ackermann *et al.*, *Nucl. Phys. A* **559** (1993) 61.
84. S. Singh and B. Singh, *Nucl. Data Sheets* **130** (2015) 127.
85. P. W. Zhao, J. Peng, H. Z. Liang, P. Ring and J. Meng, *Phys. Rev. C* **85** (2012) 054310.

www.ingramcontent.com/pod-product-compliance
Lightning Source LLC
LaVergne TN
LVHW010331220926
848484LV00004BB/881